MINNESOTA STUDIES IN THE PHILOSOPHY OF SCIENCE

Minnesota Studies in the
PHILOSOPHY OF SCIENCE

RONALD N. GIERE, GENERAL EDITOR

HERBERT FEIGL, FOUNDING EDITOR

VOLUME XIV

Scientific Theories

EDITED BY

C. WADE SAVAGE

UNIVERSITY OF MINNESOTA PRESS, MINNEAPOLIS

Published by the University of Minnesota Press
2037 University Avenue Southeast, Minneapolis, MN 55414.
Printed in the United States of America.

Library of Congress Cataloging-in-Publication Data
Scientific theories/edited by C. Wade Savage.
p.   cm. — (Minnesota studies in the philosophy of science : v. 14)
Some of these papers were originally presented
at an institute conducted by the Minnesota Center
for Philosophy of Science from 1985 to 1987.
ISBN 0-8166-1801-1
1. Science—Philosophy.   2. Science—Methodology.   3. Science—Evaluation.
I. Savage, C. Wade.   II. Minnesota Center for Philosophy of Science.   III. Series.
Q175.M64   vol. 14
[Q175.55]
501—dc20   90-32932
                                                                                        CIP

The University of Minnesota is
an equal-opportunity educator and employer.

Contents

Preface

From the fall of 1985 through the spring of 1987 the Minnesota Center for Philosophy of Science conducted an institute—an ongoing conference—whose focal question was: Is there a new consensus in the philosophy of science? The old consensus was of course logical empiricism, or positivism, the position forged during the first half of the century by Russell, Schlick, Carnap, Feigl, Reichenbach, Hempel, Nagel, and others, and dissolved during the third quarter by the criticisms of Quine, Hanson, Feyerabend, Kuhn, and others. The explicit purpose of the conference was to ascertain whether, in the wake of this criticism and the resulting loss of focus and direction, some new consensus was emerging that might come to provide the same sort of structure and direction for the field as had the old consensus. For this purpose the field was initially divided into three traditional areas: scientific explanation, scientific theories, and scientific justification or evaluation. The present volume's predecessor and companion—*Scientific Explanation,* volume 13, edited by Philip Kitcher and Wesley Salmon—contains papers in the first area. The present volume contains papers in the second and third areas. It is not a conference proceedings in the strict sense, for some participants did not write papers and some wrote them afterward, and some papers were additionally solicited. Furthermore, at later stages two other areas were added to the agenda: the relation between history and philosophy of science, and recent developments in the philosophy of cognitive science. With few exceptions, these additional symposia are not represented in the two volumes.

What then is our result? Is a new consensus emerging in the philosophy of science, either in general, or in the special areas represented by the two volumes? Comments here will be limited to the nature and acceptance of scientific theories. (Comments on scientific explanation can be found in volume 13.) The answer to the special question would seem to be negative. The *syntactic* view that a theory is an axiomatized collection of sentences has been challenged by the *semantic* view that a theory is a collection of nonlinguistic models, and both are challenged by the view that a theory is an amorphous entity consisting perhaps of sentences and models, but just as importantly of exemplars, problems, standards, skills, practices, and tendencies. Similarly, several views of theory confirmation compete for allegiance. The best confirmed theory is variously held to be the one with the greatest number and variety of observed consequences, the one with the highest degree of confirmation (probability) on the observed evidence, or the one that best explains the observed evidence. Some theorists regard a decision-theoretic approach as superior to those above. Still others hold that theories of confirmation beg the question of whether theories are, can be, or should be "confirmed" and recommend replacing them by accounts of how theories are discovered, accepted, and developed.

The answer to the general question is more complex, partly because it applies to more areas than the traditional three above, and partly because it includes second-order questions about the nature and scope of philosophy of science (which naturally complicate the first-order questions). Nonetheless, the answer here also seems to be negative. The once dominant conception of philosophy of science as the logical analysis and reconstruction of science is generally regarded to be moribund, and no comparably general conception has replaced it. Philosophy of science has become exceedingly broad and diverse. It now attends to practical and experimental science in addition to theoretical science, and it examines a virtually unrestricted range of sciences and scientific practices. It is no longer simply the logic and methodology of science, but involves in addition the history, sociology, and psychology of science. These developments have been accompanied by the growing view that philosophy of science should be scientific, naturalistic. This view regards previous theories of the structure and acceptance of scientific theories as idealizations, and recommends replacing them by accounts of what science is actually, in its natural psychosocial setting. Such studies have led some to conclude that science is not an objective, rule-governed, rational activity, and that its development is not a rational process. On one suggestion, the development of science is a process comparable to Darwinian natural selection. Another view is that, although scientific rationality cannot be equated with logicality, science is rational in a sense philosophers are currently attempting to explicate.

The welter of conflicting views has suggested to many that the only consensus in philosophy of science is that there is no consensus. Indeed, the pluralistic ideology currently in favor is unsympathetic to attempts to achieve or even locate consensus, for fear that some new and equally stifling dogma will replace the old. It seems that to seek now for anything so definite as *consensus* in philosophy of science—some set of doctrines to rival the positivist consensus of earlier years—is premature at best. The field seems to contain too much diversity and too much ferment to permit it.

Most of the essays presented here reflect the concerns and themes of current philosophy of science, either by pursuing them, or by criticizing them, or by attempting to harmonize them with others, occasionally with some of the old consensus. Although they do not constitute a new consensus, and perhaps do not even point toward one, they do provide a bridge between the old and the new.

One goal of our institute was to survey current philosophy in a manner that would be useful to a general academic audience as well as to specialists in the field. As a consequence, most of the essays are relatively accessible, and the introduction has been designed to increase their accessibility and usefulness to those who wish to sample the field. The volume should therefore be suitable for beginning graduate and advanced undergraduate courses, as well as the more advanced contexts.

We wish to acknowledge the support of institutions and individuals who made our institute on consensus possible. A major grant from the National Endowment of the Humanities provided most of the funds for the conference. (Philip Kitcher and Wade Savage were the principal investigators.) The College of Liberal Arts and the Office of the President of the University of Minnesota provided a substantial supplementary grant. We thank them all. We also thank the faculty who staffed the institute; their names appear at the end of the volume. Finally, we thank the students and faculty who attended the lectures and contributed to the discussions. It was a stimulating experience for everyone involved.

SCIENTIFIC THEORIES

Introduction

The essays presented here are arranged in three groups. Essays in the *first group* treat issues arising from special sciences. Although they continue the classical tradition of detailed attention to science, some depart from its tendencies toward instrumentalism, foundationalism, and a linguistic view of theories. Caplan examines non-epistemic factors involved in the development of medical technology and the categorization of medical experimentation and research. Grünbaum contends that although Freudian psychoanalysis is a scientific theory it has not yet been experimentally confirmed. Churchland proposes a radically new way of representing theories and their acquisition in the terms of connectionist neuroscience. Nelson suggests that the low success of microeconomics is due to its antirealist tendencies and its failure to isolate the natural kinds of its domain. Sklar shows that, although empiricist foundationalism is philosophically controversial, it was an essential research strategy in the development of relativistic physics.

Essays in the *second group* deal with the confirmation and choice of scientific theories. All recommend probabilist methods, Bayesian in most cases, thus illustrating the current dominance of the probabilist approach. Kyburg outlines a method for adding statements to the body of scientific knowledge according as the decision raises or lowers the statistical probability of others. Salmon devises a Bayesian scheme for comparing theories that employs some of Kuhn's criteria in its assignment of prior probabilities. Eells offers a solution to the problem in using "old," previously known evidence in a Bayesian system of confirmation. Howson recommends the use of prior probabilities to discriminate ad hoc from genuine theories in Bayesian confirmation. Skyrms explains why Bayesian decision theory assigns a greater value to knowledge than to ignorance.

Essays in the *third group* fall into an area of interconnected topics much discussed in current philosophy of science: "rationality, revolution, and realism," we may call it. Laudan argues that the doctrine that theories are underdetermined by the evidence does not in any of its true versions entail that theory choice is irrational. Kuhn develops his earlier thesis that scientific theories separated by a revolution are semantically incommensurable with one another and defends it from objections flowing from the causal theory of reference. Worrall criticizes the view — which he attributes to Kuhn — that theory choice is not a completely rational process with the case of scientists who stubbornly refused to give up a theory. Boyd defends his argument that a realist view of scientific theories is required to explain their empirical success from a number of objections, one of which is that history does not always exhibit realist theoretical progress. Sober criticizes Boyd's argument and develops an alternative realist theory of inference that will permit legitimate scientific inferences while excluding those of Boyd.

Arthur Caplan — Seek and Ye Might Find

Caplan acknowledges that philosophy of science has expanded. It now examines a large number and variety of sciences, concentrating less than before on the physical sciences. It has come to value the results of history and sociology of science. And its units of analysis have expanded, from theories to paradigms, problems, research strategies, etc. Nonetheless, he finds its thematic focus still too narrow. For instance, philosophers of biology tend to concentrate on abstract, theoretical topics, such as genetics and evolutionary biology; to the neglect of concrete, practical topics in such areas as immunology, paleontology, horticulture, and veterinary science. He supports his contention with a richly detailed case study from nephrology: the development of dialysis machines to treat chronic kidney disease. The case reveals features of scientific development that the traditional narrow focus overlooks. The process of scientific development (of dialysis, for example) — from discovery to acceptance to mastery — is complex: it often involves advertising and political manipulation, and it is often affected by administrative decisions, availability of funds and subject pools, and other non-epistemic factors, which operate in conjunction with epistemic factors. Such non-epistemic factors are often neglected. He concludes with a discussion of the effect on medical development of categorizing a procedure as research, experiment, or theory; and he argues that these categories are partly defined by practical and social considerations.

Caplan applauds the current tendency to add new interests, specialities, and even divisions to philosophy of science, but finds it still too narrow. Ironically, expansion may well have the practical effect of increasing the trend toward specialization and further narrowing the focus of specialists, even as the field as a whole broadens its perspective. In any event, divisions between theoretical and practical studies and between developmental, structural, and evaluational (confirmational) studies will most likely persist. Caplan's example is in the practical developmental division. The narrowest workers are in the divisions of structural theoretical and confirmational studies. The importance of developments in one area to workers in another will no doubt vary from case to case. Where a technological development (dialysis, say) affects a scientific theory (of the kidney, for example), it may become important to philosophers mainly concerned with the structure of the theory. For example, it might encourage them to think of a theory as a Kuhnian paradigm (matrix of exemplars, problems, standards, etc.), rather than a set of sentences.

Adolf Grünbaum — The Psychoanalytic Enterprise in Scientific Perspective

Grünbaum examines the evidence for some of Freud's hypotheses concerning the causation of pathological behavior by unconscious processes and finds it wanting. The first of these is that repressed (often sexual) memories cause neurotic symptoms, such as those of hysteria. Freud's evidence was that reduction (lifting) of the repression resulted in removal or reduction of the symptoms. The repression was lifted by the patient's recalling the memory, with the aid of hypnosis or free association about the neurotic symptoms. But there is a strongly competing, "placebo" hypothesis: symptom reduction is the effect of the patient's "recalling" an event suggested by the therapist together with the expectation that the symptoms will thus be reduced. To reply that the true cause is revealed by free association recall is in effect to beg the question, and hypnotic recall is notoriously unreliable. To decide between the competitors requires moving outside the clinical setting to conduct a controlled experiment: one group of subjects receives psychological treatment that does not repress memories, another group receives treatment that does. (Apparently the experiment has not been conducted.) Freud's second hypothesis is that repressed wishes cause dreams, and repressed unpleasant thoughts cause slips (of memory, the tongue, etc.). Grünbaum claims that this hypothesis is a rash and unjustified extrapolation from the first, etiologic, hypothesis, which inherits all its weakness. In addition, he offers two sorts of grounds for deeming Freud's dream theory to be false, rather than just unsupported. The reliability of free association as a method of validating causal hypotheses in psychoanalysis is thus undermined. He concludes with a brief review of experimental results, none of which alters his previous evaluation. His final conclusion is that the evidence for the psychoanalytic hypotheses in question is insufficient to support their widespread acceptance in our culture.

Grünbaum does not argue that psychoanalysis is not a science. Indeed, here and elsewhere he resists attempts of the hermeneuticists to locate it among the humanities, and elsewhere he has defended it from Popper's charge that it is unfalsifiable. He notes that rival placebo hypotheses are a species of rival causal hypotheses. One could add that it is typical for causal hypotheses in science to have strong causal rivals. The observation applies equally to hypotheses about unconscious mechanisms of perception, speech comprehension, and other behavior, which are numerous in current cognitive psychology. In general, it would seem to be no more difficult to design experiments for deciding between theories of the psychoanalytic unconscious than to design them for deciding between theories of the cognitive unconscious. Nonetheless, it may be difficult to design an experiment that will eliminate the placebo rival to Freud's.

Paul Churchland—On the Nature of Theories: A Neuocomputational Perspective

Churchland brings recent connectionist neuropsychology to bear on several central questions in philosophy of science: the nature of theories, foundationalism, etc. Connectionist theories of cognition model information processes by means of layered networks of cells and their weighted connections (inhibitory and excitatory). They are recommended because of their similarity to systems of neurons and associated dendritic and axonic synapses, which leads one to hope they can be implemented by brains. One of Churchland's examples is a simple network that discriminates naval mines from rocks by processing the intensity distribution of their sonar echoes. The network has three layers: an input layer of n cells, connected to an intermediate layer of m cells, connected to an output layer of two cells. The n stimulus intensities produce a vector (set) of n excitations in the input cells, which is transformed through cell connections into an intermediate vector of m excitations, which in turn is transformed into an output vector of two excitations. If the first element of the two-vector is high compared to the second, the sound was produced by a mine and not a rock. The network is trained to discriminate mines from rocks by subjecting it to stimuli from both and correcting it when it errs by adjusting the connection weights according to some learning rule (algorithm). Churchland recommends that we naturalize epistemology by reconceptualizing its fundamental elements in terms of such connectionist models. For example, the network's perception is its production and transformation of vectors of excitation, its percepts (representations) of objects are vectors of excitation, its knowledge of its world is the set of weights on connections among its cells, and its learning is the alteration of those weights. (Some of these identifications are only implied in Churchland's paper.) One important conclusion is that theories (representations) are not the linguistic sentences that the syntacticist epistemologists and philosophers of science have taken them to be. They are even less like the abstract models that some of the semanticists take them to be. They are more like Kuhn's paradigms. Another conclusion is that percepts and other representations are theory laden, in the sense in which an organism's theory is its knowledge of the world, for every representation depends on the set of connection weights that produce it.

Churchland's essay is a rich interdisciplinary mix of neuropsychology, psychology, (naturalistic) epistemology, and philosophy of science. The current literature in these areas suggests that it may provoke an equally rich set of objections. Cognitive psychologists who take percepts and other representations to be symbolic entities have already complained that connectionists try to model these at too low a level, the level of hardware implementation; and that such representations are functional entities that "emerge" from neural processes, much as sentences in a computer emerge from digital electrical processes in the circuits of its chips. Among philosophers of science, the syntacticists may say that, however ordinary percepts and theories may be modeled, scientific theories are sets of sentences; and they may add that because humans do in fact produce, process, and store sentences, the human brain must be capable of implementing the processes. The semanticists may make a similar point to defend their alternative view that theories are models: humans do construct and use models, and so these processes must be implementable by their brains. Some epistemologists may add that the (neural) dependence of percepts on the (connectionist) conditions of their production should not be equated with the dependence of percepts on the theories they support, a dependence that many deride as circular.

Alan Nelson—Are Economic Kinds Natural?

Nelson compares the General Equilibrium Theory (GET) of microeconomics with classical mechanics in an effort to explain why physical theories are more successful than theories in the social sciences. GET models an economic system as a collection of consumers (agents with preferences and commodity endowments) who transfer commodities (goods, labor, etc.) according to prices (relative rates of exchange) in such a way as to maximize utility. The goal of the theory is to determine the conditions of equilibrium, a state in which consumers maximizing utility in their exchange of commodities have no incentive to change their behavior. He judges GET to have relatively low explanatory and predictive power, even when applied to the capitalist economies that inspired it, clearly lower than classical mechanics. The explanation is not that it is theoretically less sophisticated, coherent, or mathematical than mechanics. Nor is it that economic concepts are less capable of theoretical refinement than concepts of mechanics— such as body, nor that they are (or are not) derived from concepts of folk science, i.e., common sense. Rather, it is that the fundamental concepts of economics do not (well) represent natural kinds, do not "divide nature at its joints," in contrast with those of mechanics. One indication is that some concepts of GET do not (fully) apply to societies without private property, and so are not universal. Nelson wonders whether the same weakness may not afflict all the social sciences. Another weakness he finds in economics may be specific to the science: namely, its tendency to treat utility as a pattern of economic behavior, rather than as part of a psychological mechanism causing individual behavior. To treat it thus deprives it of connections with noneconomic behavior that might be used to evaluate and improve theories about it. In an aside, he notes that a sophisticated philosophical instrumentalist may hold that scientists should act as if unobservables such as utility exist, and so may consistently recommend treating utility in a psychologically realistic manner.

Nelson's critique will disappoint economists, and may make psychologists uneasy. And so they may question whether universality is an appropriate criterion of a natural kind, even in physics, or whether the notion is so theory dependent that the failure of a theory to incorporate a natural kind shows only that it is not a natural kind (relative to that theory). Perhaps the intuition behind Nelson's critique has less to do with the naturalness of economic concepts than with their detachment from the behavior and psychological processes of individual human beings. Perhaps his feeling is that economics should be unified with psychology. That would seem as reasonable as recommending in the nineteenth century that phenomenological thermodynamics be unified with statistical mechanics, or as recommending in the twentieth that atomic theory be unified with quark theory. On the other hand, if the intuition is that economic phenomena are artificial because created by humans, and that artificial phenomena are incapable of scientific treatment, then it may threaten psychology as well as economics. For human perceptions and thoughts are created by humans, though perhaps less consciously than economic phenomena. Psychology too is arguably one of the "sciences of the artificial."

Lawrence Sklar — Foundational Physics and Empiricist Critique

Sklar acknowledges that foundational empiricism is problematic: for there seem to be no pure observation concepts or statements uncontaminated by theory, nor any a priori guaranteed-reliable rules for inferring objective statements from them. (He distinguishes the foundationalist view from two other types of empiricism: *realist naturalism,* which holds that empirical science tells us what the observables are and what rules of inference are reliable; and *pragmatism,* which adds that ordinary and scientific practice are the basis for epistemological norms.) Nonetheless, he contends, foundationalism is the basis of an empiricist critique that physicists conduct to analyze old theories and replace them with new ones. His first example is Einstein's special theory of relativity, which at one stage was in competition with the compensatory ether theory to explain the surprising result of Michelson and Morley that the velocity of light relative to the measuring apparatus is independent of the motion of the apparatus. The compensatory ether theory retained the intuitive view of space and time and hypothesized that clock rates change and measuring rods contract in the direction of their motion. In contrast, Einstein noted that the time interval between events in different locations is unobservable and therefore theoretically underdetermined by observation, and he proposed the counterintuitive hypothesis that the time interval is different in different reference frames. Einstein's theory employs the observation/theory distinction, defines distant simultaneity in terms of observables, and reduces the number of theoretical entities by abandoning absolute space — all traditional empiricist devices. Sklar's other examples are from the general theory of relativity, the causal theory of space-time, and quantum mechanics. In all these, competition between empirically equivalent theories tends to lead to a reductionist thinning of theory down to observational data, again a traditional empiricist device. His conclusion is that although foundational empiricism is problematic, there are stages in the process of theory development in the physical sciences where it is the best scientific methodology.

A number of philosophers of science now hold that the observation/theory distinction is theory relative, that is to say, what is observable is determined by the scientific theory in question. One wonders whether Sklar's results could be combined with this view to argue (though Sklar does not) that according to physics, observations of local simultaneity are uncontaminated by theory and hence are foundational in the classical sense! One possible objection would insist that the distinction between observable and nonobservable is established by the sciences in concert, not by physics alone. For physics is not the basic science (even if unifiable with others), and it has no special authority concerning observability. To consider one historical example, conflicting telescopic observations of stellar parallax were finally explained by differences in the speed of the observers' optic nerve transmissions, which implies that the basic observables are neurological events or sensations associated with these. Here neuropsychology corrects physics. It has been suggested that in a similar way neuropsychology corrects the assumption of relativity theory that local simultaneity is observable. If physicists conclude (as Russell did at one point) that the basically observable time interval is between sensations in the individual observer's private space-time, then they will have difficulty distinguishing physics from introspective psychology (as did Russell). The general point is that in the deepest foundational empiricism, observation is a perceptual psychological process and observability a psychological property; and so psychology should have even more authority on the question of what is observable than physics.

Henry Kyburg — Theories as Mere Conventions

Kyburg follows Quine (for some distance) in holding that whether statements in one's theoretical network are regarded as indispensable ("analytic") is a matter of convention: not arbitrary convention motivated by non-epistemic grounds such as convenience and simplicity, but nonarbitrary, "mere" convention motivated by epistemic grounds of predictive power. In Kyburg's epistemology, theoretical statements are meaning postulates assigned probability 1, and they are accepted or rejected together with the entire theory, or language, as he prefers to call it.

Acceptance consists not in determining how well confirmed the theory is on the basis of the observational evidence, but in selecting the theory in preference to competitors. A theory containing "All ravens are black" as a meaning postulate is preferable to one without the postulate if the quantity and distribution of acceptable observation statements (those whose probability exceeds some minimum value) are greater and better in the former than in the latter. For example, in a theory containing the postulate, the observation that joe is a raven and the observation that joe is not black cannot both be correct, and so the postulate will affect the quantity and distribution of such observation statements. Observation reports ("Otto observed that joe is a raven") are treated as incorrigible, but observation statements ("joe is a raven") are not. The observation vocabulary is not permanently fixed, but is specified by the degree of probability that an observation report bestows upon its associated observation statement.

Kyburg's epistemology can be viewed as an implementation of Quine's wholist view that "our statements about the external world face the tribunal of sense experience not individually but as a corporate body," which is normally but pehaps incorrectly taken to be anti-empiricist. Or it can be viewed as a way of refitting the ship of empiricism to enable it to withstand the winds of contemporary criticism that beached the classical model. Observation statements are acknowledged to be corrigible, the distinction between theoretical and observational statements is relativized to the theory (language), theories are not confirmed in the classical sense but selected from competitors, and they are selected as wholes on the basis of global criteria—features that seem to avoid the central objections to classical empiricism. Kyburg even makes bold to suggest that these features reveal the proper sense in which theories are "incommensurable." His view provides an objective, rational method of assessing theories on the basis of observation, while qualifying as an empiricist theory of knowledge under a general characterization of the view, and yet avoids the usual criticisms. However, if it is necessary to distinguish observation reports from reports of premonitions, intuitions, conjectures, etc., then the view may encounter a problem.

Wesley Salmon—Rationality and Objectivity in Science, *or* Tom Kuhn Meets Tom Bayes

Some philosophers have inferred from Kuhn's claim, that the choice between scientific theories (paradigms) "cannot be resolved by proof" and requires "persuasion," that the selection of theories is not a rational, objective process and that consequently genuine scientific knowledge is impossible. Salmon attempts to defeat the inference by combining elements of Kuhn's own view with a probabilistic system of theory evaluation based on Bayes's theorem. The simplest version of this theorem is $Pr(T/E\&B) = Pr(T/B)Pr(E/T\&B) \div Pr(E/B)$. Salmon shows how to avoid the difficulties in obtaining the probabilities on the right-hand side of the equation by formulating the theorem as an algorithm for computing the relative probability of rival theories evaluated by the same body of evidence. In the case where the theories do not entail the evidence and the likelihood, Pr(E/H&B), is less than 1, he recommends the addition of plausible auxiliary hypotheses to boost the likelihood to 1. Where the theories entail the evidence and the likelihood is 1, only the prior probabilities of the competing theories, Pr(T/B), are required. Salmon proposes that these can be obtained by employing some of the criteria mentioned by Kuhn: accuracy, consistency, scope, simplicity, and fruitfulness. Thus he hopes to reconcile Kuhn's approach to theory acceptance with a probabilist approach that appears to be objective and rational. In this connection it is important to note that Salmon subscribes to an *objectivist* (truth-frequentist) interpretation of probabilities and a *realist,* truth-based view of scientific theories.

One wonders whether Kuhn will welcome the suggested rapprochement. Undoubtedly some of his subjectivist followers will not. They will object that the algorithm merely reduces the subjectivity in theory choice or postpones it or both, since there is no objective, rational method for selecting criteria; and even if given criteria, there is no objective, rational method of weighting them or applying them. Still, as the logical empiricists might have said, reduction of disagreement may well be the essence of scientific rationality. It is important to note that Salmon's proposal does not in present form apply to the case where rival theories are based on different bodies of evidence, the case both Kuhn and Feyerabend use to argue that theories are incomparable or "incommensurable." Consequently, a pressing question is whether Salmon can extend his proposal to this case.

Ellery Eells — Bayesian Problems of Old Evidence

Eells defends Bayesian confirmation theory by clarifying and improving Garber's solution to the problem of "old evidence." Bayesian confirmation theory holds that evidence, E, confirms a theory, T, if and only if $Pr(T/E) > Pr(T)$, where the left hand term is computed by Bayes' theorem: $Pr(T/E) = Pr(T)Pr(E/T) \div Pr(E)$ (Note: the theorem is used here in its categorical form, rather than the conditional form employed by Salmon.) In the case where E has been learned prior to the time at which T is formulated (is old evidence) — so that $Pr(E) = 1 - E$ does not confirm T, since $Pr(T/E) = Pr(T)$. Garber has proposed the following solution. Where T was formulated to explain E (is "ad hoc," as some would put it), E indeed does not confirm T, in accordance both with Bayesian confirmation theory and our intuition. But in the case where T was not formulated to explain E, E in conjunction with the discovery that T explains E does confirm T. For until the moment of discovery that T explains E, $Pr(T \text{explains} E) < 1$, from which it follows on apparently reasonable assumptions that $Pr(T/T \text{explains} E \text{ \& } E) >$

$Pr(T)$. Because Garber takes the explanation relation to be a "logical" relation, like that of logical implication, he must explain how the agent can initially fail to know it obtains and thus be less than the logically omniscient agent presupposed by traditional Bayesian confirmation theory. His explanation in effect is that logical relations can be hidden by the coarseness of the language of the agent, in the way that propositional logic hides relations in predicate logic. Jeffrey takes a different tack, and attempts to characterize the explanation relation by means of conditions on the probability function of the explainer. Eells finds several difficulties in Garber's and Jeffrey's approaches. He argues that it is complexity, and not coarseness of the language, that is crucial to understanding logical omniscience, and that Jeffrey's conditions can be replaced with the main condition itself, i.e., $Pr(T/TexplainsE) > Pr(T)$, which can be regarded as a postulate.

It may seem that the Garber-Eells solution postpones the question at issue. For if the agent's personal probability function satisfies the main condition, then the agent believes that the probability of T is boosted by the fact that T explains E. In that case the agent must have already discovered the fact that T explains E; consequently, the fact is old knowledge, and by Bayes's theorem does not in itself boost the probability of T. On this perspective the efforts of Garber and Eells bring the problem of old evidence into even higher relief. For it appears that there are numerous cases in the history of science where evidence that E and knowledge that T explains E are both old for A, having been previously obtained and temporarily forgotten, and yet our strong intuition is that $TexplainsE$ & E confirms T for A. Eells could of course reply that at the moment of confirmation, the knowledge was rediscovered by A and hence was not old. In this vein, he distinguishes various problems of old evidence: what he calls the problems of "old new," "old old," and "new old" evidence. The resolution of this dispute seems to require further clarification of the notion of old knowledge and various concepts involved in its analysis, i.e., discovery (whether it involves possession as well as acquisition of belief), evidence (whether it is a psychological relation), and probability or degree of belief (whether values are assigned to inactive beliefs). Perhaps cognitive science has a contribution to make here. It may be noted that the problem of old evidence is one of those Salmon attempts to finesse in his paper by recommending a Bayesian selection—rather than confirmation—theory.

Colin Howson—Fitting Your Theory to the Facts: Probably Not Such a Bad Thing After All

Howson treats a problem for confirmation theory that may be called the problem of ad hoc theories (which is related to, though distinct from, both the problem of old evidence and the problem of underdetermination). Howson introduces the problem by criticizing various arguments for what he calls the *null-support thesis*: the thesis that ad hoc theories — theories contrived specifically to explain some evidence in hand (old evidence) — are not as strongly confirmed by the evidence as others. The thesis, if true, creates an apparent problem for deductivist views of confirmation, for example, the simple version on which E confirms T if and only if T logically entails E. Consider a theory T_1 that entails E_1 and other evidence E_2 but not evidence E_3. Now consider the distinct theory $T_2 = T_2' \& (T_2' \rightarrow (E_1 \& \sim E_2 \& E_3))$, for arbitrary T_2'. By increasing the number of bodies of evidence we can construct any number of theories, T_i, that entail E_1 yet disagree on other evidence. (This example is a generalization of the mathematical example Howson employs throughout his paper.) Suppose the available evidence is E_1. Informed scientific intuition will consider the second theory ad hoc and less highly confirmed than the first; but the deductivist view must hold that they are equally confirmed, since both entail the available evidence. Howson shows with several examples that science is full of respectable theories contrived to explain previously obtained evidence, which cannot properly be regarded as ad hoc. Nonetheless, theories like T_2 are properly so regarded, and these seem to present a serious problem. He maintains that a solution to the problem is available on the Bayesian view of confirmation, since the prior probability of a theory contributes to its confirmation as defined in terms of Bayes's theorem. The theorem reads: $\Pr(T/E) = \Pr(T)\Pr(E/T) \div \Pr(E)$. Now $\Pr(E/T)$ and $\Pr(E)$ are the same for theories T_1 and T_2; but $\Pr(T_2) < \Pr(T_1)$, because T_2 is ad hoc and T_1 is not; consequently $\Pr(T_2/E) < \Pr(T_1/E)$. Furthermore, Howson notes, because an ad hoc theory has an infinite number of competitors, none of which has any initial plausibility, the prior probability of each competitor is zero.

Howson's solution, in brief, is that because the ad hoc T_2 was formulated solely to explain E, we may reasonably assume initially that the probability of its other consequences being observed (its prior probability) is less than that of T_1. The larger question is whether the availability of his solution is a reason to adopt Bayesian confirmation theory. Proponents of the view that theories are best confirmed by evidence they best explain may not concede that it is. For they can argue, in the manner of defenders of the null-support thesis, that T_2 is less highly confirmed than T_1 because it does not explain E as well. Howson can of course reply that in such a case the theory's likelihood, $\Pr(E/T_2)$, is lower. Deductivist confirmation theorists also may refuse to withdraw from the competition. For they can argue that although any of an infinite set of theories is as highly confirmed as T_1 by evidence E, the next body of evidence obtained will decide in favor of just one member of the set, its members having been constructed to entail incompatible evidential statements. Where the infinite set is so constructed that its members entail the same evidential statements, the problem is not that of ad hoc theories, but the problem of underdetermination of theories.

Brian Skyrms — The Value of Knowledge

The ("Bayesian") statistical decision theory of Ramsey, L. J. Savage, and de Finetti recommends that we choose actions — (A_i) — on the basis of their expected *utility* (EU) — defined as the sum of products of their utilities (desirabilities) in various circumstances — $U(A_i \ \& \ C_j)$ — and the probability of the circumstances — $Pr(C_j)$. That is, EU (A_i) = $\Sigma_j \ U(A_i \ \& \ C_j) \ Pr(C_j)$. (Note that circumstances may include causal effects of actions.) The recommended rule is to choose the act with the greatest expected utility. Now suppose the agent is a Bayesian decision maker and changes her probabilities in the face of new evidence, E, according to the rule, $Pr_f(S) = Pr_i(S/E)$, where Pr_i is the initial (old) probability and Pr_f is the final (new) probability, the right-hand term being computed by Bayes's theorem. I. J. Good has proved that under these conditions, decisions made by the above rules that use cost-free information will have at least as much utility as those that do not use it. That is, in Bayesian decision theory knowledge has at least as much value as ignorance. In the case where the agent is a gambler trying to select the shell with the most money under it, Good's theorem harmonizes Bayesian decision theory with our intuitions: of course the gambler could only benefit from a quick look under the shells. Skyrms reviews results showing that the theorem can be extended to the case where the information obtained is uncertain, and (surprisingly) to the case where the agent does not know what sort of information will be obtained but only that its receipt will alter the probabilities. Both extensions require a principle of belief consistency or efficacy, $Pr_i(q/Pr_f = Pr^*) = Pr^*(q)$, which in words says that the agent believes that (currently assigns (initial) probability such that) the final probability is the result of accurate learning. Ulysses did not subscribe to this principle as he anticipated his encounter with the sirens, for he believed that they would produce in him false navigational beliefs (probabilities) that would lead his ship onto the rocks. Skyrms employs this principle as a condition for *deliberational* decision making, that is decision making in which an initial decision is altered by the receipt of new knowledge. Where the principle is not satisfied, deliberational decision making can lead to decisional impasses and paradoxes if the agent receives information about how rewards were arranged to influence his choices (Newcomb's paradox arises in such a case). According to Skyrms, *causal* decision theory — which includes the causal effect of the agent's action in the computation of expected utility — avoids these paradoxes, while *evidential* decision theory succumbs to them. He shows that Good's theorem holds in causal decision theory, but fails in evidential, producing thus another argument for the former.

At first sight decision theory seems to have little if any relevance to the issue of confirmation or scientific theory choice. However, it can be viewed as a general theory capable of handling both epistemic theory choice and more complex sorts of theory choice emphasized by critics of the classical approach.

Epistemic theory choice is the special case in which actions are believings of theories and circumstances are identical with actions; thus the expression for expected utility has just one term. Epistemic theory choice divides into two subcases: (i) that in which actions (believings) are assigned equal utilities on the ground that any other assignment begs the question of which theory is most likely true, (ii) that in which actions (believings) are assigned their prior or posterior probabilities, as utilities, which are not generally equal. These cases consider only epistemic factors. Non-epistemic factors would include the need for practical knowledge (by sick patients, a nation at war, etc.), the desire of scientists for fame, the cost of experimentation, etc. Most of these can be represented in the decision situation as circumstances (causal or otherwise). Other allegedly neglected aspects of science can be represented by taking actions to be the public, perhaps social, process of adopting theories. In all these cases Good's theorem that knowledge is at least as valuable as ignorance should hold, whether nonepistemic considerations are involved or not. It therefore deserves the close attention of those who think science cannot be shown to be a rational process.

Larry Laudan – Demystifying Underdetermination

Laudan attempts to neutralize the doctrine that the choice of scientific theories is underdetermined by rational methods, a doctrine used by many writers as a premise for the conclusion that theory choice is partly determined by social, psychological, and other non-epistemic factors, and is therefore relative, subjective, nonrational. The outline of his argument is that the true versions of the doctrine do not have undesirable consequences, and that versions that do have such consequences are false or unproved. He identifies two main versions. The *non-uniqueness* version holds that for every theory and every rational method of selection there are (perhaps infinitely many) rival theories between which and the original the method fails to decide. The *egalitarian* version holds that for every rational method of selection and every pair of rival theories the method fails to decide between them. (His alternative formulation – every theory is as well supported by the evidence as any of its rivals – may in fact describe an intermediate version.) Laudan concedes that some rules of theory selection are underdeterminate in the non-uniqueness sense; for example, the deductivist rule: choose the theory that entails the observed evidence. (One proof of this fact employs the procedure for constructing ad hoc theories described in the comment on Howson. Another proof is suggested by Goodman: given that "all emeralds are green" entails the observed evidence, so does "all emeralds are grue" where *grue* means "green and observed or blue and unobserved.") But, he argues, the concession does not lead to nonrationalism. In the first place, other methods may not be indeterminate; for instance, Bayesian confirmation theory, which can distinguish between rival theories by their prior probabilities (compare Howson). Second, even if every method of selection were underdeterminate in the non-uniqueness sense, subjectivism and nonrationalism would not follow. For even the deductivist rule chooses between the class of theories that entail the observed evidence and the class that does not, and so may choose between rival theories that are under consideration at a given stage of science. Laudan then proceeds to argue that subjectivism and nonrationalism follow only from egalitarian underdetermination, and that although this strong form has been embraced by Quine, Kuhn, Hesse, and members of the sociological school of epistemology, it has been not been established by anyone. Furthermore, he says, it is false, even with respect to epistemic methods of selection employed in the most successful sciences, as an examination of cases shows.

As the essays here illustrate, Kuhn's position is capable of many interpretations. Laudan ascribes to him the position that no objective, rational method determinately selects among theories. Salmon on the other hand concentrates on his reference to the prospect of "probabilistic algorithms" in science and uses some of the criteria he mentions for evaluating theories in such an algorithm. In any event, the fundamental issue is the doctrine that selection of theories is underdetermined by every *rational* method, and Laudan seems to have shown, without any analysis of the difficult notion of rationality, that no established version of the doctrine leads to the conclusion that theory selection is irrational. It is an important result, but it may seem a surprising one.

Thomas Kuhn—Dubbing and Redubbing: The Vulnerability of Rigid Designation

In his *Structure of Scientific Revolutions* (chapter 12) Kuhn advanced the thesis that two scientific theories separated by a scientific revolution are "incommensurable": in problems and standards, in terms and concepts, and in points of view. He used the thesis to buttress his claim that the revolutionary transition from one scientific theory, or "paradigm," to another is a "conversion experience" that "cannot be justified by proof," as well as his claim that science cannot be said to achieve the goal of getting closer to the truth. In the present essay he develops the second, semantic part of his incommensurability thesis and defends it from the challenge of a causal theory of meaning. Precisely stated, this part of the thesis is that a theory and its revolutionary successor are not translatable into one another. Its basis is that the meanings of terms are given by the accepted statements (laws, for instance) in which they appear and the examples through which they are introduced to learners of the language. The terms *force, mass,* and *weight* thus have different meanings in Newtonian and in relativistic mechanics, since they appear in different laws and are introduced to students in different ways; and they are not fully translatable in the terms of the other theory. Because Kuhn bases his thesis on the (Wittgensteinian) theory that the meaning of a term is determined by the contexts of its use, he sees a potential challenge to it in the opposed causal, referential theory of meaning championed by Putnam and Kripke. The causal theory holds that the meaning of a referring expression (name or noun) is the individual or substance that was originally dubbed with the expression, and consequently implies that the designation of the expression is rigid, and cannot undergo the postulated change of meaning through change of contexts of use. Kuhn's response is that designation in science is only temporarily rigid, and that the meanings of scientific expressions such as *force, mass,* and *weight* are changed in the process of scientific revolution by *redubbing* new instances with them. Thus his thesis of semantic incommensurability is preserved, together with its implication that scientific progress cannot be conceived as a cumulative process in which the successor theory describes the same objects as the former, only better or more truly. Strictly speaking, the theories do not talk about the same things, and the new theory gives its adherents access to possible worlds that cannot even be described by the displaced theory.

Although Kuhn maintains that theories separated by a scientific revolution are not intertranslatable, he acknowledges that it is possible for an exponent of the one theory to comprehend the other. Such a scientist is comparable to a speaker fluent in both French and English who finds that sentences of the one language do not have exact translations in the other. Clearly there could be cultures sufficiently unlike one another for their languages to be thus untranslatable, and perhaps some existing cultures are reasonable approximations to the case. So the question remaining is whether scientific theories separated by revolutions are in fact such a case. For those inclined to doubt it, Kuhn provides detailed treatments of the semantics of *force, mass,* and *weight,* and their (alleged) change of meaning under the development of the theory of relativity. As to the question of whether standards are incommensurable, Laudan and others have pointed out that it is possible to characterize the problems theories are trying to solve (e.g., "They are trying to predict the exact local time of the next solar eclipse") and compare their success in a manner that does not depend on the theories, even if the theories themselves cannot do it.

John Worrall—Scientific Revolutions and Scientific Rationality: The Case of the Elderly Holdout

The "elderly holdout" is the British physicist, Sir David Brewster, who refused to join the early nineteenth-century scientific revolution that replaced the Newtonian particle (emission) theory of light with the empirically more successful wave theory of light, and continued to believe that the older theory would ultimately be improved and come to enjoy even more empirical success than the then-current challenger. Worrall uses the case to evaluate Kuhn's claim that competent scientists who hold out for the old theory cannot be pronounced irrational, and also a more general claim attributed to Kuhn that theory choice is partly determined by subjective factors and consequently is not a completely rational process. Kuhn acknowledges that theory choice is partly determined by the criteria of accuracy, scope, consistency, simplicity, and fruitfulness. At the same time Worrall finds him arguing that different scientists often arrive at different evaluations of theories by emphasizing favored criteria and by applying the same criterion in different ways. Worrall replies that when Kuhn's criteria are analyzed — so that empirical adequacy is subdivided into predictive empirical success and empirical content, and both are distinguished from fruitfulness (capacity for modification that yields predictive empirical success) — it is seen that competent scientists in fact do not disagree on the empirical adequacy of theories but on their fruitfulness and other less important qualities. But Kuhn also argues, according to Worrall, that even where scientists agree on the empirical adequacy of theories, they may accept different theories without being pronounced irrational. Worrall rebuts this argument by proposing an objective Lakatosian principle of rationality on which such hold-outs as Brewster would be classified as irrational in the sense of "unscientific," though rational in the sense of "logical." The principle is: abandon research programs that have shown themselves to be degenerate in favor of progressive programs. He cautions that this principle has induction as its warrant, and is therefore subject to Humean scepticism about induction.

Worrall's critique raises important questions of interpretation. He takes Kuhn to hold that a satisfactory principle of rationality must specify objectively applicable criteria for evaluating the theoretical truth of theories as well as their empirical adequacy, and that neither requirement can be satisfied. Worrall's view — which is anti-realist (or "structural realist," as he prefers to call it) — is that the first requirement should be dropped and the second can be satisfied. However, Kuhn's position is often interpreted as anti-realist (instrumentalist), and on this interpretation his position may be close to Worrall's. If it were identical, Kuhn would hold that empirical adequacy is the ability of a theory to predict objective empirical facts; and there is much in his writing, even in the present essay, to suggest that he holds there are no such facts. Whatever the correct interpretation, Worrall has provided a challenge to those who argue on historical grounds that the growth of science is a nonrational process. At the same time, his historical example provides yet another countercase to the view of Boyd and others that the best explanation of empirical progress in science is that its theories are getting ever closer to the truth. Worrall notes that the sequence of theories developed in nineteenth century optics — particle emission theory, then wave theory, then photon quantum theory — exemplifies increasing empirical success but not convergence to a true theory. And with this interpretation Kuhn would probably agree.

Richard Boyd—Realism, Approximate Truth, and Philosophical Method

Scientific realism for Boyd is the view that empirically successful scientific theories, in both their theoretical and their observational components, characteristically refer to real entities and are approximately true in significant respects. He notes that this view need not claim that in their historical development scientific theories converge to some exact limit, but only that they successively approximate the truth. He contrasts the view both with empiricism—which holds that theories are true only in their observational parts, and with constructivism—which holds that the "real" world of theoretical entities is constructed by the knower. Over a period of years Boyd has developed a distinctive argument for a scientific realist view: that scientific theories are true is the best explanation both for their empirical success (their success in predicting observational data) and for the success of the scientific methods by means of which they were devised and adopted. In the present essay he defends the argument from four objections, the first three scientific, the last philosophical. In preparation he surveys important elements of the epistemological view that buttresses his argument for realism. For example, a realist epistemology employs a causal definition of reference, and maintains that terms denoting kinds are not conventionally defined, but instead are based on property clusters that naturally co-occur and are discovered and refined in the process of theory development. Causation is taken in a realistic, non-Humean sense. And respects of approximation are defined in terms of respects of similarity and difference between actual causal situations and certain possible ones. The first objection Boyd considers to his argument for realism is that history contains examples of false (i.e., ultimately rejected) but empirically successful theories. Boyd's reply is that these are cases of weak theories where neither they nor their methods were extendable, and that even so they were true in *some* relevant respects. To the counter that every theory is true in some at least contrived respect (the second and third objections), Boyd replies that the respect is not contrived if it is essential to a successful theory and that any respect that argues for realism is relevant. The final objection is that his argument for realism assumes that realism is true and thus begs the question. He replies that when realism is taken to be part of a package that includes an epistemology and a comprehensive scientific method, and is contrasted with its competitor packages, the argument is seen to be noncircular.

Boyd's argument for realism is the "miracle argument" criticized by Sober in his essay. Sober's objection is that when the appropriate restrictions are placed on inference to the best explanation, it is seen that the inference to realism is no better than an inference to empiricism. To counter this objection Boyd can point to particular respects, described here and in earlier papers, that make his the better inference. He might also point out that the notion of successive approximation to truth by a sequence of theories is theory dependent and must be understood by detailed analysis of the particular science in question. One of his examples in this connection is the convergence to classical mechanics by relativity theory in the limit of zero velocity. This example raises the question whether approximation to truth can be assimilated to approximation to a theory, and, if not, whether the notion is capable of analysis.

Elliott Sober — Contrastive Empiricism

Sober develops the view of his title by analyzing *empiricism* (which holds that knowledge cannot go beyond experience) and *realism* (which holds that it can) and combining elements of each into a new synthesis. He finds traditional empiricism too restrictive: it distinguishes between observation statements and theoretical statements and stipulates that only observation statements may legitimately be inferred from observed evidence (whether by deduction, induction, or abduction), thus disallowing inferences to nonobservables. He finds traditional realism too permissive: it permits abductive inference (inference to the best explanation) not only to decide between the theory and contrasting, competing theories, but also to select single theories in isolation, which is not generally permissible. He illustrates the fallacy in noncontrastive abductive inferences with the "miracle argument" for realism, endorsed by Putnam (at one stage) and Boyd. According to this argument the realist hypothesis that scientific theories about nonobservable entities are usually approximately true of those entities is the best explanation for the empirical predictive adequacy of the theory. Sober objects that when the realist hypothesis is contrasted with the anti-realist hypothesis that the theories say nothing true about nonobservables, and a reasonable (for example, Bayesian) analysis of explanation is employed, the realist hypothesis is seen to be no better supported than the anti-realist. On the other hand, abductive inference to the best explanation can be employed to choose between two *scientific* (as opposed to philosophical), theories about nonobservables, as long as the theories are not empirically equivalent. The restrictions he places on such inferences disallow the use of traditional simplicity and parsimony to distinguish the prior probabilities of competing theories in Bayesian confirmation. The result is that on his contrastive empiricist view of confirmation, knowledge can in a safely restricted manner "go beyond experience." How the restrictions have this effect is, however, puzzling; for he ultimately abandons the distinction between observation statements and theoretical statements in favor of a distinction between sensory stimulation and thought or language.

Sober's proposal contrasts sharply with Salmon's. According to Salmon, if two theories explain the available evidence equally well (i.e., their likelihoods in Bayes's theorem are the same), one may still be more highly confirmed than the other by having a higher prior probability in virtue of satisfying criteria such as scope, consistency, fruitfulness, etc. But according to Sober, theories obtain their virtues solely from the observational evidence, and empirically equivalent theories are equally confirmed; otherwise knowledge would go beyond experience. Sober's abandonment of the distinction between observation statements and theoretical statements seems to have the curious effect that there are no statements describing "experience" that other statements might "go beyond." However, he attributes his new distinction between sensory stimulation and thought/language to Quine, whom he could follow even further by relativizing observation to the speaker and defining the speaker's observation statements as those whose acceptance and rejection remain unchanged when sensory stimulation remains unchanged.

Arthur L. Caplan

Seek and Ye Might Find

I. Looking for Answers in Only a Few Places

There is an old joke about a man who has lost a key, which reveals much about the analysis of theories and their evolution in the philosophy of science. In the joke, a man is walking his dog down a dark street one night when he encounters his neighbor. The neighbor is on his knees under a street lamp, obviously hunting around for something on the ground. "What are you looking for?" the dogwalker asks. "I lost my house key and I am trying to find it" responds the neighbor. "Why are you looking there?" the inquisitive dogwalker asks. "Because," the neighbor responds, "that is where the light is."

I believe that philosophers of science have spent the better part of the twentieth century looking where the light is. An endless parade of philosophers of science have set out to understand the process of conceptual evolution in the sciences by looking only where the light is. They have taken as their data base almost exactly the same range of concepts, theories, and research paradigms that those in previous generations have sought to examine.

But, it might be objected, the claim that the scope of the philosophy of science has remained static cannot possibly be true. The philosophy of science has undergone a rapid expansion in the variety of scientific theories (Suppe 1977) subjected to philosophical analysis in the past few decades.

Philosophers of science once confined their analyses of conceptual evolution almost exclusively to the theories of the physical sciences. The writings of the postwar positivists devoted an extraordinary amount of attention to mechanics, thermodynamics and quantum theory (Feigl and Brodbeck 1953; Hempel 1952; Nagel 1961; Popper 1959). But today there are any number of philosophers of science interested in examining conceptual change in the biological sciences, the geological sciences, and in the various social sciences. The observational domain of the philosophy of science in the 1980s is far more extensive than was true even in the 1960s and 1970s.

Moreover, defenders of the catholicism of the present generation of philoso-

22

phers of science might add, the philosophy of science has discovered the history and sociology of science. Even though this discovery has spawned all manner of interdisciplinary wrangling as to which students of science have the right set of tools in hand for understanding conceptual change, the fact is that the debate has been joined (Engelhardt and Caplan 1987; Giere 1988).

The debate has forced philosophers of science to ruminate upon a much broader range of theories within the sciences than they might if left to their own devices. Since sociologists and historians have been less prone to develop obsessions with the physical sciences, the philosophy of science, if only to preserve its intellectual autonomy, has had to fish in all manner of new scientific waters.

Moreover, there has been an expansion of the units of analysis where conceptual evolution is concerned. Where theories were once seen as the sole units of conceptual evolution, it is now widely understood that science organizes its inquiries around concepts, problems, paradigms, themata, research strategies, controversies, and exemplars (Englehardt and Caplan 1987)

Those excited by the expansion of the exemplary domain of the philosophy of science are correct. There has been a great deal of growth, for a variety of reasons, in the scientific subfields and specialities that contemporary philosophers of science are willing to examine.

But, the boundaries of philosophical reflection about science were once so limited that almost any broadening of disciplinary vision appears to be significant. If one looks more closely at the specific examples that constitute the explanatory domain of the philosophy of science today, some of the catholicity is more apparent than real.

Consider the set of prominent examples in the area of the philosophy of science that has grown more quickly than any other during the past twenty years—the philosophy of biology. Even a cursory examination of the biological theories, research strategies, and paradigms that have been discussed reveals that there is a sampling strategy in use by those in this subfield that has resulted in a very peculiar picture of the biological sciences.

If one reviews the periodical literature in the philosophy of biology by browsing through the tables of contents of those books that form the basic reference points for the field (Beckner 1968; Brandon and Burian 1984; Hull 1974; Mayr 1982; Munson 1971; Rosenberg 1985; Ruse 1973; Sober 1984), scanning dissertation topics written about the biological sciences by philosophers during the past two decades, and looking at the contributions to the major journals in the philosophy of science, it quickly becomes obvious that the philosophy of biology, with few exceptions, is in actuality either the philosophy of evolutionary biology or the philosophy of population genetics.

Darwinism and the subsequent disputes that have followed in its wake constitute one primary focus of philosophical attention. Most of the rest of the philosophy of biology's attention span is riveted on understanding the nature of the relationship that exists between three theories: Mendelian genetics, population genetics, and molecular genetics.

Admittedly evolution and genetics are significant areas of inquiry within biology. But there is more going on in biology than disputes over the plausibility of a punctuated equilibrium approach in explaining macroevolution or whether a molecular explanation can be given for either Mendel's laws or the Hardy-Weinberg law. Almost nothing has been written by philosophers of biology about ideas, theories, research programs, exemplars, and the like in such areas of the biological sciences as biochemistry, botany, developmental biology, ecology, physiology, immunology, behavioral biology, paleontology, or biogeography.

The data sample of the philosophy of biology is surprisingly sparse. But matters are even worse than this. For the angle of the street lamp that has been determining where philosophers of biology should look in order to understand conceptual evolution is very definitely skewed.

The choice of examples in the literature of the field is drawn almost exclusively from the most abstract concerns of biology. To read the examples of theories cited by philosophers of biology, one would come away with the very distinct impression that most biologists spend most of their work days driven by conceptual questions of a very general nature. In the world of biology, as presented in the periodical literature of the philosophy of biology, most biologists spend no time with living organisms, rarely venture outside the boundaries of their office or laboratory, and are driven only by a desire to concoct a single overarching theory that can answer fundamental but very abstract questions.

This is a distorted view of the state of affairs in the biological sciences. The overwhelming majority of biologists spend their time wrestling with theories, concepts, research programs, and problems in the service of practical and applied problems.

The majority of biological scientists work in schools of forestry, horticulture, veterinary science, agriculture, pharmacy, or medicine. The majority of those who think of themselves as biologists and who recognize others as doing work in biology spend their days thinking about problems, constructing theories, and refining models in conceptual terrain where nary a philosopher of biology has deigned to tread.

The avoidance of the applied in science is not a prejudice that is confined to philosophers of biology. Most other specialty fields of the philosophy of science do not pay attention to the practical or applied domains of science.

Most areas of engineering, chemistry, psychology, dentistry, nursing, nutrition, geography, architecture, pharmacy, oceanography, soil science, and metallurgy, to name some of the most obvious, remain unremarked upon by philosophers of science. The scope of the philosophy of science, while expanding, is still focused on only a tiny fraction of the domain of science in which most of those who call themselves scientists, who are funded to do science, and whom society recognizes as scientists work.

The lack of attention to applied areas of science by philosophers of science is, to put it bluntly, systematic and thorough. Why this is so is a topic worthy of inquiry in its own right. That it is so is obvious.

This paper has a grandiose goal—to try and highlight the fact that the domain of examples governing philosophical reflection about science is highly determinative of how the conceptual evolution and change is understood and explained. The importance attributed to paradigms or exemplars in theory construction (Kuhn 1970, Kitcher 1983a, 1983b; Schaffner 1986), the pace at which theories are thought to evolve in science (Laudan 1977), the extent to which theories can be axiomatized, and the deductive or nondeductive nature of relationships between theoretical statements are closely tied to the examples of scientific theories that are selected for analysis. Similarly, the adequacy of evolutionary models of theory change, attempts to ascertain the logic of discovery (Nickles 1980), Nagelian models of theoretical reduction (Caplan 1981; Schaffner 1967; Robinson 1986) or of various views of the role of crucial experiments in science (Feyerabend 1981), all pivot on the adequacy of the data base that is used by those doing the philosophizing.

And those doing the philosophizing have shown no interest in the practical, pragmatic side of science. I believe this orientation has come at great cost in terms of a valid understanding of conceptual change.

The evidence cited in support of this claim will itself rest on a rather flimsy evidential base. I shall discuss a single example of inquiry in medicine, one drawn from the field of nephrology. The example, the development of a treatment for chronic renal failure, reveals an interesting pattern of theoretical and practical efforts to solve a pressing problem in one tiny subdomain of applied science.

The example may or may not be typical of how ideas evolve in medicine. It may or may not be typical of how ideas evolve in applied domains of science. And it may or may not be illustrative of how theory and practice coevolve to produce both stasis and change in the evolution of ideas.

But the example is certainly a valid instance of how ideas in one portion of science have evolved. And, as such, it reveals how narrow the present scope of philosophical reflection about science is and what sorts of insights might await those willing to broaden their perspective by broadening the domain of their analytical efforts.

Conceptual evolution may not be uniform in all areas of science. I remain agnostic as to whether the dream of a theory of "unified science" (Feigl and Brodbeck 1953) can be discovered. But no one will ever know unless a complete and comprehensive range of theories, concepts, research programs, paradigms, and problem-solving strategies from all domains of science is utilized in constructing the subject matter of the philosophy of science.

The philosophy of science in general, and those interested in the problem of conceptual change in the biomedical sciences in particular, are not well served by confining their attention to those places where "the street lamp is shining." Unless an argument is mounted, and to date none has been advanced, that certain areas of science are more representative, typical, paradigmatic, or illustrative of science than are others, then philosophers will have no excuse for continuing to generate analyses of conceptual change that ignore most of what is funded as, published as, and rewarded as science. The only plausible research strategy for arriving at an understanding of the dynamics of scientific change is to look at science in its entirety, not simply at those areas of science that have departments with the closest locations, physically, historically and spiritually, to philosophy departments.

II. Kidneys and Diseases of the Kidneys

The kidney is an organ that sometimes makes a cameo appearance in philosophical discussions of the life sciences. Sometimes it appears in discussions of teleological explanation, as a refutation of the claim that teleological claims can be translated without remainder into claims about causally necessary conditions for achieving a goal. Occassionally it winds up lumped with hearts, as illustrative of the kind of products natural selection has produced in various species.

But from the point of view of most scientists in biology and medicine, the kidney represents much much more than a readily comprehensible example in the ongoing dispute about the irreducibility of teleology in biological explanations. In particular, those in medicine and veterinary science have a keen interest in the function of kidneys, since when they do not function the consequences for the organism possessing them are fatal.

Death has a way of riveting the attention of those interested in physiology. Moreover, preventing death is a goal that fuels a great deal of inquiry in the biomedical sciences. The ability to restore the function of failing kidneys, or to provide a substitute for kidneys whose function has been irreversibly lost, has been a major preoccupation of biomedicine in the years since the end of the Second World War.

The kidneys' primary function is to remove waste materials from the blood. The normal metabolism of fat and carbohydrates in the human body produces on average about 70 mEq/kg each day of nonvolatile acid. An additional 13,000 mEq of carbonic acid is produced as well. If these acids are not removed, coma, shock, and heart failure will result (Fishman, et. al. 1981).

The kidney also serves important regulatory functions in terms of the maintenance of total body fluid and electrolytes. The body's pH levels are also maintained by the kidney. Since the ability of mitochondria in each cell to generate ATP through oxidative phosphorylation is a function of the pH gradient present across the mitochondrial membrane, renal function is an absolute necessity for life.

The primary business of the kidney is performed in units called nephrons. Each kidney contains about one million. A nephron consists of two major parts.

The glomerulus is a ball of thin-walled capillaries. Blood enters the glomerulus from the renal artery. The small size of the vessels composing the glomerulus raises the pressure of flow, resulting in the filtration of fluid and solutes through the thin walls of the capillaries. This filtrate then travels through a long looping tube, the Loop of Henle, to a third element known as a tubule. Here various solutes are reabsorbed back into the blood stream. The remaining fluids and heavier solutes, now constituting urine, drain from the tubules into the bladder.

The nephrons of the kidney constitute an enormous countercurrent system. The differential permeability of the glomerulus, the Loop of Henle, and the tubule creates a cycle in which sodium ions are first actively pumped out of the blood and then, as water diffuses out, return and are excreted as concentrated salts in the form we know as urine.

A sudden loss of kidney function is termed acute renal failure. The most common cause of this condition is traumatic injury to the kidneys. But kidney failure can also result from the consumption of various toxic agents such as heavy metals, aminoglycoside antibiotics, or alchohol.

Acute renal failure can sometimes be reversed. For example, it may be possible to reverse the effects of a traumatic injury through surgical intervention. When this is not possible, a new diagnosis is applied—chronic renal failure.

The most common cause of chronic renal failure is not sudden injury or insult to the kidney. There are a number of progressive degenerative diseases that strike the kidneys and cause them to slowly fail over a period of months or years. These diseases can be classified into two forms: glomerular and interstitial.

In glomerular renal failure, the glomeruli of the kidney cannot admit blood into the kidney. This can result from hypertension, which destroys the ability of the capillaries to generate sufficient pressure to filter the blood. Or it can result from streptococcal infections, autoimmune diseases, or myelomas.

In interstitial diseases, the tubules of the kidney become inflamed and suffer fibrosis. The tubules are then incapable of permitting the exchange of salts and water between the nephron and the bloodstream. Diabetes, sickle-cell anemia, and the long-term ingestion of analgesics can cause this form of chronic renal failure.

For the first half of the twentieth century, medicine had no treatments to offer those who suffered acute or chronic renal failure. However, physiologists realized as early as the 1910s that it might be possible to mimic the function of the kidneys by creating a machine that could pump blood through a tube composed of a semipermeable membrane. If a salt solution of the proper concentration was placed on the other side of a membrane of the right thickness and composition, waste materials should diffuse across the membrane, duplicating, to some extent, the processes by which urine is formed and removed from the body. The first "artificial" kidney was built in 1914. But the membrane used was so fragile that it could not stand up to the volumes and pressures necessary for use in human beings.

III. The Early Evolution of Hemodialysis — Solving the Problem of Circulatory Access

The first machines capable of substituting for the function of the kidneys were invented during World War II. The key breakthrough was the invention of cellophane, which was both permeable enough and sturdy enough to stand up to prolonged blood flows.

Willem Kolff built a somewhat primitive artificial kidney and performed the first "hemodialysis" in a human being in 1943. In 1956 Kolff and Watschinger introduced a new model of artificial kidney, the twin-coil dialyzer, a design still evident in many of the machines used in dialysis during the next 25 years (Czaczkes and Kaplan De-Nour 1978).

By the late 1950s several American and European medical centers had artificial kidney machines that were capable of "resting" the kidneys of persons afflicted with acute renal failure. The clinical approach was to temporarily substitute an artificial kidney for the natural kidneys in the hope that the damaged kidneys might regain their functional capacities.

The use of these machines required the insertion of tubes or cannulas into a patient's artery and vein in order to bring blood from the patient to the machine and return it to the patient. However, the amount of blood involved required the use of needles with such large bores that each treatment required the use of a new artery and vein. Physicians quickly ran out of accessible sites to a patient's circulatory system.

Physicians in the early fifties had a treatment that was efficacious for many forms of acute renal failure. But the problems associated with gaining access to the circulatory system made it impossible to use the first generation of artificial kidney machines for those with chronic renal failure.

The problem of how to gain continuous access to the circulatory system without destroying it dominated the theoretical and clinical efforts of nephrologists. In 1959 Scribner and Quinton of the University of Washington discovered a solution.

Scribner and Quinton realized (Fox and Swazey 1978) that by implanting a permanent tube, or shunt, between an artery and a vein, blood circulation in the vessels could be maintained. Yet a surgically implanted tube would permit repeated access to the circulatory system without damaging blood vessels.

The original version of the Scribner-Quinton shunt was a T-shaped plastic tube. One end of each bar of the shunt was sewn directly onto an artery and a vein, usually in the patient's arm. When access to the circulatory system was necessary, the remaining portion of the T could be removed and the ends fed to tubes leading directly to an artificial kidney machine. By permanently short-circuiting the circulation in an artery and a vein, a permanent access site to the circulatory system could be maintained.

There was a key breakthrough in materials science that facilitated the invention of the Scribner-Quinton shunt. A new inert material, Teflon, had recently become available. Though not promoted by the manufacturer for its medical applications, Scribner realized that the smooth surface of the new substance made it unlikely that blood cells would be damaged in passing over a Teflon surface. The fact that it was inert made it a plausible material to attempt to use in the human body, since Teflon would not trigger an immunological reaction.

The Scribner-Quinton shunt, and the subsequent versions constructed of Silastic, a more flexible inert material, made it possible to treat those with chronic renal failure. With a mechanism available for maintaining access to the blood that did not destroy access sites in the process, existing artificial kidneys could be used for persons suffering from all forms of chronic renal failure.

Scribner and Quinton had made an important breakthrough with respect to renal disease. But, the discovery immediately created two new problems.

IV. Selling Is the Daughter of Invention

A problem facing those who wanted to use the shunt to try and treat patients in chronic renal failure was that no one knew what the effects would be of long-term exposure to an artificial kidney machine. No one knew how often or how long patients ought to be hemodialyzed on the artificial kidney. Nephrologists did not know whether the machines would continue to be effective over time.

Nor did they know what the side effects would be of long-term treatment using an artificial kidney. Would unfiltered impurities enter the bloodstream? Would the machine remove too many vital salts or cause other problems in patients with renal failure?

Another, more practical, problem confronted the inventors of the shunt. Now that they had concocted a solution to the problem of allowing constant access to the circulatory system, they needed to inform other doctors of their discovery.

Certainly publication was a critical element in communicating the new breakthrough and Scribner and Quinton published a number of papers on the subject of the shunt (Scribner, et. al. 1960). But publication was not enough. Physicians and especially surgeons needed to see and handle the new shunt in order to understand how it could be implanted and what opportunities it might create. Having invented a shunt, the inventors had to let others see and manipulate the invention in order to demonstrate its practicality and utility.

Scribner and Quinton spent roughly three years, from 1959 to 1962, disseminating their ideas concerning the opportunities offered by a permanently implanted shunt to other nephrologists. Scribner made presentations at many medical meetings. He brought samples of his shunt with him to show to his peers sometimes during coffee breaks, or in his hotel room.

Many physicians, particularly younger doctors interested in renal failure, showed an interest in the shunt. A number of physicians went to Seattle to do residencies with Scribner in order to learn how to implant the shunt (Rettig 1976).

V. Moving from Skepticism to Acceptance

The sales efforts and training programs of Scribner and his colleagues paid dividends. In 1963 the Veterans Administration (VA) hospital system decided to create artificial kidney units, or as they are now called, dialysis centers, in thirty hospitals around the United States. The VA system had many patients dying from chronic renal failure as a result of hypertension, diabetes, and injuries and was desperate to locate a technique that might help these patients.

VA patients were well suited to dialysis treatment, since many were permanently institutionalized with serious diseases or injuries that made it unlikely they would ever be discharged from the hospital. There were many patients readily available in a relatively small number of locations. This made it possible to bring the artificial kidney to the patients, rather than the patients to the artifical kidney. This was an especially important consideration for a treatment that physicians were beginning to realize required at least three sessions a week of four to five hours each to have any hope of efficacy.

The interest of the VA was crucial as a vehicle for disseminating Scribner's invention and the concept of using dialysis to treat chronic renal failure throughout the American hospital system. A large number of physicians receive their advanced training within the VA system and, thus, were exposed to Scribner's shunt and the concept of chronic hemodialysis early on in their careers.

By 1968 about 1,000 patients were receiving long-term renal dialysis for chronic renal failure in the United States (Rettig 1976; Russell 1979). A small number of physicians had acquired some experience with the treatment. And a small number of facilities capable of carrying out the procedure had been created.

It was about this time that the first of what eventually became a torrent of articles appeared discussing the ethical consequences of a shortage of dialysis machines available for those with chronic renal failure (Abram and Wadlington 1968; Beecher 1969; Rescher 1969). There were far fewer centers and doctors available to do the procedure then there were patients dying from chronic renal failure.

About 5,000 patients were dying each year in the United States from chronic renal failure. The solution of the problem of how to gain permanent access to the circulatory system created a new problem of how to determine access to the small number of centers capable of performing the procedure.

Matters were made more complex by the fact that the answer to the question of whether long-term dialysis for chronic renal failure was safe and efficacious were slow in coming. Scribner initially obtained good results using his shunt on a fairly broad cross-section of patients with chronic renal failure. His initial success was so impressive that it became hard for him and other physicians to obtain funds for research on chronic hemodialysis since it looked as if chronic hemodialysis, had sprung into existence as a full-fledged medical cure.

But as the number of doctors and centers using the technique began to grow, physicians began to have trouble duplicating Scribner's low morbidity and mortality rates. Some thought the results of chronic hemodialysis were so poor that the procedure ought be abandoned. Others felt that chronic hemodialysis was still very experimental and ought only be allowed to spread slowly into the general hospital system.

By its very nature the effects of chronic hemodialysis on the human body were difficult to study. Long periods of time were required both to determine efficacy and to ascertain whether any adverse side effects might be associated with long-term exposure to hemodialysis.

After an initial period of success, many doctors became skeptical of the promise of chronic hemodialysis. But after a few more years, roughly by the mid-1960s, other centers began to achieve better success in their morbidity and mortality rates. However some hospitals continued to have patients who suffered many severe side effects or even died while on chronic hemodialysis. This mixed bag of results led to a dispute within nephrology as to what the proper course of action should be with respect to the expansion of dialysis services.

Matters were made even more contentious as a result of the emergence during the early 1960s of a new form of treatment for chronic renal failure – transplantation. Renal transplants from either living or cadaver sources were seen by some as less costly and more enhancing of a patient's quality of life than thrice-weekly dialysis treatments. Chronic hemodialysis was competing for the same therapeutic niche as kidney transplants, and each evolving treatment had its advocates and detractors.

The dispute over which technique was more efficacious, involving as it did the VA hospital system, caught the attention of federal officials. They needed to know if the VA was wasting its money on a losing form of treatment. The VA was also under increasing pressure to expand services, both within the VA system and in other federally funded health care programs to minimize the need to ration access to dialysis. The federal government created a commission headed by Dr. C. W. Gottschalk to examine the controversy over chronic hemodialysis.

The commission issued a report in 1967. The report gave a strong endorsement to both chronic dialysis and kidney transplantation. The report noted that,

. . . transplantation and dialysis techniques are sufficiently perfected at present to warrant launching a national treatment program and urges this course of action. (Report of the Committee on Chronic Kidney Disease, 1967)

The committee was composed primarily of people who made their livings performing either dialysis or transplantation. So in one sense it is not surprising that the commission took the Solomonic course of weighing the available evidence and then deciding to bless both approaches to renal failure as therapies.

V. What Can the Philosophy of Science Learn from an Examination of Early Efforts to Treat Chronic Renal Failure?

The case of the evolution of a clinical treatment for chronic renal failure poses a number of challenges to existing models of conceptual change in the philosophy of science. Philosophers rarely acknowledge that, occasionally, disputes in science are resolved by procedural means. In the case of hemodialysis, a committee was formed, a study undertaken, a vote taken, and a conclusive finding was issued.

Using a committee to resolve a dispute is a long way from designing a crucial experiment to test conflicting hypotheses. But committees and other bureaucratic structures have important roles to play in determining the direction and course of inquiry in many areas of science (Englehardt and Caplan 1987)

Less dramatic but nonetheless of interest is the role played by practical success in determining the pace and course of scientific work. Deaths had a discouraging effect on the willingness of scientists to believe that long-term hemodialysis using a Scribner-Quinton shunt was a valid therapeutic approach. Theoretical questions about the degree to which a combination of a plastic shunt and plastic membrane could duplicate the function of the nephron of the kidney were almost irrelevant in understanding the evolution of clinical strategies for treating renal failure. What mattered was how many patients were alive and healthy and how many were not, after prolonged exposure to the artificial kidney.

The assessment of the ideas and hypotheses advanced in the pursuit of a treatment for chronic renal failure was complicated by the presence of another therapeutic option. The fact that two approaches to treatment were in competition had a very real impact on how the success rates of dialysis and transplantation were evaluated. As we shall shortly see, the competition between plausible research strategies for solving the problem of how to treat renal failure had a distinct impact on the composition of the subject pool that was afforded access to chronic hemodialysis and the interpretation of evidence concerning efficacy.

Of great import is the pattern of development manifest in the early history of chronic hemodialysis. Discovery in the biomedical realm is often only the first in a long sequence of events, not the final stage of a lengthy inquiry. Those who discover new drugs, devices, or materials must convince their peers that they have found an answer to a problem. But the proof requires more than publication. It requires active promotion, an active interaction between doctor and invention, good results initially, good results in the long run, and results that are better than other available options.

Technological solutions, at least in the case of chronic hemodialysis, follow a course of evolution that moves from the recognition of a problem to the formulation of a solution to selling others on the merits of the purported solution. A discovery is almost always followed by a long period of what might best be termed "advertising" during which agnostics, skeptics, and devotees of alternative approaches must all be persuaded to accept the merits of the purported breakthrough (Banta 1984).

This phase is then followed by a stage that might be termed "acceptance," during which evidence of efficacy must be accumulated and evaluated. Sometimes formal certification of the sort that can best be provided by a blue-ribbon panel of experts is the only mechanism by which disputes concerning the acceptability of a new treatment can be resolved.

Philosophers of science have not given sufficient emphasis to the identification of the developmental phases that lead from discovery to acceptance. Discovery, even by recognized authorities in a given field, does not always lead to acceptance. Analyses of theoretical evolution must remain alert to the sequence of events following on the heels of a discovery, which may or may not result in acceptance.

VI. From Acceptance to Mastery

In 1968 about 1,000 Americans were receiving renal dialysis treatment. By 1978 more than 35,000 were having their lives extended by dialysis. And by 1988, more than 98,000 patients were receiving dialysis treatment (HCFA 1988).

These numbers raise a number of interesting and important questions relevant to understanding the evolution of theories in medicine. Why has the number of patients receiving dialysis continued to grow twenty years after the medical profession decreed that dialysis was in fact a legitimate form of medical therapy? What kinds of factors played a role in decisions made by physicians to control the numbers of persons with fatal forms of renal failure who had access to dialysis treatment?

Some of the variables influencing the rate of growth manifest by dialysis throughout the general population are easy to identify. The number of centers, kidney machines, and trained personnel available to deliver dialysis was much smaller in 1968 than was the case in 1978 or 1988. The cost of performing thrice-weekly regimens of dialysis, ranging from $20,000 to $30,000 per year, were large enough to discourage some patients dying of renal failure from seeking care.

The growth in the number of patients receiving dialysis also reflects the fact that some persons who started dialysis in 1968 were still receiving care in 1988. The growth in the number of patients is accumulative, since the overall patient pool reflects both new patients added each year plus ongoing cases of renal failure from previous years.

But it is important to realize that the composition of the patient population receiving dialysis changed drastically from 1968 to 1988. In the late 1960s the majority of dialysis recipients were young, 25 to 45, middle or upper class, married males with no other significant illnesses. By 1988 the majority of patients being dialyzed were 45 years of age or older. Dialysis had been attempted on newborns as well as on those over 90. Current recipients included men and women from all walks of life. Many had signficant complicating illnesses such as diabetes, arthritis, alchoholism, depression, and even cancer.

As one physician observed in 1968,

> We had what was in many ways an idealized population. A large fraction of the patients were living in a productive period of their lives. They were young and had little else wrong with them. (Kolata 1980)

By 1988 the patient pool barely resembled that of this early period.

The most common reasons cited in the literature of medicine and the history of medicine for the change in the composition of the pool of patients receiving hemodialysis are money and bias. Some commentators argue that physicians were simply biased against those from backgrounds different then their own and when faced with a shortage of resources acted to penalize those they disliked (Fox and Swazey 1978; Plough 1986).

Others argue that it was the decision by the federal government to pay for the costs of dialysis therapy for all patients in renal failure in 1972 that opened the floodgates to this treatment modality. When the federal government created a special program, the End Stage Renal Dialysis Program, to cover nearly all of the costs of dialysis, money was no longer an obstacle to access.

Explanations such as these are likely to delight sociologists and historians who are sensitive to the nonepistemic, or external factors that drive the evolution of science (Englehardt and Caplan 1987). But, while doctors did show bias, especially with respect to race and sex, in their selection of potential recipients, and while money played an important role in enabling more patients to receive care then would have been possible without governmental assistance, there is more going on in the expansion of the dialysis patient pool than is captured by explanations of change confined to only prejudice, money, or both.

The easiest demonstration that more than money was involved in the sudden expansion of the pool of dialysis patients comes from the British experience with renal dialysis. The United Kingdom had a national health insurance system in place that dated back to the end of the Second World War. Money was not an obstacle for offering dialysis to those with renal failure.

Yet, British nephrologists were as selective as their American counterparts in controlling the access of patients with renal failure to dialysis. During the 1960s four out of five potential candidates for dialysis were rejected. Patients were usually young, male, married, working, and homeowners. Yet by the 1980s the pool of patients receiving dialysis care had demonstrated enormous growth.

The explanation of the selective policies regarding access to renal dialysis that prevailed in the United States and England well into the 1970s lies in what are sometimes termed internal or epistemic considerations (Englehardt and Caplan 1987). Victor Parsons, a British nephrologist, puts his finger on the methodological considerations that influenced the selection of patients for dialysis in both countries:

> Individual desire to go on living or the desire to be treated was not considered. . . . very often the patients were unaware they were up for selection. This enabled the renal units to achieve high survival rates, and quite rightly since in the early stages it was important that the treatment should be seen in its best possible light. To have adopted a totally non-selective policy at the outset would have led the technique into disrepute as being nothing but a technical exercise in the prolongation of a very poor quality of life. (Parsons 1978)

Those involved with dialysis had to demonstrate both its safety and its efficacy. Moreover, they had to demonstrate its efficacy relative to an alternative therapy—renal transplants. The way to accomplish these goals was to utilize patients who would permit physicians to observe adverse side effects. It was also to use patients who were likely to respond well to the treatment. Those who were relatively well-off, who had no other disabling diseases, who were young, and who had supportive families were the most likely to fulfill the need to demonstrate efficacy, safety, and technical superiority over alternative approaches.

It is important to realize that selectivity in the pool of patients receiving dialysis remained a fact of life in both the United States and Britain long after the Gottschalk commission had declared renal dialysis to be a therapy and long after the American government had created a fund to pay for the costs of dialysis care. By the late 1960s doctors had learned to perform chronic renal dialysis with success. But part of the reason behind their success was the care with which patients were selected for treatment.

During the next twenty years nephrologists learned to utilize dialysis on a broad range of patients. They mastered the technique by treating the very young and the very old, those with many serious medical complications, and those whose psychosocial environment was less than optimal.

The process by which health care professionals learn to master a therapy should be of great interest to those interested in theory change in the sciences. Long after a new innovation is recognized as therapeutic, sustained efforts are directed toward learning the proper management of the therapy across the entire spectrum of potential patients. Much the same process is currently underway with respect to cardiac transplants, liver transplants, balloon angioplasty, and lithotripsy. Competition and money, as well as prestige and concerns for safety and efficacy, all play roles in determining the boundaries of the patient population for which doctors believe a therapy is indicated.

VII. Forces Driving the Evolution of Experiments into Therapies

One of the most interesting questions to emerge from a consideration of the recent history of chronic hemodialysis is: Why did dialysis move so quickly from the status of experimentation to that of therapy? After all, despite the fact that dialysis was declared a therapy in 1967, it took another twenty years for physicians to feel comfortable enough with the technique to offer it to all persons suffering from renal failure.

The desire of biomedical scientists who do human experimentation to quickly shed the label of experimentation has interesting analogues in other areas of medical innovation. Those involved in attempting the first artificial heart implant and who conducted the first xenograft of a heart to a young infant were quick to declare their efforts therapeutic and nonexperimental (Caplan 1985a, 1985b). The therapeutic status of these interventions was seen as established by the demonstration of the mere feasibility of the undertaking.

There are many reasons why those involved in the development of new medical interventions such as dialysis wish to dispose of the label "experimental" as soon as they can. Experimentation carries with it connotations of the unknown, the risky, and the especially dangerous. Talk of experimentation can make it difficult to recruit willing subjects. Such connotations are unfortunate since there are many experiments that are relatively safe and risk free, and many therapies that are precisely the opposite.

Another reason for the rapid transmutation of experiments into therapies is hope. Those who treat the sick and those who are sick or disabled naturally want to have a cure. Talk of "therapy" is far more conducive to optimism than is talk of "experimentation" or "research." Researchers involved in the development of new drugs to treat terminal cancers or to help those suffering from AIDS find it much easier to offer comfort to the dying using the language of therapy rather than research.

Another factor influencing the speed with which experimentation becomes therapy is the assignment of credit for discoveries. Few scientists or physicians get credit or public acclaim for being the first to conduct experiments. Credit goes to those who find cures, who discover therapies.

If the first use of an artificial kidney or heart is experimental, or, if the first baby born as a result of IVF constitutes a successful outcome of research, then others may try to claim credit for the first truly "successful" application of a new technology or technique. And as the long evolution of hemodialysis shows, there are many steps that must be made to move from discovery to mastery. The race in biomedicine for fame, fortune and celebrity goes to those who find cures. Priority is no small matter in a scientific world of fierce competition for grants, fame, and recognition.

There is another important force driving new techniques, drugs, and devices down the continuum from research to therapy—money. Third-party payers, whether public or private, do not want to pay for experimentation. Those giving grants for basic research do not want to fund therapy.

There are many reasons for physicians to push new treatments across the spectrum from discoveries to therapies. But as the history of chronic hemodialysis reveals, there are also reasons for wondering whether these two categories are adequate for understanding the evolution of treatments in medicine or products in other areas of applied science.

VIII. What Criteria Ought Govern the Language of Experimentation and Therapy?

It is interesting to see exactly how existing regulations governing human experimentation define research. The so-called Belmont Report, published in 1979, which played a key role in the formation of existing federal guidelines concerning human experimentation, used the following definitions:

> practice—"interventions designed solely to enhance the well-being of an individual patient that have a reasonable expectation of success."

> research—"an activity designed to test a hypothesis, permit conclusions to be drawn and thereby to develop or contribute to generalizable knowledge."

Existing regulations governing institutional review boards (IRBs) reflect the considerations raised in the Belmont report definitions. Research is defined as, "a systematic investigation designed to develop or contribute to generalizable knowledge."

These definitions place great weight on intent. If a biomedical scientist believes that what he or she is doing is done solely to help benefit a patient with some chance of success, then what is done is a part of practice. It is therapy. If the goal or intent is to produce generalizable knowledge, then what is done is research.

But these definitions leave too much to intentions. While the goals of therapy and research clearly are different, it is odd to make the distinction entirely contingent upon the aim of the health care professional. Even the most honest and forthright clinician is going to have a hard time describing what he or she is doing as research if it means adverse and often disastrous fiscal consequences for patients plus a loss of hope for those suffering from fatal illness.

Whatever the defining characteristics of research, they must go beyond the subjective intent of the health care provider or scientist. Two characteristics that would appear to be especially relevant are the state of knowledge prevailing about the underlying mechanisms or processes that produce a particular result, and the efficacy associated with a particular activity in terms of the probability that it will produce its intended outcome.

If a physician can produce a cure in a patient but does not have any idea why the cure comes about, then, in a key respect, the intervention is experimental. The reason this is so is that it may not be clear exactly what intervention or contributing factor is responsible for producing the outcome that is sought.

For example, dermatologists have long known that a combination of ultra‾ ‾)-let light and coal tar helps many persons suffering from psoriasis. The cure rate associated with this regimen is quite high. Nonetheless, physicians remain uncertain exactly what frequencies of ultraviolet light, what components of the coal tar and what combinations of light and coal tar, are responsible for symptomatic relief. The treatment is certainly useful. But, it is still experimental. Why it works is at least as relevant to the description of what dermatologists do as is the fact that it has benefits for those who receive the ministrations.

Nephrologists in the 1960s felt they had a treatment that could duplicate the function of the normal kidneys. But, in fact, it took twenty years of clinical and experimental research to establish the biochemical and biophysical similarities and differences between chronic hemodialysis and the naturally occurring process of filtration and osmosis in the kidney.

But theoretical knowledge is not sufficient in medicine for establishing a treatment as therapeutic. Interventions that fail to extend life or that have disastrous side effects are still seen as experimental, even if the underlying mechanisms of action are well understood.

Oncologists may describe various drugs as therapeutic for lung cancer, cancer of the pancreas, or cancer of the liver, but the fact is, no available drug regimens are efficacious against these forms of cancer. While an oncologist may believe he or she is providing therapy to those with cancers of this sort, they often refer to their efforts in talking to one another as experimentation.

Background knowledge concerning causal mechanisms and efficacy are both key factors that must be considered in assigning a particular intervention a place on the experimentation-to-therapy continuum. Intentions in and of themselves are not sufficient to distinguish research from therapy.

Unfortunately there is very little discussion of the degrees of efficacy and safety that ought constitute adequate evidence for establishing the therapeutic status of an intervention. And, as the history of chronic hemodialysis should make plain, there are many opportunities presented for the drawing of lines on the path from discovery to mastery.

The problem becomes even more pressing when it is made clear how the provision of new treatments is closely linked to the need to establish therapeutic efficacy and safety, as well as to competitive and economic factors. Values and facts mix freely in fueling the course of thinking in clinical medicine. Those seeking examples of such interactions would be well advised to look in this domain of science. And those who would advance overarching theories to explain conceptual change in the sciences must look in this domain to see whether the theoretical and the applied are miscible to a degree compatible with a univocal approach to understanding scientific change.

References

Abram, H., and Wadlington, W. 1968. Selection of Patients for Artificial and Transplanted Organs *Annals of Internal Medicine* 59: 615–20.

Banta, H. 1984. "Embracing or Rejecting Innovations: Clinical Diffusion of Health Care Technology." In *The Machine at the Bedside*, eds. S. Reiser and M. Anbar. New York: Cambridge University Press: 65–94.

Beckner, M. 1968. *The Biological Way of Thought*. Berkeley: University of California Press.

Beecher, H. 1969. Scarce Resources and Medical Advancement. *Daedalus* 98: 275–313.

Brandon, R. and Burian R., eds. 1984. *Genes, Organisms, Populations*. Cambridge: MIT Press.

Caplan, A. 1982. "Babies, Bathwater and Derivational Reduction," in PSA–1978, Vol.II, eds. P. Asquith and I. Hacking. East Lansing, Michigan: PSA: 357–70.

——. 1985a. Ethical Issues Raised by Research Involving xenografts. *Journal of the American Medical Association* 254: 3339–43.

——. 1985b. Good Intentions Are Not Enough: The Case of Baby Fae. *Transplantation Today* 2: 4–7.

Committee on Chronic Kidney Disease. 1967. Final Report. Washington, D.C.: U. S. G. P. O.

Czaczkes, J. and Kaplan De Nour, A. 1978. *Chronic Hemodialysis as a Way of Life*. New York: Brunner/Mazel.

Englelhardt, T., and Caplan, A., eds. 1987. *Scientific Controversies*. New York: Cambridge University Press.

Feigl, H., and Brodbeck, eds. 1953. *Readings in the Philosophy of Science*. New York: Appleton-Century-Crofts.

Feyerabend, P.K. 1981. *Realism, Rationalism and Scientific Method*. Cambridge: Cambridge University Press.

Fishman, M., Hoffman, A., Klausner, R., Rockson, S., Thaler, M. 1981. *Medicine*. Philadelphia: J.B. Lippincott.

Fox, R., and Swazey, J. 1978. *The Courage to Fail*, 2d. ed. Chicago: University of Chicago Press.

Giere R.N. 1988. *Explaining Science*. Chicago: University of Chicago Press.

HCFA (Health Care Financing Adminstration). 1988. *Nephrology News and Issues*, October): 7.

Hempel, H. 1952. *Fundamentals of Concept Formation*. Chicago: University of Chicago Press.

Hull, D. L. 1974. *Philosophy of Biological Science*. Englewood Cliffs, N. J.: Prentice-Hall.

——. 1988. Science as a Process. Chicago: University of Chicago Press.

Laudan, L. 1977. *Progress and its Problems*. Berkeley: University of California Press.

Kitcher, P. 1983a. 1953 and All That, a Tale of Two Sciences. *Philosophical Review* 93: 335–73.

——. 1983b. *The Nature of Mathematical Knowledge*. New York: Oxford University Press.

Kolata, G. 1980. Dialysis after Nearly a Decade. *Science* 208: 473–76.

Kuhn, T.S. 1970. *The Structure of Scientific Revolutions*, 2d ed. Chicago: University of Chicago Press.

Mayr, E. 1982. *The Growth of Biological Thought*. Cambridge: Harvard University Press.

Munson, R., ed. 1971. *Man and Nature*. New York: Dell.

Nagel, E. 1961. *The Structure of Science*. New York: Harcourt, Brace.

Nickles, T., ed. 1980. *Scientific Discovery, Logic and Rationality*. Dordrecht: Reidel.

Parsons, V. 1978. The Ethical Challenges of Dialysis and Transplantation. *The Practitioner*, 220: 872–77.

Plough, A. 1986. *Borrowed Time: Artificial Organs and the Politics of Extending Lives*. Philadelphia: Temple University Press.

Popper, K.R. 1959. *The Logic of Scientific Discovery*. New York: Basic Books.

Rescher, N. 1969. The Allocation of Exotic Medical Lifesaving Therapy. *Ethics* 79: 173–86.

Rettig, R. 1976. Health Care Technology: Lessons Learned from the End-Stage Renal Disease Experience. The Rand Paper Series: P-5820.

Robinson, J. 1986. Reduction, Explanation and the Quests of Biological Research. *Philosophy of Science* 53: 333–53.

Rosenberg, A. 1985. *The Structure of Biological Science*. Cambridge: Cambridge University Press.

Ruse, M. 1973. *The Philosophy of Biology*. London: Hutchinson.

Russell, L. 1979. *Technology in Hospitals*. Washington, D.C.: Brookings.

Schaffner, K. 1967. Approaches to Reduction. *Philosophy of Science* 34: 137–47.

Schaffner, K. 1986. Exemplar Reasoning about Biological Models and Diseases. *Journal of Medicine and Philosophy* 11: 55–72.

Scribner, B., Buri, R., Caner, J., Hegstrom, R., and Burnell, J. 1960. The Treatment of Chronic Uremia by Means of Intermittent Hemodialysis: a Preliminary Report. *Transactions of the American Society for Artificial Organs* 6: 144–149.

Sober, E., ed. 1984. *Conceptual Issues in Evolutionary Biology*. Cambridge: MIT Press.

Suppe, F. ed. 1977. *The Structure of Scientific Theories*, 2d. ed. Urbana: University of Illinois Press.

Adolf Grünbaum

The Psychoanalytic Enterprise in Scientific Perspective

1. Introduction

It is well known that Freud was hardly the first thinker who postulated *unconscious* processes in order to explain much conscious psychic life and overt conduct. In Plato's dialogue the *Meno,* we encounter a slave boy who had never studied geometry. Yet by just showing him a diagram and asking him appropriate questions, his interlocutor was able to elicit geometric truths from him. This phenomenon is then used by Plato to interpret the acquisition of conscious knowledge as the recall of information, which the soul had unconsciously stored in the meantime.

In the early nineteenth century, the German philosopher Johann Herbart, who died fifteen years before Freud was born, taught that conscious mental life is affected by subliminal processes, which function like ideas, except for being beneath the threshold of focal awareness (Fancher 1973, 12). Moreover, Arthur Schopenhauer had claimed, *before* Freud, that consciousness resists the intrusion of unpleasant thoughts and perceptions. Indeed, Schopenhauer was Freud's precursor in the field of psychopathology even to the extent of enunciating, in very general terms, that repressed ideation is pathogenic (Ellenberger 1970, 209)! No wonder that Thomas Mann had a sense of déjà vu when he read Freud, after delving into Schopenhauer. Furthermore, Eduard von Hartmann published his *Philosophy of the Unconscious* when Freud was thirteen years of age. There, von Hartmann gave wide explanatory scope to unconscious processes (Ellenberger 1970, 210).

The founding father of psychoanalysis died in 1939. Several decades thereafter, new experimental findings have prompted cognitive psychologists to conclude that unconscious ideation plays a cognitive role in mental life that even Freud had not envisioned. In fact, despite some similarity between the *cognitive* unconscious of recent psychology and Freud's *dynamic* unconscious, there are also important differences between them that should not be glossed over. For example, the psychoanalytic unconscious is *affect*-laden, and its contents are deemed to be recoverable by lifting their repression. By contrast, the implicit problem-solving capabilities of the cognitive unconscious are neither repressed nor conscious. Thus, therapists who have used subliminal techniques to increase self-esteem, or to induce weight loss, are feeling reassured by the new recognition that a substantial portion of *cognitive* activity is, in fact, unconscious.

But one factor that may have made psychoanalytic theory so extraordinarily influential in some segments of our culture was Freud's particular articulation of the assumed *causal* role of unconscious processes. It is a measure of this influence, at least in the United States, that the Science section of the *New York Times* (January 24, 1984), no less than the *New Yorker, Time,* and other such magazines, gave wide publicity to Jeffrey Masson's book (1984) about Freud, which purports to contain highly derogatory revelations about his lack of intellectual integrity. To boot, Janet Malcolm (1981) gave prominence to Peter Swales' allegation that one of Freud's paradigmatic examples of a memory lapse pertained to a shady episode involving Freud himself, who had supposedly impregnated his own sister-in-law, and had then taken her to Italy for an abortion. Indeed, the intellectual and cultural historian Peter Gay (1985, 1988)—to name only one— invokes psychoanalytic theory *uncritically,* as if it were holy writ. And he applies it to generate so-called psychohistory (Lifton 1980).

Yet, in my view, Freud's massive elaboration of *clinical observations* into his own doctrine of hidden motives still cries out for further careful scrutiny. By the same token, I contend, such scrutiny is at least equally imperative in the case of the various *post*-Freudian, revisionist versions of psychoanalysis. Although the specific content of their theories of psychic conflict is more or less different, they also rely on Freud's clinical methods of validating causal inferences (Eagle, 1983; Grünbaum, 1984, chapter 7). I shall be challenging just these causal inferences. And it will be an immediate corollary of my challenge that it applies not only to Freud's own original hypotheses, but also to any and all post-Freudian versions of psychoanalysis that rely on his clinical methods of justifying causal claims. After all, the changes made by post-Freudians in the specific content of the founding father's theory of psychic conflict (repression) hardly make the *validation* of the revisionist versions more secure!

Therefore, as Morris Eagle documented in a recent publication (1983), those analysts who have objected to my critique as anachronistic have simply not come to grips with it. For example, such inadequate engagement is present, in my view, in the recent Freud Anniversary Lecture "Psychoanalysis as a Science: A Response to the New Challenges," given by Robert Wallerstein (1986), the current president of the International Psychoanalytical Association. As he tells us (1988, 6, n.1), "The Freud Anniversary Lecture was intended primarily as response to Grünbaum." Yet he does not come to grips at all with the gravamen of my challenge: *Even if clinical data could be taken at face value as being uncontaminated epistemically,* the inability of the psychoanalytic method of clinical investigation by free association to warrant causal inferences leaves the major pillars of the clinical theory of repression ill-supported.

The heart of Freud's distinctive theory of repression is not just that we harbor repressed memories, thoughts, desires, and feelings. Instead, it is that sexual repressions are the *crucial pathogens* of mental disorders, that repressed infantile wishes are the generators of our dreams, and that various sorts of repressed, unpleasant thoughts *engender* our slips of memory, of the tongue, the ear, the eye, the pen, etc.

Freud referred to these various sorts of unsuccessful, bungled actions collectively in German as *Fehlleistungen,* or misbegotten performances. And James Strachey, the principal translator of Freud's psychological works into English, coined the new term *parapraxes* to denote them *generically* in English. But, the Vienna-born American psychoanalyst Bruno Bettelheim (1982), deplored this translation. The German word *Fehlleistungen,* he tells us, has a familiar, mellifluously humanistic ring, even smacking of poetry. By contrast, according to Bettelheim, the technical term *parapraxes* allegedly has a coldly scientific tang. Thus, as he would have it, the Englishman Strachey has misportrayed the psychoanalytic enterprise to the English-speaking public by giving a misleading scientific twist to it. Freud, we are asked to believe, wanted psychoanalysis to be a branch of the humanities, but deplorably Strachey made it appear as if Freud worshiped the natural sciences and idolatrously intended psychoanalysis to be a natural science! Strachey created this allegedly false impression by insidiously mistranslating Freud's key German vocabulary. Let me just say very briefly that Bettelheim's indictment of Strachey is, I believe, an unfortunate exegetical fabrication which, alas, grasps at straws. For example, one need only read the German original of Freud's 1933 lecture "Über eine Weltanschauung," which appeared only six years before he died, to see that Bettelheim's complaint is a hermeneutic red herring. In that lecture, Freud declared that "psychoanalysis has a special right to speak for the scientific *Weltanschauung*" (S.E. 1933, 22:159),[1] the word *scientific* being intended in the sense of the *natural* sciences.

The contemporary philosophic spokesmen for the so-called hermeneutic reconstruction of psychoanalysis have even gone further than Bettleheim by condemning Freud for an alleged "scientistic" misunderstanding of his own clinical theory. As we know, the word *scientism* is a derogatory term, used to refer to a misguidedly utopian, intellectually imperialistic worship of science. Thus the hermeneutic philosophers, such as Jürgen Habermas and Paul Ricoeur, claim that even Freud's *aspiration* to build a scientific depth psychology was misguided from the outset. In my book (Grünbaum 1984), I argued that these hermeneutic criticisms are based on misconceptions of both the content and the methods of the natural sciences. Furthermore, I have contended (1988) that Karl Jaspers and the hermeneutic philosophers have mishandled so-called "meaning connections" between mental events vis-à-vis causal connections between such states. As against their claim that Freud assigned much too little explanatory significance to thematic kinships ("meaning connections"), I maintain that he fallaciously gave much too much explanatory weight to them. Therefore, here, I shall try to appraise Freud's own principal arguments for the *cornerstone* of his entire edifice: the theory of repression, or psychic conflict. His theory of psychopathology was the logical and historical foundation of his entire theory of repression. Thus, I shall now turn to it.

2. The Theory of Psychopathology

The central causal and explanatory significance of unconscious ideation *throughout* the psychoanalytic theoretical edifice rests, I claim, on two cardinal inductive inferences. They were drawn by Freud in collaboration with his senior mentor, Josef Breuer. As we are told in their joint "Preliminary Communication" of 1893 (S.E. 1893, 2:6–7), they began with an observation made after having administered their cathartic treatment by hypnosis to patients suffering from various symptoms of hysteria. In the course of such treatment, it had turned out that, for each distinct symptom S afflicting such a neurotic, the victim had *repressed* the memory of a trauma that had closely preceded the onset of S and was thematically cognate to this particular symptom. Besides repressing this traumatic memory, the patient had also strangulated the affect induced by the trauma. In the case of each symptom, our two therapists tried to lift the ongoing repression of the pertinent traumatic experience, and to effect a release of the pent-up affect by expressing the previously suppressed feelings verbally. When their technique succeeded in implementing this cognitive and cathartic objective, they reportedly observed the dramatic disappearance of the given symptom. Furthermore, the symptom removal *seemed* to be durable, although the presumed cures later turned out to be ephemeral, especially after Freud had begun to practice without Breuer.

Impressed by the positive treatment outcome while it lasted, Breuer and Freud drew their first momentous *causal* inference. Thus they enunciated the following fundamental therapeutic hypothesis: The dramatic improvements observed *after* treatment were produced by none other than the cathartic lifting of the pertinent repressions. But before the founders of psychoanalysis credited the undoing of repressions with remedial efficacy, they had been keenly alert to a rival hypothesis, which derived at least prima facie credibility from the known achievements of the admittedly suggestive therapies. On that alternative explanation of the positive outcome after cathartic treatment, its therapeutic benefit was actually wrought by the patient's credulous expectation of symptom relief, or at any rate *not* by achieving insight into his or her repressions. In this perspective, the quest for such insight is only a particular treatment ritual, which serves to fortify the patient's therapeutic expectations. But Breuer and Freud believed that they could *rule out* such a contrary account of the treatment gains, a challenge to which I shall refer as "the hypothesis of *placebo* effect" (Grünbaum 1986a). In an attempt to counter it, they pointed out that the distinct symptoms had been removed separately — one at a time — such that any one symptom disappeared only after lifting a *particular* repression (S.E. 1893, 2:7).

But, even such separate removals, I submit, may not be due at all to the lifting of repressions; instead they may be a *placebo effect* after all, generated by the patient's awareness that the therapist was intent upon uncovering a thematically particular episode *E* when focusing attention upon the initial appearance of the distinct symptom *S*. Thus, it was presumably communicated to the patient that his or her doctor attached potential therapeutic significance to the recall of *E* with respect to *S*. Indeed, Breuer and Freud do not tell us why the likelihood of placebo effect should be deemed to be lower when *several* symptoms are wiped out *seriatim* (one at a time), than in the case of getting rid of only one symptom. To discredit the hypothesis of placebo effect, it would have been essential to have comparisons with treatment outcome from a suitable control group whose repressions were *not* lifted. If that control group were to fare equally well therapeutically, treatment gains from psychoanalysis would then presumably be placebo effects, since such a result would then *not* have been wrought by psychoanalytic insight. Hence the attribution of remedial efficacy to the cathartic lifting of repressions was devoid of adequate evidential warrant.

Let me be clear on what I understand here by a "placebo effect" (Grünbaum 1986a). I do not rely on the vague notion of suggestion at all to characterize such an effect. Instead, I say essentially the following: A treatment gain is a "placebo effect" with respect to a particular target disorder, and also with respect to a particular therapeutic theory, just when that positive effect is produced by treatment factors *other than* those designated as the efficacious ones by the given theory.

At the time, Breuer and Freud believed that their therapeutic results *had* ruled out the dangerous competing hypothesis of placebo effect. Thus they credited the gains made by their hysterics to the resurrection of buried painful memories. And they thought furthermore that this supposed therapeutic potency of lifting repressions spelled a paramount etiologic moral as follows: A coexisting ongoing repression is *causally necessary* for the *maintenance* of a neurosis *N*, and an original act of repression was the causal *sine qua non* for the origination of *N*. This second groundbreaking causal inference was animated by the fact that the inferred etiology then yielded a *deductive* explanation of the supposed remedial efficacy of cathartic recall. How so? Clearly, if an ongoing repression *R* is causally *necessary* for the pathogenesis *and* persistence of a neurosis *N*, then it follows that the removal of *R* will actually engender the disappearance of *N*. As we can see, it was not the observed therapeutic gain itself that had prompted the inference of the repression etiology. Instead, it was only the *causal attribution* of the patient's gain to the lifting of repressions *in particular* that had furnished this motivation. Thus, without reliance on the presumed dynamics of their therapeutic results,

Breuer and Freud could never have propelled their clinical data into repression etiologies. Freud often put this very briefly by saying that the attempt to unravel the *cause* of a disorder by means of free association was *simultaneously* a therapeutic maneuver (S.E. 1893, 3:35). Or he made the same point by declaring that therapy and etiologic research coincide in psychoanalysis (S.E. 1909, 10: 104–5; 1926, 20: 256).

Freud saw hypnosis, and then free association, as means for the authentic restoration of forgotten traumatic experiences to conscious awareness. Hence one might wonder how he thought he could rule out the *fancied* recall of events that never occurred. Thus, the question is how *pseudo*memories could be distinguishable from genuine ones among the emerging associations. I shall not pursue this issue here. Suffice it to say that it derives added poignancy from recent experimental studies (Dywan & Bowers 1983), which have cast much doubt on the authenticity of hypnotically enhanced remembering. These investigations have shown that hypnotically increased recall is achieved largely at the expense of reliability: When highly hypnotizable subjects recalled twice as many new items hypnotically as the control subjects, they introduced three times as many *new* errors, despite their intense conviction that the new memories were trustworthy! Thus, hypnotic memory is less, not more, reliable than normal prehypnotic recall. It would seem that, by being more suggestible, the hypnotized person translates some of his or her own beliefs and/or those of the hypnotist into *pseudo*-memories.[2] Yet in the 1918 case history of the Wolf Man, who suffered from obsessional neurosis (S.E. 1918, 17:57–60, 95–97), Freud (S.E. 1918, 17: 33–4) gave credence to the adult patient's report of a dream that had occurred between the ages of three and five, which Freud then used to retrodict the following event: When the Wolf Man was a mere infant of eighteen months, he witnessed his parents engage in sexual intercourse *a tergo*. Alleged hypnotic findings have also figured in contexts *removed* from psychoanalysis.

More recently, we have it on the authority of illustrious Hollywood actors such as Glenn Ford and Shirley MacLaine that, under hypnosis, they discovered having each gone through several prior incarnations. For example, it was reported[3] that Glenn Ford was once a Christian martyr who was eaten by a lion. *Mirabile dictu,* Shirley MacLaine reports having been beheaded by Louis XV, and claims to have watched her own head rolling on the floor, but landing face up, a big tear coming out of one eye. This experience, she explains, cured her of stage fright during her current reincarnation. Moreover, her own daughter was her mother in one life, and her sister in another. Recently, the Phil Donahue show on television featured a hypnotherapist from California who blithely reported the following statistic: Two-thirds of his patients retrieve memories from their *current* incarnation, but fully *one-third* recall experiences from *several prior incarnations*. In each case, personal identity remains stunningly intact, unencumbered by the death of the brain.

Even police departments in Los Angeles and elsewhere have made use of hypnotists as detectives in "Svengali squads." But recently, a number of state supreme courts in the United States have declared hypnotically induced testimony inadmissible in criminal trials as inherently unreliable, because a hypnotized person is as likely to concoct wild hallucinations as to achieve veridical recall.[4] I do *not* say that free association should be simply equated with hypnosis. I do claim that the pitfalls of hypnosis at least spell a sobering caveat for the credibility of recall under free association.

But Freud went much beyond crediting free association with the authentic restoration of memories when he drew a major methodological inference from his repression-etiology. As we saw, the outcome and presumed dynamics of successful therapy had been the original evidential basis for the pathogenic role of repression. And once Freud had thus convinced himself of the etiologic role of (unsuccessful) repression on *therapeutic* grounds, he drew a crucial, momentous *investigative* lesson: Free association is a tool of etiologic research into pathogenesis, precisely because it excavates repressions (psychic conflict). In sum, it was only the therapeutically inferred etiology *itself* that gave the license for supposing that when previous repressions are uncovered by free associations, then these emerging repressions are indeed pathogens (S.E. 1900, 5: 528). Hence, at least as far as clinical evidence goes, the credibility of free association as a means of indentifying pathogens depends crucially on whether the therapeutic results do, in fact, support the alleged pathogenic role of repression.

In fact, though it is widely overlooked, *the attribution of therapeutic success to the removal of repressions not only was but, to this day, remains the sole epistemological underwriter of the purported ability of the patient's free associations to ascertain causes.* Analysts such as Strachey (S.E. 1955, 2:xvi) — the translator and editor of the *Standard Edition* of Freud's works — and Kurt Eissler (1969, 461) have hailed free association as an instrument comparable to the microscope and the telescope. More recently, it has been compared to x-ray tomography or CAT scanning. And it is asserted to be a trustworthy means of etiologic inquiry in the sense of licensing a specific major *causal inference.* To state the inference, consider a set of previously repressed traumatic memories, wishes, or fantasies, whose associative emergence is triggered by one of the patient's neurotic symptoms; then, it is claimed, precisely this surfacing in the causal chain of ensuing associations is strong evidence that the prior ongoing repressions of these memories or wishes contributed *pathogenically* to the formation of the symptom.

Whereas all Freudians presumably champion this causal inference, a number of influential ones have explicitly renounced its legitimation by the presumed *therapeutic dynamics* of undoing repressions. To them I say: Without this therapeutic vindication, or some as yet unknown other warrant, no rational person should believe that free associations can ascertain pathogens or any other causes! For without the stated *therapeutic* foundation, this epistemic tribute to free associations rests on nothing to date but a glaring causal fallacy of causal inversion (Grünbaum 1984, 186–87 and 233–34; 1986b, 277). And even that therapeutic foundation will turn out to be quite flimsy. Therefore, it is unavailing to extol the method of clinical investigation by free association as a trustworthy resource of etiologic inquiry, while issuing a disclaimer as to the therapeutic efficacy of psychoanalytic treatment. In brief, those who have made it fashionable nowadays to dissociate the clinical credentials of Freud's theory of personality — the so-called science — from the merits of psychoanalytic therapy are stepping on thin ice indeed.

But let me caution against a possible serious misunderstanding. I do claim that insofar as the credentials of psychoanalytic hypotheses avowedly rest on the purported ability of free associations to ascertain causes, these credentials are ultimately predicated on the therapeutic efficacy of lifting repressions. In concert with Freud himself, the vast majority of his followers continue to maintain that the use of free association as a touchstone of causal certification is crucial for the validation of their *etiologic* hypotheses, as well as of the psychoanalytic theory of dreams and of slips. Hence I contend that these advocates cannot dispense with a therapeutic foundation. But, for my own part, I abjure as unwarranted the invocation of free association as a hallmark of causal certification. Thus, unlike these Freudians, who insist on clinical validation via free association, I am not at all constrained to grant the consequences of such an insistence. Insofar as I can envisage potentially cogent tests of psychoanalytic hypotheses, I see no reason to assign any special or privileged role to therapeutic results. For example, I can envision an epidemiologic test of one of Freud's etiologies such that therapeutic results are irrelevant to the test (Grünbaum 1984, 38–39, 110–11).

Within the clinical confines of traditional psychoanalysis, the foundational role of the presumed dynamics of the therapy to which I have called attention has an important corollary: Namely, that the whole structure of clinical hypotheses based on the therapeutic foundation is in serious jeopardy from the ominous threat posed by the hypothesis of placebo effect. Mind you, the thus endangered hypotheses comprise not only the asserted therapeutic efficacy of lifting repressions, and the repression etiologies of the psychoneuroses, but also the theory of dreams and of sundry sorts of "slips," which Freud extrapolated from his etiologic theory, as we shall see.

Even before he had developed his theory of dreams, and of slips, the menace of placebogenic gains in the therapy was driven home to him. By his own account, soon after he had begun to practice without Breuer, it became devastatingly plain that both of them had been all too hasty in rejecting the rival hypothesis of placebo effect. The remissions achieved by additional patients whom Freud himself treated cathartically turned out *not* to be durable. And these symptom relapses showed him that his treatment had *not* uprooted the cause of the symptoms. Indeed, the ensuing pattern of relapses, additional treatment, ephemeral remissions, and further relapses undermined the attribution of therapeutic credit to the lifting of those repressions that Freud had uncovered. Ironically, he began to be haunted by the triumph of the hypothesis of placebo effect over the fundamental therapeutic tenet that Breuer and he had originally enunciated. As he recognized, the vicissitudes of his personal relations to the patient were highly correlated with the pattern of symptom relapses and intermittent remissions. And, in his own view, this correlation "proved that the personal emotional relation between doctor and patient was after all stronger than the whole cathartic process" (S.E. 1925, 20:27). But, once the repression etiology was thus bereft of therapeutic support, the very cornerstone of psychoanalysis had been completely undermined. Hence at that point, the new clinical psychoanalytic structure tumbled down and lay in shambles. So also then did free association as a method of etiologic certification!

Nonetheless Freud was undaunted. He took courage in 1896, because he thought that, in a new *sexual* version going back to childhood, the repression etiology could be rehabilitated on secure therapeutic foundations after all. And his strenuous effort to achieve such a rehabilitation culminated in his 1917 paper on "Analytic Therapy" (S.E. 1917, 16:448–63). There he tried to offer an explicitly *therapeutic* vindication of the psychoanalytic method and theory of personality, including its specific etiologies of the psychoneuroses, and even its general hypotheses about psychosexual development. But, as I have argued in detail in my book (1984, chapter 2), his attempt to vindicate the rehabilitated etiology therapeutically in 1917 fared no better empirically than his and Breuer's original reliance on cathartic success in the mid-1890s. In particular, this 1917 endeavor foundered, if only because Freud again failed to rule out the opposing placebo hypothesis. Indeed, to this day, psychoanalytic treatment process research has failed to discredit the altogether reasonable challenge posed by this rival account of gains from its therapy. Hence I claim that the whole of the clinical psychoanalytic enterprise continues to be haunted by it.

Moreover, no empirically viable surrogate for Freud's discredited 1917 effort seems to be even remotely in sight. Nor, to my knowledge, are there any other cogent *therapeutic* defenses of the sexual repression etiology of the neuroses. Hence this etiology should now be regarded as devoid of significant therapeutic support, just like Breuer's nonsexual cathartic etiologies, which Freud himself had disavowed as clinically dubious.

But that is not all. As I have already pointed out, the credibility of free association as a method of etiologic certification has rested entirely on the cogency of Freud's *therapeutic* argument for his theory of pathogenesis (S.E. 1900, 5: 528). Hence the collapse of precisely that argument, even in its mature version of 1917, also undermines the etiologic credibility of the fundamental rule of free association. After all, as we saw, Freud had enunciated this rule as a maxim of research in psychopathology, because he thought—on therapeutic grounds—that associations governed by it can reliably certify the unconscious pathogens of the neuroses.

In fact, *it was too good to be true at the outset, I think, that psychoanalysis should have been able to make reliable etiologic determinations just by having someone lie on a couch and associate freely,* even for years. At least prima facie, this skeptical attitude appears justified, if only because the validation of etiologic hypotheses in somatic medicine, for example, normally requires controls of one sort or another. And there is typically no counterpart to such controls in Freud's fundamental rule of free association. The well-known psychoanalyst Erik Erikson seems to be cognizant of this doubt. Thus, he speaks very defensively, when he claims "the necessity to abandon well-established methods of sober investigation (invented to find out a few things exactly and safely to overlook the rest) for a method of self-revelation apt to open the floodgates of the unconscious" (Erikson 1954, 54). This brings us to the theory of dreams and of slips.

3. The Dream Theory and the Theory of Slips

Freud did not limit his investigative esteem for free associations to etiologic research. When he found that his patients reported their dreams while freely associating to their neurotic symptoms, he drew a very weighty but highly risky conclusion as follows: Manifest dream contents can be causally *assimilated* to the status of neurotic symptoms, and thus are likewise presumed to be generated by repressions. And he saw neurotic symptoms, in turn, as *substitutive* gratifications and outlets, or as "compromises between the demands of a repressed impulse and the resistance of a censoring force in the ego" (S.E. 1925, 20:45). And when he extended this notion of compromise formation to manifest dream content, he carried out a bold extrapolation of his repression etiology from neurotic symptoms to manifest dream content. By the same token, he felt entitled to enlarge the scope of free association from being a method of etiologic research *aimed at therapy,* to serving likewise as an avenue for finding and certifying the purported unconscious causes of dreaming (S.E. 1900, 4:101; 5:528).

Mutatis mutandis, he also assimilated slips and bungled actions (*Fehlleistungen*) to his compromise conception of neurotic symptoms. And, by parity with his previous reasoning, he again resorted to free association not merely heuristically, as a means of generating causal explanations of slips, but also probatively, as a basis for validating the explanatory hypotheses. For example, he conceived of a slip of the tongue as a compromise between a repressed motive that crops out in the form of a disturbance, on the one hand, and the conscious intention to make a certain utterance, on the other.

In short, once Freud had postulated *by sheer extrapolation* that dreams and slips are indeed compromise formations, no less than neurotic symptoms themselves, it seemed evident that dream production and slip generation are also due to repressed motives. By the same token, *if* it be granted that the method of free association can show repressions to be the pathogens of neuroses, that method seems eminently capable of reliably ascertaining as well the unconscious motives of other purported compromise formations, notably of dreams and slips.

But Freud's assimilation of dreams and slips to compromise formations without ado was an audacious, if not just foolhardy, extrapolation from the etiologic role that he had attributed to repression in psychopathology. The more so, since just that purported etiologic role is itself in grave jeopardy—as I have argued— from lack of cogent therapeutic support.

Indeed, I contend that even if the original *therapeutic* defense of the repression etiology of neuroses had actually turned out to be empirically viable, Freud's compromise models of parapraxes and of manifest dream content would still be *misextrapolations* of that etiology, precisely because they lacked any corresponding *therapeutic* base at the outset. For example, he did not adduce any evidence that the permanent lifting of a repression to which he had attributed a slip will be "therapeutic" in the sense of enabling the person himself or herself to correct the slip and/or to avoid its repetition in the future. Thus, it turns out that whereas he claimed to have therapeutic evidence for postulating the pathogenic role of repressions, he never produced any *independent* clinical support for his two daredevil extrapolations of the compromise model to dreams and to slips—nor for believing that free associations can ascertain the *causes* of dreams and of slips. Let me emphasize that when I speak of evidence as "clinical," I adopt its technical usage as referring to the observations made by psychoanalysts of their patients' productions in the treatment setting of the couch.

Hence, within the confines of his clinical evidence, where free association is invoked as a hallmark of causal validation, the epistemic fortunes of Freud's extrapolations of the compromise model are dependent on those of his theory of psychopathology. As a consequence of just this epistemic dependence in the clinical context, the ravages from the clinical collapse of Freudian psychopathology, and of free association as a tool of etiologic certification, turn out to extend, with a vengeance, to the psychoanalytic theory of dreams and of sundry sorts of slips. Mind you, when I speak of the clinical collapse, I mean the clinical unfoundedness, not the clinical refutation.

Thus, for instance, I *allow* that there may be slips that are engendered by repressions. But I maintain that if there are such slips, Freud did not give us any good reason to think that his clinical methods can identify and validate their causes as such, no matter how interesting the elicited "free" associations might otherwise be.

Thus, the failure of his clinical arguments for the protean causal role of repressions does not itself betoken the falsity of his theory of psychopathology, or of his dream theory, or yet of his account of slips. But my discreditation of his arguments so far does, I claim, undermine the foundations of his theoretical edifice, since he rested it on just the clinical reasoning I have challenged. And the vast majority of his disciples nowadays also rest their case on *clinical* evidence from the analytic treatment setting.

But even if we confine ourselves solely to clinical considerations and disregard experimental ones (Hobson 1988), there are additional good grounds for deeming Freud's dream theory to be false rather than just ill supported. These further grounds for claiming falsity are of two sorts.

1. As I shall now argue, in the context of the remainder of Freud's theory of repression and psychoanalytic therapy, his dream theory predicts a reduction in the *frequency* of dreaming among extensively psychoanalyzed patients. But there is no evidence for such a reduction. More precisely, Freud's assimilation of manifest dream content to minineurotic symptoms has the consequence that *either* extensively analyzed patients should be "cured" of dreaming, or free association fails as a means of lifting presumably repressed infantile wishes.

It will be recalled that just as sexual repressions, in particular, are postulated to be causally necessary for neurosogenesis, so also sundry sorts of repressed infantile wishes are hypothesized to be the causal sine qua non of dream generation. Breuer and Freud had told us in 1893 that if particular repressions are, in fact, causally necessary for psychopathology, then it follows that the lifting (undoing) of these pathogenic repressions or conflicts will issue in the conquest of the patient's affliction. By parity of reasoning, if *repressed* infantile wishes are the sine qua non of dream formation, the patient's achievement of conscious awareness of these wishes will rob them of their previous causal role as dream generators. It emerges, therefore, that in proportion as the patient's free associations do succeed in bringing his or her buried infantile wishes to light, the analysand should experience—and presumably exhibit neurophysiologically—a noticeable reduction in dream formation. Evidently this reduction should be a diminution in the *frequency* of dream generation, as distinct from a mere change in dream content. But whereas changes in dream content are commonplace as a function of the thematic content of analytic sessions, even protractedly analyzed patients do not report any remarkable subjective diminution of their recalled dream experiences. Nor, to my knowledge, have analysts been aware that the theoretical expectation of a reduction in the frequency of dreaming is a consequence of Freud's dream theory, which makes the assumption that free associations of sufficient duration normally retrieve at least some buried infantile wishes, at least among extensively analyzed patients.

We now see that if neurophysiological indicators (perhaps REM sleep) bear out that, among such psychoanalytic patients, the expected decline in dream activity *fails* to materialize, then an important indictment would seem to follow: Either their free associations are chronically unsuccessful in retrieving their buried infantile wishes, or, if there is such retrieval, then Freud's account of dream generation is false. But if free association were to fail chronically even in just lifting repressions, that would be a much greater threat to the psychoanalytic enterprise than the mere demise of Freud's dream theory.

In response to my development of this argument, Philip Holzman suggested that I comment on the retort that no reduction of dream frequency is to be expected, because the impulse behind the emerging wishes remains undiminished in the unconscious. I reply that this retort cannot obviate the discreditation of Freud's account of dream formation, precisely because of the parity of the reasoning in my argument with the basic rationale of psychoanalytic therapy. If we grant Freud's assimilation of manifest dream content to his compromise model of neurotic symptoms, then I ask: Why shoud the *therapeutic* import of this model not hold alike for dream production and ordinary symptom formation? What is sauce for the goose is sauce for the gander. If lifting (and working through) the sexual repressions that are deemed pathogenic more or less cures the neuroses, then lifting (and working through) the repressions of infantile wishes should "cure" dreaming to the same extent, as it were. On the other hand, if the impulse behind the previously repressed wishes generates ever new ones that, in turn, engender dreams even as the patient becomes conscious of the earlier ones, why does the pathogenic action of sexual repressions not also remain equally undiminished after *they* are lifted? If psychoanalytic theory is taken to assert that we have an inexhaustible store of the impulses that beget dreams, how can it claim that we do not also have a like store of pathogenic impulses? By the same token, if psychoanalytic therapy is not doomed to fail at the outset in the case of neuroses, then the recourse to undiminished dream generation, even as repressed wishes are made conscious, is impermissibly ad hoc.

2. I contend that, contrary to Freud (1900, pp. 151–59), the so-called "counter-wish dreams" cannot be reconciled with his wish-fulfillment theory. Freud claims compatibility on the grounds that the contents of these dreams do fulfill one of two wishes: (1) the purported wish to prove his psychoanalytic theory wrong, or (2) the masochistic wish for humiliation and mental torture. As shown by his examples, Freud makes the sound assumption that imputations of the desire to prove him wrong or of a masochistic disposition require *independent* evidence, which does not rely on the de facto occurrence of counter-wish dreams.

As Freud (p. 151) reports, after he had explained his wish-fulfillment theory to "the cleverest of all my dreamers," she dreamt that "she was traveling down with her mother-in-law to the place in the country where they were to spend their holidays together." Yet he "knew that she had violently rebelled against the idea of spending the summer near her mother-in-law and that a few days earlier she had successfully avoided the propinquity she dreaded by engaging rooms in a far distant resort." Thus, "now her dream had undone the solution she had wished for."

To deal with this seemingly recalcitrant finding, Freud (pp. 151–52) points out that, at the time of the dream, the patient had been rejecting his inference as to the occurrence of certain events in her life that had presumably been pathogenic, but which she could not recall. This resistance had been prompted by her "well-justified wish that the events of which she was then becoming aware for the first time might never have occurred." By the same token, this wish supposedly also engendered her broader intellectual desire that Freud's theory be generally wrong. Finally, we learn, the latter desire "was tranformed into her dream of spending her holidays with her mother-in-law" (p. 152).

Freud (p. 151) pays tribute to her cunning unconscious, which purportedly has a perspicacious appreciation of the dreams' logical consequence: "it was only necessary to follow the dream's logical consequence in order to arrive at its interpretation. The dream showed that I was wrong. *Thus it was her wish that I might be wrong, and her dream showed that wish fulfilled"* (italics in original). In the German original, the second of these two conjuncts says literally that the dream *showed her* the fulfillment of this wish (Freud, 1940–52, 2/3, p. 157).

Though Freud speaks of counter-wish dreams as "very frequent" (p. 157), he would have us believe that they are *confined* to people who either have a wish to prove him wrong or who, qua "mental masochists," have the self-punitive wish to suffer mental torture. The former group includes (a) patients who become aware of his dream theory while "in a state of resistance" to him during psychoanalytic treatment, and (b) others who are exposed to his writings or lectures on his dream theory but are unfavorably disposed toward it (p. 158). By thus confining the counter-wish dreamers, he hopes to sustain his thesis that their dreams conform to the pattern "the non-fulfilment of one wish meant the fulfilment of another." But this confinement is untenable, unless he could demonstrate that people, children, or animals (e.g., monkeys) who give no *independent* evidence of a significant masochistic disposition and who are in no position to harbor a wish to disprove Freud's dream theory have at least incomparably fewer counter-wish dreams than patients who are in a resistance phase of their analysis or are otherwise hostile to Freudian ideas.

It does not even seem to have occurred to Freud that without the provision of a baseline as to the incidence of counter-wish dreams in the former huge class, there is at best no cogent reason for attributing such dreams causally to hostile motives toward his theory.

Indeed, it would seem that there just is no difference in the incidence of counter-wish dreams in the two classes. For example, there are presumably lots of students who are devoid of a masochistic disposition and who have nightmarish examination dreams before they ever hear of the content of Freud's theory. Moreover, even if a desire to prove Freud wrong were to motivate some counter-wish dreams, that wish fails to satisfy his requirement of being a *repressed infantile* wish! Besides, what of the counter-wish dreams of *ardent* believers in psychoanalysis who are not masochists? And, finally, why should Freud assume that *unanalyzed* educated people exposed to his wish-fulfillment theory will often have a motive for resisting it? After all, the notion that *some* dreams are wish fulfilling is a commonplace in folk (commonsense) psychology.

Recall Freud's wish-fulfillment claim: The dream "*content was the fulfilment of a wish*" [first conjunct] *and its motive was a wish* [second conjunct]" (p. 119; italics in original). My criticisms have shown, I believe, that Freud's *causal* attribution of nonmasochistic counter-wish dreams to the desire to prove him wrong is probably false. So much for the considerations that, in my view, suggest the falsity of Freud's dream theory, rather than merely its clinical ill foundedness.

All the same, there is some plausibility, I believe, in the psychoanalytic claim that there exist such defense mechanisms as repression, denial, rationalization (in Ernest Jones's sense), reaction formation, and projection. But, as I have emphasized, psychoanalytic theory goes far beyond asserting that repression operates in these defense mechanisms. Besides claiming the bare operation of a mechanism of repression, the theory assigns a crucial causal role to it in pathogenesis, dream formation, and in the generation of slips. And, I claim, just that causal role is questionable.

4. Experimental Results

So far, I have made no mention of the results obtained in actual attempts to test some parts of psychoanalytic theory experimentally. Such laboratory tests of psychoanalysis are discussed in a book by the *pro*-Freudian English psychologist Paul Kline (1981). In this work, Kline tries to refute the extremely skeptical conclusions offered by Hans Eysenck and Glenn Wilson (1973). Interestingly enough, Freud himself told us in 1933 that "In psycho[analysis] however, we have to do without the assistance afforded to [other] research by experiment" (S.E. 1933, 22:174).

In my view, the debate between Eysenck and Kline yields essentially the same verdict as the one I have drawn from the *clinical* evidence offered by Freudians. So far, the evidence available from the laboratory has provided no significant support for any of the major hypotheses of psychoanalytic theory or therapy. Let me give two brief but important illustrative examples, which will also convey what I mean by the term *major hypothesis* in this context.

Freud hypothesized that strongly *repressed* homosexual desires are the causal sine qua non for the pathogenesis of paranoid delusions. The experimenter Harold Zamansky (1958) attempted to test this etiologic hypothesis (Grünbaum 1986b, 269–70). But this investigator himself makes only the following quite modest claim for his findings: "Though the present experiment has demonstrated a greater degree of [repressed] homosexuality in men with paranoid delusions than in nonparanoid individuals, these results *tell us nothing* about the role which homosexuality plays in the development of these delusions" (quoted from Zamansky's study, as reprinted in Eysenck and Wilson [1973, p. 308]; italics added). In the same vein, I claim more generally: It is not enough to provide evidence for the existence of a *mechanism* of repression, which was already postulated before Freud by Herbart, among others. The point is that there is no good experimental evidence for Freud's much stronger claim that specific sorts of sexual repressions are *causally necessary* for the production of designated kinds of mental disorders. In fact, Kline (1981) admits as much in the following disclaimer: "Many of the Freudian claims concerning the neuroses have simply never been put to the objective test . . . Thus, the hypotheses which have been put to the test are usually those where convenient measures are at hand rather than those most crucial to the theory" (437).

The psychoanalytic theory of dream production by repressed infantile wishes furnishes a second illustration of my thesis that there is a poverty of experimental support for Freud's *cardinal* postulates. His theory of dream interpretation is appraised by the psychoanalytically oriented psychologists Fisher & Greenberg (1977). On the construal they deem most reasonable, its central thesis is that dreams express (vent) not only wishes, but sundry drives or impulses that originate in the unconscious (46–47). But, besides opting for some revisions, they make two sobering points: (1) Even if one can show that a dream does, indeed, express a certain impulse, there are, at present, still no scientifically reliable means of warranting that the impulse originated in the unconscious sector of the psyche (47); and (2) the available "findings are *congruent* with Freud's venting model. But . . . they do not specifically document the model" (53; italics in original). Moreover, "the data . . . are encouraging but not definitively validating with respect to Freud's venting model" (63).

Yet, despite issuing these vital admonitions, Fisher and Greenberg permitted themselves to *begin* their summary by declaring that Freud's model "seems to be supported by the scientific evidence that can be mustered" (63). By thus sliding from compatibility ("congruence") with the theory into *"support"* for it, these friends of psychoanalysis have lent substance to Popper's complaint against the methodological behavior of some of Freud's sympathizers, although Popper's critique of psychoanalytic theory is largely unsound (Grünbaum 1984, ch. 1; 1986b, 266–70; 1989).

5. Conclusion

But my primary concern here has been with the *clinical* rather than with the *experimental* evidence. After all, it is the *clinical* evidence on which most psychoanalysts rest their case (Grünbaum 1988). And I conclude that despite their appeal to such evidence, the operation of hidden motives in Freud's sense has yet to be cogently tested on an adequate scale. And until it is, the widespread belief in psychoanalytic theory in some segments of our culture is ill founded.[5] For just the reasons I recapitulated in the *Introduction* (§1) from Eagle (1983) and Grünbaum (1984, chapter 7), this unfavorable verdict applies alike to Freud's original formulations and to any and all post-Freudian, revisionist versions of psychoanalysis that rely on the founding father's clinical methods of justifying causal inferences.

Notes

1. Sigmund Freud, "The Question of a Weltanschauung," in *Standard Edition of the Complete Psychological Works of Sigmund Freud,* vol. 22, trans. J. Strachey et al. (London: Hogarth Press, 1955), p. 159. This paper first appeared in 1933. Hereafter any references to Freud's writings in English will be to this *Standard Edition* under its acronym "S.E." followed by the year of first appearance, volume number, and page(s). Thus the 1933 paper just cited in full would be cited within the text as: S.E. 1933, 22:159.

2. Compare also the summary of Loftus's work (1980) in my book (1984, 243–44).

3. "I Was Beheaded in the 1700s," *Time,* September 10, 1984, 68.

4. Compare "Breaking the Spell of Hypnosis," *Time,* September 17, 1984, 62.

5. For a further discussion of the issues raised in this essay, the reader is referred to my "Author's Response" (1986b) to nearly forty commentaries on my (1984) book, *The Foundations of Psychoanalysis: A Philosophical Critique.*

References

Bettelheim, B. 1982. *Freud and Man's Soul.* New York: Random House.

Dywan, J. and Bowers, K. 1983. The Use of Hypnosis to Enhance Recall. *Science* 222 (October 14):184–85.

Eagle, M. 1983. "*The Epistemological Status of Recent Developments in Psychoanalytic Theory.*" In: *Physics, Philosophy and Psychoanalysis,* eds., R.S. Cohen and L. Laudan. Boston: Reidel.

Eissler, K. R. 1969. Irreverent Remarks About the Present and the Future of Psychoanalysis. *International Journal of PsychoAnalysis* 50:461–71.

Ellenberger, H. F. 1970. *The Discovery of the Unconscious.* New York: Basic Books.

Erikson, E. H. 1954. The Dream Specimen of Psychoanalysis. *Journal of the American Psychoanalytic Association* 2:5–56.

Eysenck, H., and Wilson, G. D. 1973. *The Experimental Study of Freudian Theories.* London: Methuen.

Fancher, R. E. 1973. *Psychoanalytic Psychology.* New York: Norton.

Fisher, S., and Greenberg, R. D. 1977. *The Scientific Credibility of Freud's Theories and Therapy.* New York: Basic Books.

Freud, S. 1955. *Standard Edition of the Complete Psychological Works of Sigmund Freud,* trans. J. Strachey, et al. London: Hogarth Press.

Gay, P. 1985. *Freud for Historians.* New York: Oxford University Press.

———. 1988. *Freud.* New York: W. W. Norton.

Grünbaum, A. 1984. *The Foundations of Psychoanalysis: A Philosophical Critique.* Berkeley, London: University of California Press.

———. 1986a. The Placebo Concept in Medicine and Psychiatry. *Psychological Medicine* 16:19–38.

———. 1986b. Précis of *The Foundations of Psychoanalysis: A Philosophical Critique,* and Author's Response: Is Freud's Theory Well-Founded? *The Behavioral and Brain Sciences* 9:217–28 and 266–84.

———. 1988. The Role of the Case Study Method in the Foundations of Psychoanalysis. *Canadian Journal of Philosophy* 18: 623–58.

———. 1989."The Degeneration of Popper's Theory of Demarcation." In *Freedom and Rationality*, eds. F. D'Agostino and I. C. Jarvie. Boston: Reidel, 141–61.

Hobson, A. J. 1988. *The Dreaming Brain*. New York: Basic Books.

Kline, P. 1981. *Fact and Fantasy in Freudian Theory,* 2d ed. London: Methuen.

Lifton, R. J. 1980. "Psychohistory." In: *Comprehensive Textbook of Psychiatry,* 3d ed., vol. 3, eds. A. M. Freedman, H. T. Kaplan, and B. J. Sadock, 3104–12.

Loftus, E. 1980. *Memory*. Reading, Mass.: Addison-Wesley.

Malcolm, J. 1981. *Psychoanalysis: The Impossible Profession*. New York: Knopf.

Masson, J. M. 1984. *The Assault on Truth*. New York: Farrar, Straus & Giroux.

S.E. (See Freud, S., above.)

Wallerstein, R. S. 1986. Psychoanalysis as a Science: A Response to the New Challenges (Freud Anniversary Lecture). *Psychoanalytic Quarterly* 55: 414–51.

———. 1988. Psychoanalysis, Psychoanalytic Science, and Psychoanalytic Research – 1986. *Journal of the American Psychoanalytic Association* 36: 3–30.

Zamansky, H. 1958. An Investigation of the Psychoanalytic Theory of Paranoid Delusions. *Journal of Personality* 26:410–25.

On the Nature of Theories: A Neurocomputational Perspective

I. The Classical View of Theories

Not long ago, we all knew what a theory was: it was a set of sentences or propositions, expressible in the first-order predicate calculus. And we had what seemed to be excellent reasons for that view. Surely any theory had to be stat*able*. And after it had been fully stated, as a set of sentences, what residue remained? Furthermore, the sentential view made systematic sense of how theories could perform the primary business of theories, namely, prediction, explanation, and intertheoretic reduction. It was basically a matter of first-order deduction from the sentences of the theory conjoined with relevant premises about the domain at hand.

Equally important, the sentential view promised an account of the nature of learning, and of rationality. Required was a set of formal rules to dictate appropriate changes or updates in the overall set of believed sentences as a function of new beliefs supplied by observation. Of course there was substantial disagreement about which rules were appropriate. Inductivists, falsificationists, hypothetico-deductivists, and Bayesian subjectivists each proposed a different account of them. But the general approach seemed clearly correct. Rationality would be captured as the proper set of formal rules emerged from logical investigation.

Finally, if theories are just sentences, then the ultimate virtue of a theory is truth. And it was widely expected that an adequate account of rational methodol-

Several pages in section 4 are based on material from an earlier paper, "Reductionism, Connectionism, and the Plasticity of Human Consciousness," *Cultural Dynamics*, 1 (1): 1988. Three pages in section 5 are drawn from "Simplicity: The View from the Neuronal Level." In *Aesthetic Factors in Natural Science*, ed. N. Rescher (forthcoming, 1990). My thanks to the editors for permission to use that material here. For many useful discussions, thanks also to Terry Sejnowski, Patricia Churchland, David Zipser, Dave Rumelhart, Francis Crick, Stephen Stich, and Philip Kitcher.

ogy would reveal why humans must tend, in the long run, toward theories that are true.

Hardly anyone will now deny that there are serious problems with every element of the preceding picture—difficulties we shall discuss below. Yet the majority of the profession is not yet willing to regard them as fatal. I profess myself among the minority that does so regard them. In urging the poverty of 'sentential epistemologies' for over a decade now (Churchland 1975, 1979, 1981, 1986), I have been motivated primarily by the *pattern* of the failures displayed by that approach. Those failures suggest to me that what is defective in the classical approach is its fundamental assumption that languagelike structures of some kind constitute the basic or most important form of representation in cognitive creatures, and the correlative assumption that cognition consists in the manipulation of those representations by means of structure-sensitive rules.

To be sure, not everyone saw the same pattern of failure, nor were they prepared to draw such a strong conclusion even if they did. For any research program has difficulties, and so long as we lack a comparably compelling *alternative* conception of representation and computation, it may be best to stick with the familiar research program of sentences and rules for their manipulation.

However, it is no longer true that we lack a comparably compelling alternative approach. Within the last five years, there have been some striking theoretical developments and experimental results within cognitive neurobiology and 'connectionist' AI (artificial intelligence). These have provided us with a powerful and fertile framework with which to address problems of cognition, a framework that owes nothing to the sentential paradigm of the classical view. My main purpose in this essay is to make the rudiments of that framework available to a wider audience, and to explore its far-reaching consequences for traditional issues in the philosophy of science. Before turning to this task, let me prepare the stage by briefly summarizing the principal failures of the classical view, and the most prominent responses to them.

II. Problems and Alternative Approaches

The depiction of learning as the rule-governed updating of a system of sentences or propositional attitudes encountered a wide range of failures. For starters, even the best of the rules proposed failed to reproduce reliably our preanalytic judgments of credibility, even in the artificially restricted or 'toy' situations in which they were asked to function. Paradoxes of confirmation plagued the H-D accounts (Hempel 1965; Scheffler 1963). The indeterminacy of falsification plagued the Popperian accounts (Lakatos 1970; Feyerabend 1970; Churchland 1975). Laws were assigned negligible credibility on Carnapian accounts (Salmon, 1966). Bayesian accounts, like Carnapian ones, presupposed a given probability space as the epistemic playground within which learning takes place, and they could not account for the rationality of major shifts from one probability space to another, which is what the most interesting and important cases of learning amount to. The rationality of large-scale *conceptual change*, accordingly, seemed beyond the reach of such approaches. Furthermore, simplicity emerged as a major determinant of theoretical credibility on most accounts, but none of them could provide an adequate definition of simplicity in syntactic terms, or give a convincing explanation of why it was relevant to truth or credibility in any case. One could begin to question whether the basic factors relevant to learning were to be found at the linguistic level at all.

Beyond these annoyances, the initial resources ascribed to a learning subject by the sentential approach plainly presupposed the successful completion of a good deal of sophisticated learning on the part of that subject already. For example, reliable observation judgments do not just appear out of nowhere. Living subjects have to *learn* to make the complex perceptual discriminations that make perceptual judgments possible. And they also have to *learn* the linguistic or propositional system within which their beliefs are to be constituted. Plainly, both cases of learning will have to involve some procedure quite distinct from that of the classical account. For that account presupposes antecedent possession of both a determinate propositional system and a capacity for determinate perceptual judgment, which is precisely what, prior to extensive learning, the human infant lacks. Accordingly, the classical story cannot possibly account for all cases of learning. There must exist a type of learning that is prior to and more basic than the process of sentence manipulation at issue.

Thus are we led rather swiftly to the idea that there is a level of representation *beneath* the level of the sentential or propositional attitudes, and to the correlative idea that there is a learning dynamic that operates primarily on sublinguistic factors. This idea is reinforced by reflection on the problem of cognition and learning in nonhuman animals, none of which appear to have the benefit of language, either the external speech or the internal structures, but all of which engage in sophisticated cognition. Perhaps their cognition proceeds entirely without benefit of any system for processing sentencelike representations.

Even in the human case, the depiction of one's knowledge as an immense set of individually stored 'sentences' raises a severe problem concerning the relevant retrieval or application of those internal representations. How is it one is able to retrieve, from the millions of sentences stored, exactly the handful that is relevant to one's current predictive or explanatory problem, and how is it one is generally able to do this in a few tenths of a second? This is known as the "Frame Problem" in AI, and it arises because, from the point of view of fast and relevant retrieval, a long list of sentences is an appallingly inefficient way to store information. And the more information a creature has, the worse its application problem becomes.

A further problem with the classical view of learning is that it finds no essential connection whatever between the learning of *facts* and the learning of *skills*. This is a problem in itself, since one might have hoped for a unified account of learning, but it is doubly a problem when one realizes that so much of the business of understanding a theory and being a scientist is a matter of the skills one has acquired. Memorizing a set of sentences is not remotely sufficient: one must learn to *recognize* the often quite various instances of the terms they contain; one must learn to *manipulate* the peculiar formalism in which they may be embedded; one must learn to *apply* the formalism to novel situations; one must learn to *control* the instruments that typically produce or monitor the phenomena at issue. As T. S. Kuhn first made clear (Kuhn 1962), these dimensions of the scientific trade are only artificially separable from one's understanding of its current theories. It begins to appear that even if we do harbor internal sentences, they capture only a small part of human knowledge.

These failures of the classical view over the full range of learning, both in humans and in nonhuman animals, are the more suspicious given the classical view's total disconnection from any theory concerning the structure of the biological *brain*, and the manner in which it might *implement* the kind of representations and computations proposed. Making acceptable contact with neurophysiological theory is a long-term constraint on any epistemology: a scheme of representation and computation that cannot be implemented in the machinery of the human brain cannot be an adequate account of human cognitive activities.

The situation on this score used to be much better than it now is: it was clear that the classical account of representation and learning could easily be realized in typical digital computers, and it was thought that the human brain would turn out to be relevantly like a digital computer. But quite aside from the fact that computer implementations of sentential learning chronically produced disappointing results, it has become increasingly clear that the brain is organized along computational lines radically different from those employed in conventional digital computers. The brain, as we shall see below, is a massively parallel processor, and it performs computational tasks of the classical kind at issue only very slowly and comparatively badly. To speak loosely, it does not appear to be designed to perform the tasks the classical view assigns to it.

I conclude this survey by returning to specifically philosophical matters. A final problem with the classical approach has been the failure of all attempts to explain why the learning process must tend, at least in the long run, to lead us toward *true* theories. Surprisingly, and perhaps distressingly, this Panglossean hope has proved very resistant to vindication (Van Fraassen 1980; Laudan 1981). Although the history of human intellectual endeavor does support the view that, over the centuries, our theories have become dramatically *better* in many dimensions, it is quite problematic whether they are successively 'closer' to 'truth'. Indeed, the notion of truth itself has recently come in for critical scrutiny (Putnam 1981; Churchland 1985; Stich 1990). It is no longer clear that there *is* any unique and unitary relation that virtuous belief systems must bear to the nonlinguistic world. Which leaves us free to reconsider the great many different dimensions of epistemic and pragmatic virtue that a cognitive system can display.

The problems of the preceding pages have not usually been presented in concert, and they are not usually regarded as conveying a unitary lesson. A few philosophers, however, have been moved by them, or by some subset of them, to suggest significant modifications in the classical framework. One approach that has captured some adherents is the 'semantic view' of theories (Suppe 1974; Van Fraassen 1980; Giere 1988). This approach attempts to drive a wedge between a theory and its possibly quite various linguistic formulations by characterizing a theory as a *set of models*, those that will make a first-order linguistic statement of the theory come out *true* under the relevant assignments. The models in the set all share a common abstract structure, and that structure is what is important about any theory, according to the semantic view, not any of its idiosyncratic linguistic expressions. A theory is true, on this view, just in case it includes the actual world, or some part of it, as one of the models in the set.

This view buys us some advantages, perhaps, but I find it to be a relatively narrow response to the panoply of problems addressed above. In particular, I think it strange that we should be asked, at this stage of the debate, to embrace an account of theories that has absolutely nothing to do with the question of how real physical systems might embody representations of the world, and how they might execute principled computations on those representations in such a fashion as to learn. Prima facie, at least, the semantic approach takes theories even farther into Plato's Heaven, and away from the buzzing brains that use them, than did the view that a theory is a set of sentences. This complaint does not do justice to the positive virtues of the semantic approach (see especially Giere, whose version does make some contact with current cognitive psychology). But it is clear that the semantic approach is a response to only a small subset of the extant difficulties.

A more celebrated response is embodied in Kuhn's *The Structure of Scientific Revolutions* (1962). Kuhn centers our attention not on sets of sentences, nor on sets of models, but on what he calls paradigms or exemplars, which are specific *applications* of our conceptual, mathematical, and instrumental resources. Mastering a theory, on this view, is more a matter of being able to perform in various ways, of being able to solve a certain class of problems, of being able to recognize diverse situations as relevantly similar to that of the original or paradigmatic application. Kuhn's view brings to the fore the historical, the sociological, and the psychological factors that structure our theoretical cognition. Of central importance is the manner in which one comes to perceive the world as one internalizes a theory. The perceptual world is redivided into new categories, and while the theory may be able to provide necessary and sufficient conditions for being an instance of any of its categories, the perceptual recognition of any instance of a category does not generally proceed by reference to those conditions, which often transcend perceptual experience. Rather, perceptual recognition proceeds by some inarticulable process that registers *similarity* to one or more perceptual *prototypes* of the category at issue. The recognition of new applications of the apparatus of the entire theory displays a similar dynamic. In all, a successful theory provides a prototypical beachhead that one attempts to expand by analogical extensions to new domains.

Reaction to this view has been deeply divided. Some applaud Kuhn's move toward naturalism, toward a performance conception of knowledge, and away from the notion of truth as the guiding compass of cognitive activity (Munevar 1981; Stich 1990). Others deplore his neglect of normative issues, his instrumentalism and relativism, and his alleged exaggeration of certain lessons from perceptual and developmental psychology (Fodor 1984). We shall address these issues later in the paper.

A third and less visible reaction to the classical difficulties has simply rejected the sentential or propositional attitudes as the most important form of representation used by cognitive creatures, and has insisted on the necessity of empirical and theoretical research into *brain* function in order to answer the question of what *are* the most important forms of representation and computation within cognitive creatures. Early statements can be found in Churchland 1975 and Hooker 1975; extended arguments appear in Churchland 1979 and 1981; and further arguments appear in Churchland, P.S., 1980 and 1986, and in Hooker 1987.

While the antisentential diagnosis could be given some considerable support, as the opening summary of this section illustrates, neuroscience as the recommended cure was always more difficult to sell, given the functional opacity of the biological brain. Recently, however, this has changed dramatically. We now have some provisional insight into the functional significance of the brain's microstructure, and some idea of how it represents and computes. What has been discovered so far appears to vindicate the claims of philosophical relevance and the expectations of fertility in this area, and it appears to vindicate some central elements in Kuhn's perspective as well. This neurofunctional framework promises to sustain wholly new directions of cognitive research. In the sections to follow I shall try to outline the elements of this framework and its applications to some familiar problems in the philosophy of science. I begin with the physical structure and the basic activities of the brainlike systems at issue.

III. Elementary Brainlike Networks

The functional atoms of the brain are cells called neurons (figure 1). These have a natural or default level of activity that can, however, be modulated up or down by external influences. From each neuron there extends a long, thin output fiber called an *axon*, which typically branches at the far end so as to make a large number of *synaptic connections* with either the central cell body or the bushy *den-*

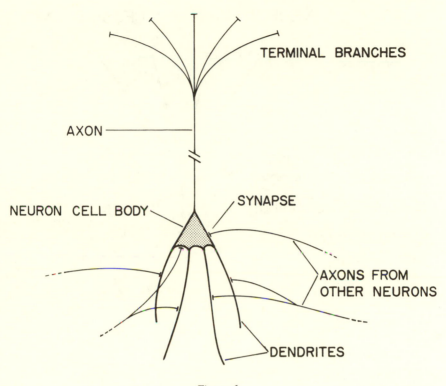

Figure 1.

drites of other neurons. Each neuron thus receives inputs from a great many other neurons, which inputs tend to excite (or to inhibit, depending on the type of synaptic connection) its normal or default level of activation. The level of activation induced is a function of the *number* of connections, of their size or *weight*, of their *polarity* (stimulatory or inhibitory), and of the *strength* of the incoming signals. Furthermore, each neuron is constantly emitting an output signal along its own axon, a signal whose strength is a direct function of the overall level of activation in the originating cell body. That signal is a train of pulses or *spikes*, as they are called, which are propagated swiftly along the axon. A typical cell can emit spikes along its axon at anything between zero and perhaps 200 Hz. Neurons, if you like, are humming to one another, in basso notes of varying frequency.

The networks to be explored attempt to simulate natural neurons with artifical units of the kind depicted in figure 2. These units admit of various levels of activation, which we shall assume to vary between 0 and 1. Each unit receives input signals from other units via 'synaptic' connections of various weights and polari-

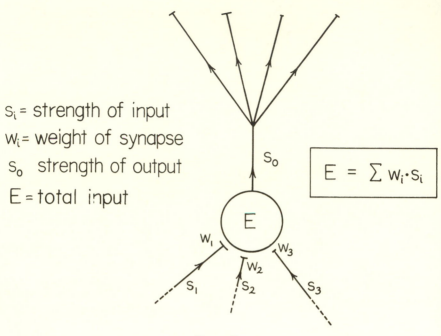

NEURON–LIKE PROCESSING UNIT

s_i = strength of input
w_i = weight of synapse
s_o strength of output
E = total input

$$E = \sum w_i \cdot s_i$$

Figure 2.

ties. These are represented in the diagram as small end-plates of various sizes. For simplicity's sake, we dispense with dendritic trees: the axonal end branches from other units all make connections directly to the 'cell body' of the receiving unit. The total modulating effect E impacting on that unit is just the sum of the contributions made by each of the connections. The contribution of a single connection is just the product of its weight w_i times the strength s_i of the signal arriving at that connection. Let me emphasize that if for some reason the connection weights were to change over time, then the unit would receive a quite different level of overall excitation or inhibition in response to the very same configuration of input signals.

Turn now to the output side of things. As a function of the total input E, the unit modulates its activity level and emits an output signal of a certain strength s_o along its 'axonal' output fiber. But s_o is not a direct or *linear* function of E. Rather, it is an S-shaped function as in figure 3. The reasons for this small wrinkle will emerge later. I mention it here because its inclusion completes the

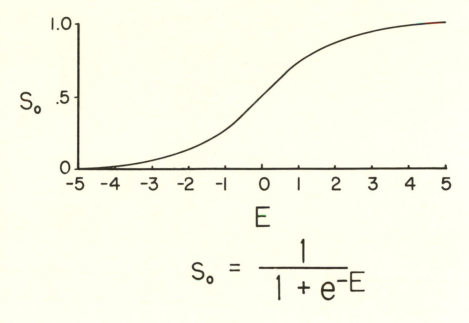

$$S_0 = \frac{1}{1 + e^{-E}}$$

Figure 3.

story of the elementary units. Of their intrinsic properties, there is nothing left to tell. They are very simple indeed.

It remains to arrange them into networks. In the brain, neurons frequently consitute a population, all of which send their axons to the site of a second population of neurons, where each arriving axon divides into terminal end branches in order to make synaptic connections with many different cells within the target population. Axons from cells in this second population can then project to a third population of cells, and so on. This is the inspiration for the arrangement of figure 4.

The units in the bottom or input layer of the network may be thought of as 'sensory' units, since the level of activation in each is directly determined by aspects of the environment (or perhaps by the experimenter, in the process of simulating some environmental input). The activation level of a given input unit is designed to be a response to a specific aspect or dimension of the overall input stimulus that strikes the bottom layer. The assembled set of simultaneous activation levels in all of the input units is the network's *representation* of the input stimulus. We may refer to that configuration of stimulation levels as the *input vector*, since it is just an ordered set of numbers or magnitudes. For example, a given stimulus might produce the vector ⟨.5, .3, .9, .2⟩.

These input activation levels are then propagated upwards, via the output sig-

A SIMPLE NETWORK

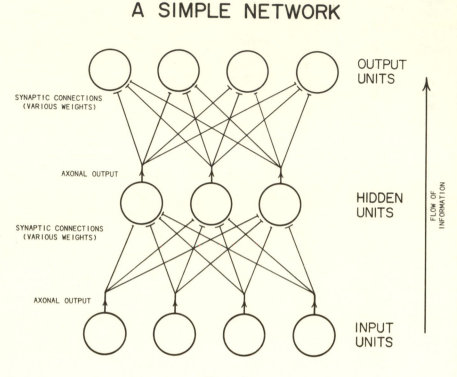

Figure 4.

nal in each unit's axon, to the middle layer of the network, to what are called the *hidden units*. As can be seen in figure 4, any unit in the input layer makes a synaptic connection of some weight or other with every unit at this intermediate layer. Each hidden unit is thus the target of several inputs, one for each cell at the input layer. The resulting activation level of a given hidden unit is essentially just the sum of all of the influences reaching it from the cells in the lower layer.

The result of this upward propagation of the input vector is a set of activation levels across the three units in the hidden layer, called the *hidden unit activation vector*. The values of that three-element vector are strictly determined by

(a) the makeup of the *input vector* at the input layer, and
(b) the various values of the *connection weights* at the ends of the terminal branches of the input units.

What this bottom half of the network does, evidently, is convert or transform one activation vector into another.

The top half of the network does exactly the same thing, in exactly the same way. The activation vector at the hidden layer is propagated upward to the output (topmost) layer of units, where an *output vector* is produced, whose character is determined by

(a) the makeup of the activation vector at the hidden layer, and
(b) the various values of the connection weights at the ends of the terminal branches of the hidden units.

Looking now at the whole network, we can see that it is just a device for transforming any given input-level activation vector into a uniquely corresponding output-level activation vector. And what determines the character of the global transformation effected is the peculiar set of values possessed by the many connection weights. This much is easy to grasp. What is not so easy to grasp, prior to exploring examples, is just how very powerful and useful those transformations can be. So let us explore some real examples.

IV. Representation and Learning in Brainlike Networks

A great many of the environmental features to which humans respond are difficult to define or characterize in terms of their purely physical properties. Even something as mundane as being the vowel sound $\bar{a}$, as in "rain," resists such characterization, for the range of acoustical variation among acceptable and recognizable $\bar{a}$s is enormous. A female child at two years and a basso male at fifty will produce quite different sorts of atmospheric excitations in pronouncing this vowel, but each sound will be easily recognized as an $\bar{a}$ by other members of the same linguistic culture.

I do not mean to suggest that the matter is utterly intractable from a physical point of view, for an examination of the acoustical power spectrum of voiced vowels begins to reveal some of the similarities that unite $\bar{a}$s. And yet the analysis continues to resist a simple list of necessary and sufficient physical conditions on being an $\bar{a}$. Instead, being an $\bar{a}$ seems to be a matter of being *close enough* to a *typical $\bar{a}$* sound along a *sufficient* number of distinct *dimensions of relevance*, where each notion in italics remains difficult to characterize in a nonarbitrary way. Moreover, some of those dimensions are highly contextual. A sound type that would not normally be counted or recognized as an $\bar{a}$ when voiced in isolation may be unproblematically so counted if it regularly occurs, in someone's modestly accented speech, in all of the phonetic places that would normally be occupied by $\bar{a}$s. Evidently, what makes something an $\bar{a}$ is in part a matter of the entire linguistic surround. In this way do we very quickly ascend to the abstract and holistic level, for even the simplest of culturally embedded properties.

What holds for phonemes holds also for a great many other important features recognizable by us—colors, faces, flowers, trees, animals, voices, smells, feelings, songs, words, meanings, and even metaphorical meanings. At the outset, the categories and resources of physics, and even neuroscience, look puny and impotent in the face of such subtlety.

And yet it is a purely physical system that recognizes such intricacies. Short of appealing to magic, or of simply refusing to confront the problem at all, we must assume that some configuration of purely physical elements is capable of grasping and manipulating these features, and by means of purely physical principles. Surprisingly, networks of the kind described in the preceding section have many of the properties needed to address precisely this problem. Let me explain.

Suppose we are submarine engineers confronted with the problem of designing a sonar system that will distinguish between the sonar echoes returned from explosive mines, such as might lie on the bottom of sensitive waterways during wartime, and the sonar echoes returned from rocks of comparable sizes that dot the same underwater landscapes. The difficulty is twofold: echoes from both objects sound indistinguishable to the casual ear, and echoes from each type show wide variation in sonic character, since both rocks and mines come in various sizes, shapes, and orientations relative to the probing sonar pulse.

Enter the network of figure 5. This one has thirteen units at the input layer, since we need to code a fairly complex stimulus. A given sonar echo is run through a frequency analyzer, and is sampled for its relative energy levels at thirteen frequencies. These thirteen values, expressed as fractions of 1, are then entered as activation levels in the respective units of the input layer, as indicated in figure 5. From here they are propagated through the network, being transformed as they go, as explained earlier. The result is a pair of activation levels in the two units at the output layer. We need only two units here, for we want the network eventually to produce an output activation vector at or near $\langle 1, 0 \rangle$ when a mine echo is entered as input, and an output activation vector at or near $\langle 0, 1 \rangle$ when a rock echo is entered as input. In a word, we want it to *distinguish* mines from rocks.

It would of course be a miracle if the network made the desired discrimination immediately, since the connection weights that determine its transformational activity are initially set at random values. At the beginning of this experiment then, the output vectors are sure to disappoint us. But we proceed to *teach* the network by means of the following procedure.

We procure a large set of recorded samples of various (genuine) mine echoes, from mines of various sizes and orientations, and a comparable set of genuine rock echoes, keeping careful track of which is which. We then feed these echoes into the network, one by one, and observe the output vector produced in each case. What interests us in each case is the amount by which the actual output vector *differs* from what would have been the 'correct' vector, given the identity of the specific echo that produced it. The details of that error, for each element of the output vector, are then fed into a special rule that computes a set of small

PERCEPTUAL RECOGNITION
WITH A
LARGE NETWORK

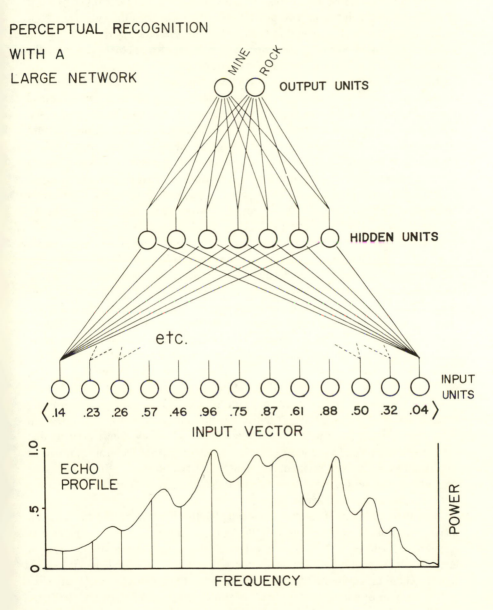

Figure 5.

changes in the values of the various synaptic weights in the system. The idea is to identify those weights most responsible for the error, and then to nudge their values in a direction that would at least reduce the amount by which the output vector is in error. The slighty modified system is then fed another echo from the training set, and the entire procedure is repeated.

This provides the network with a 'teacher'. The process is called "training up the network," and it is standardly executed by an auxiliary computer programmed to feed samples from the training set into the network, monitor its responses, and adjust the weights according to the special rule after each trial. Under the pressure of such repeated corrections, the behavior of the network slowly converges on the behavior we desire. That is to say, after several thousands of presentations of recorded echoes and subsequent adjustments, the network starts to give the right answer close to 90 percent of the time. When fed a mine echo, it generally gives something close to a $\langle 1, 0 \rangle$ output. And when fed a rock echo, it generally gives something close to a $\langle 0, 1 \rangle$.

A useful way to think of this is captured in figure 6. Think of an abstract space of many dimensions, one for each weight in the network (105 in this case), plus one dimension for representing the overall error of the output vector on any given trial. Any point in that space represents a unique configuration of weights, plus the performance error that that configuration produces. What the learning rule does is steadily nudge that configuration away from erroneous positions and toward positions that are less erroneous. The system inches its way down an 'error gradient' toward a global error minimum. Once there, it responds reliably to the relevant kinds of echoes. It even responds well to echoes that are 'similar' to mine echoes, by giving output vectors that are closer to $\langle 1, 0 \rangle$ than to $\langle 0, 1 \rangle$.

There was no guarantee the network would succeed in learning to discriminate the two kinds of echoes, because there was no guarantee that rock echoes and mine echoes would differ in any systematic or detectable way. But it turns out that mine echoes do indeed have some complex of relational or structural features that distinguishes them from rock echoes, and under the pressure of repeated error corrections, the network manages to lock onto, or become 'tuned' to, that subtle but distinctive weave of features.

We can test whether it has truly succeeded in this by now feeding the network some mine and rock echoes not included in the training set, echoes it has never encountered before. In fact, the network does almost as well classifying the new echoes as it does with the samples in its training set. The 'knowledge' it has acquired generalizes quite successfully to new cases. (This example is a highly simplified account of some striking results from Gorman and Sejnowski 1988.)

All of this is modestly amazing, because the problem is quite a difficult one, at least as difficult as learning to discriminate the phoneme $\bar{a}$. Human sonar operators, during a long tour of submarine duty, eventually learn to distinguish the two kinds of echoes with some uncertain but nontrivial regularity. But they never per-

LEARNING: GRADIENT DESCENT IN WEIGHT SPACE

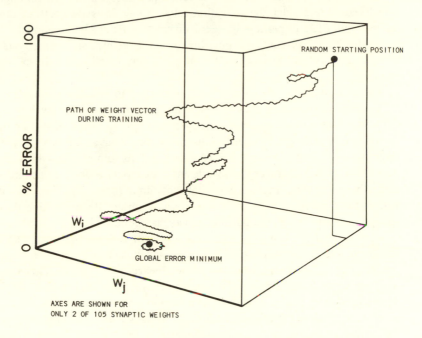

Figure 6.

form at the level of the artificial network. Spurred on by this success, work is currently underway to train up a network to distinguish the various phonemes characteristic of English speech (Zipser and Elman 1988). The idea is to produce a speech-recognition system that will not be troubled by the acoustic idiosyncracies of diverse speakers, as existing speech-recognition systems are.

The success of the mine/rock network is further intriguing because the 'knowledge' the network has acquired, concerning the distinctive character of mine echoes, consists of nothing more than a carefully orchestrated set of connection weights. And it is finally intriguing because there exists a learning algorithm—the rule for adjusting the weights as a function of the error displayed in the output vector—that will eventually produce the required set of weights, given sufficient examples on which to train the network (Rumelhart et al. 1986).

How can a set of connection weights possibly embody knowledge of the desired distinction? Think of it in the following way. Each of the thirteen input units represents one aspect or dimension of the incoming stimulus. Collectively, they give a simultaneous profile of the input echo along thirteen distinct dimensions. Now perhaps there is only one profile that is roughly characteristic of mine echoes; or perhaps there are many different profiles, united by a common relational feature (e.g., that the activation value of unit #6 is always three times the value of unit #12); or perhaps there is a disjunctive set of such relational features; and so forth. In each case, it is possible to rig the weights so that the system will respond in a typical fashion, at the output layer, to all and only the relevant profiles.

The units at the hidden layer are very important in this. If we consider the abstract space whose seven axes represent the possible activation levels of each of the seven hidden units, then what the system is searching for during the training period is a set of weights that *partitions* this space so that any mine input produces an activation vector across the hidden units that falls somewhere within one large subvolume of this abstract space, while any rock input produces a vector that falls somewhere into the complement of that subvolume (figure 7). The job of the top half of the network is then the relatively easy one of distinguishing these two subvolumes into which the abstract space has been divided.

Vectors near the center of (or along a certain path in) the mine-vector subvolume represent *prototypical* mine echoes, and these will produce an output vector very close to the desired ⟨1, 0⟩. Vectors nearer to the surface (strictly speaking, the *hyper*surface) that partitions the abstract space represent atypical or problematic mine echoes, and these produce more ambiguous output vectors such as ⟨.6, .4⟩. The network's discriminative responses are thus graded responses: the system is sensitive to *similarities* along all of the relevant dimensions, and especially to rough conjunctions of these subordinate similarities.

So we have a system that learns to discriminate hard-to-define perceptual features, and to be sensitive to similarities of a comparably diffuse but highly relevant character. And once the network is trained up, the recognitional task takes only a split second, since the system processes the input stimulus in parallel. It finally gives us a discriminatory system that performs something like a living creature, both in its speed and in its overall character.

I have explained this system in some detail, so that the reader will have a clear idea of how things work in at least one case. But the network described is only one instance of a general technique that works well in a large variety of cases. Networks can be constructed with a larger number of units at the output layer, so as to be able to express not just two, but a large number of distinct discriminations.

One network, aptly called NETtalk by its authors (Rosenberg and Sejnowski 1987), takes vector codings for seven-letter segments of printed words as inputs, and gives vector codings for phonemes as outputs. These output vectors can be fed directly into a sound synthesizer as they occur, to produce audible sounds. What this network learns to do is to transform printed words into audible speech. Though it involves no understanding of the words that it 'reads', the network's feat

LEARNED PARTITION ON
HIDDEN UNIT ACTIVATION-VECTOR SPACE

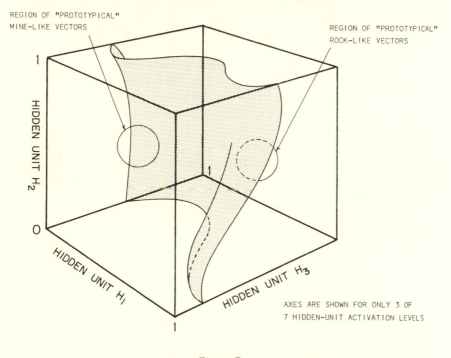

Figure 7.

is still very impressive, because it was given no rules whatever concerning the phonetic significance of standard English spelling. It began its training period by producing a stream of unintelligible babble in response to text entered as input. But in the course of many thousands of word presentations, and under the steady pressure of the weight-nudging algorithm, the set of weights slowly meanders its way to a configuration that reduces the measured error close to zero. After such training it will then produce as output, given arbitrary English text as input, perfectly intelligible speech with only rare and minor errors.

This case is significant for a number of reasons. First, the trained network makes a large number of discriminations (79, in fact), not just a binary one. Second, it contains no explicit representation of any *rules*, however much it might seem to be following a set of rules. Third, it has mastered an input/output transformation that is notoriously irregular, and it must be sensitive to lexical context in order to do so. (Specifically, the phoneme it assigns to the center or focal letter of its seven-letter input is in large part a function of the identity of the three letters on either side.) And fourth, it portrays some aspects of a 'sensori*motor*' skill, rather than a purely sensory skill: it is producing highly complex behavior.

NETtalk has some limitations, of course. Pronunciations that depend on specifically semantical or grammatical distinctions will generally elude its grasp (unless they happen to be reflected in some way in the corpus of its training words, as occasionally they are), since NETtalk knows neither meanings nor syntax. But such dependencies affect only a very small percentage of the transformations appropriate to any text, and they are in any case to be expected. To overcome them completely would require a network that actually understands the text being read. And even then mistakes would occur, for even humans occasionally misread words as a result of grammatical or semantical confusion. What is arresting about NETtalk is just how very much of the complex and irregular business of text-based pronunciation can be mastered by a simple network with only a few hundred neuronlike units.

Another rather large network (by Lehky and Sejnowski 1988a, 1988b) addresses problems in vision. It takes codings for smoothly varying gray-scale pictures as input, and after training it yields as outputs surprisingly accurate codings for the curvatures and orientations of the physical objects portrayed in the pictures. It solves a form of the 'shape from shading' problem long familiar to theorists in the field of vision. This network is of special interest because a subsequent examination of the 'receptive fields' of the trained hidden units shows them to have acquired some of the same response properties as are displayed by cells in the visual cortex of mature animals. Specifically, they show a maximum sensitivity to spots, edges, and bars in specific orientations. This finding echoes the seminal work of Hubel and Wiesel (1962), in which cells in the visual cortex were discovered to have receptive fields of this same character. Results of this kind are very important, for if we are to take these artificial networks as models for how the brain works, then they must display realistic behavior not just at the macrolevel: they must also display realistic behavior at the microlevel.

Enough examples. You have seen something of what networks of this kind can do, and of how they do it. In both respects they contrast sharply with the kinds of representational and processing strategies that philosophers of science, inductive logicians, cognitive psychologists, and AI workers have traditionally ascribed to us (namely, sentencelike representations manipulated by formal rules). You can see also why this theoretical and experimental approach has captured the interest of those who seek to understand how the microarchitecture of the biological brain produces the phenomena displayed in human and animal cognition. Let us now explore the functional properties of these networks in more detail, and see how they bear on some of the traditional issues in epistemology and the philosophy of science.

V. Some Functional Properties of Brainlike Networks

The networks described above are descended from a device called the *Percep-tron* (Rosenblatt 1959), which was essentially just a two-layer as opposed to a three-layer network. Devices of this configuration could and did learn to dis-criminate a considerable variety of input patterns. Unfortunately, having the in-put layer connected directly to the output layer imposes very severe limitations on the range of possible transformations a network can perform (Minsky and Papert 1969), and interest in Perceptron-like devices was soon eclipsed by the much faster-moving developments in standard 'program-writing' AI, which ex-ploited the high-speed general-purpose digital machines that were then starting to become widely available. Throughout the seventies, research in artificial 'neural nets' was an underground program by comparison.

It has emerged from the shadows for a number of reasons. One important fac-tor is just the troubled doldrums into which mainstream or program-writing AI has fallen. In many respects, these doldrums parallel the infertility of the classical approach to theories and learning within the philosophy of science. This is not surprising, since mainstream AI was proceeding on many of the same basic as-sumptions about cognition, and many of its attempts were just machine im-plementations of learning algorithms proposed earlier by philosophers of science and inductive logicians (Glymour 1987). The failures of mainstream AI—unrealistic learning, poor performance in complex perceptual and motor tasks, weak handling of analogies, and snaillike cognitive performance despite the use of very large and fast machines—teach us even more dramatically than do the failures of mainstream philosophy that we need to rethink the style of representa-tion and computation we have been ascribing to cognitive creatures.

Other reasons for the resurgence of interest in networks are more positive. The introduction of additional layers of intervening or 'hidden' units produced a dra-matic increase in the range of possible transformations that the network could effect. As Sejnowski et al. (1986) describe it:

. . . only the first-order statistics of the input pattern can be captured by di-rect connections between input and output units. The role of the hidden units is to capture higher-order statistical relationships and this can be accomplished if significant underlying features can be found that have strong, regular rela-tionships with the patterns on the visible units. The hard part of learning is to find the set of weights which turn the hidden units into useful feature detectors.

Equally important is the S-shaped, nonlinear response profile (figure 3) now as-signed to every unit in the network. So long as this response profile remains lin-ear, any network will be limited to computing purely linear transformations. (A transformation $f(x)$ is *linear* just in case $f(n \bullet x) = n \bullet f(x)$, and $f(x + y) = f(x) + f(y)$.) But a nonlinear response profile for each unit brings the entire range of possible nonlinear transformations within reach of three-layer networks, a dramatic expansion of their computational potential. Now there are *no* transformations that lie beyond the computational power of a large enough and suitably weighted network.

A third factor was the articulation, by Rumelhart, Hinton, and Williams (1986a), of the *generalized delta rule* (a generalization, to three-layer networks, of Rosenblatt's original teaching rule for adjusting the weights of the Perceptron), and the empirical discovery that this new rule very rarely got permanently stuck in inefficient 'local minima' on its way toward finding the best possible configuration of connection weights for a given network and a given problem. This was a major breakthrough, not so much because "learning by the back-propagation of error," as it has come to be called, was just like human learning, but because it provided us with an efficient technology for quickly training up various networks on various problems, so that we could study their properties and explore their potential.

The way the generalized delta rule works can be made fairly intuitive given the idea of an abstract weight space as represented in figure 6. Consider any output vector produced by a network with a specific configuration of weights, a configuration represented by a specific position in weight space. Suppose that this output vector is in error by various degrees in various of its elements. Consider now a single synapse at the ouput layer, and consider the effect on the output vector that a small positive or negative change in its weight would have had. Since the output vector is a determinate function of the system's weights (assuming we hold the input vector fixed), we can calculate which of these two possible changes, if either, would have made the greater improvement in the output vector. The relevant change is made accordingly. (For more detail, see Rumelhart et al. 1986b.)

If a similar calculation is performed over every synapse in the network, and the change in its weight is then made accordingly, what the resulting shift in the position of the system's overall point in weight space amounts to is a small slide *down* the steepest face of the local 'error surface'. Note that there is no guarantee that this incremental shift moves the system directly towards the global position of zero error (that is why perfection cannot be achieved in a single jump). On the contrary, the descending path to a global error minimum may be highly circuitous. Nor is there any guarantee that the system must eventually reach such a global minimum. On the contrary, the downward path from a given starting point may well lead to a merely 'local' minimum, from which only a large change in the system's weights will afford escape, a change beyond the reach of the delta rule. But in fact this happens relatively rarely, for it turns out that the more dimensions (synapses) a system has, the smaller the probability of there being an intersecting local minimum in *every one* of the available dimensions. The global point is usually able to slide down some narrow cleft in the local topography. Empirically then, the back-propagation algorithm is surprisingly effective at driving the system to the global error minimum, at least where we can identify that global minimum effectively.

The advantage this algorithm provides is easily appreciated. The possible combinations of weights in a network increases exponentially with the size of the network. Assuming conservatively that each weight admits of only ten possible values, the number of distinct positions in 'weight space' (i.e., the number of possible weight configurations) for the simple rock/mine network of figure 5 is already 10^{105}! This space is far too large to explore efficiently without something like the generalized delta rule and the back-propagation of error to do it for us. But with the delta rule, administered by an auxiliary computer, researchers have shown that networks of the simple kind described are capable of learning some quite extraordinary skills, and of displaying some highly intriguing properties. Let me now return to an exploration of these.

An important exploratory technique in cognitive and behavioral neuroscience is to record, with an implanted microelectrode, the electrical activity of a single neuron during cognition or behavior in the intact animal. This is relatively easy to do, and it does give us tantalizing bits of information about the cognitive significance of neural activity (recall the results of Hubel and Wiesel mentioned earlier). Single-cell recordings give us only isolated bits of information, however, and what we would really like to monitor are the *patterns* of simultaneous neural activation across large numbers of cells in the same subsystem. Unfortunately, effective techniques for simultaneous recording from large numbers of adjacent cells are still in their infancy. The task is extremely difficult.

By contrast, this task is extremely easy with the artificial networks we have been describing. If the network is real hardware, its units are far more accessible than the fragile and microscopic units of a living brain. And if the network is merely being simulated within a standard computer (as is usually the case), one can write the program so that the activation levels of any unit, or set of units, can be read out on command. Accordingly, once a network has been successfully trained up on some skill or other, one can then examine the collective behavior of its units during the exercise of that skill.

We have already seen the results of one such analysis in the rock/mine network. Once the weights have reached their optimum configuration, the activation vectors (i.e., the patterns of activation) at the hidden layer fall into two disjoint classes: the vector space is partitioned in two, as depicted schematically in figure 7. But a mere binary discrimination is an atypically simple case. The reader NETtalk, for example, partitions its hidden-unit vector space into fully seventy nine subspaces. The reason is simple. For each of the twenty six letters in the alphabet, there is at least one phoneme assigned to it, and for many letters there are several phonemes that might be signified, depending on the lexical context. As it happens, there are seventy nine distinct letter-to-phoneme associations to be learned if one is to master the pronunciation of English spelling, and in the successfully trained network a distinct hidden-unit activation vector occurs when each of these seventy nine possible transformations is effected.

In the case of the rock/mine network, we noted a similarity metric within each of its two hidden-unit subspaces. In the case of NETtalk, we also find a similarity metric, this time across the seventy nine functional hidden-unit vectors (by 'functional vector', I mean a vector that corresponds to one of the seventy nine desired letter-to-phoneme transformations in the trained network). Rosenberg and Sejnowski (1987) did a 'cluster analysis' of these vectors in the trained network. Roughly, their procedure was as follows. They asked, for every functional vector in that space, what other such vector is closest to it? The answers yielded about thirty vector pairs. They then constructed a secondary vector for each such pair, by averaging the two original vectors, and asked, for every such secondary vector, what other secondary vector (or so far unpaired primary vector) is closest to it? This produced a smaller set of secondary-vector pairs, on which the averaging procedure was repeated to produce a set of tertiary vectors. These were then paired in turn, and so forth. This procedure produces a hierarchy of groupings among the original transformations, and it comes to an end with a grand division of the seventy nine original vectors into two disjoint classes.

As it happens, that deepest and most fundamental division within the hidden-unit vector space corresponds to the division between the consonants and the vowels! Looking further into this hierarchy, into the consonant branch, for example, we find that there are subdivisions into the principal consonant types, and that within these branches there are further subdivisions into the most similar consonants. All of this is depicted in the tree diagram of figure 8. What the network has managed to recover, from its training set of several thousand English words, is the highly irregular phonological significance of standard English spelling, plus the hierarchical organization of the phonetic structure of English speech.

Here we have a clear illustration of two things at once. The first lesson is the capacity of an activation-vector space to embody a rich and well-structured hierarchy of categories, complete with a similarity metric embracing everything within it. And the second lesson is the capacity of such networks to embody representations of factors and patterns that are only partially or implicitly reflected in the corpus of inputs. Though I did not mention it earlier, the rock/mine network provides another example of this, in that the final partition made on its hidden-unit vector space corresponds in fact to the objective distinction between sonar targets made of *metal* and sonar targets made of *nonmetal*. That is the true uniformity that lies behind the apparently chaotic variety displayed in the inputs.

It is briefly tempting to suggest that NETtalk has the concept of a 'hard *c*', for example, and that the rock/mine network has the concept of 'metal'. But this won't really do, since the vector-space representations at issue do not play a conceptual

HIERARCHY OF PARTITIONS
ON HIDDEN-UNIT
VECTOR SPACE

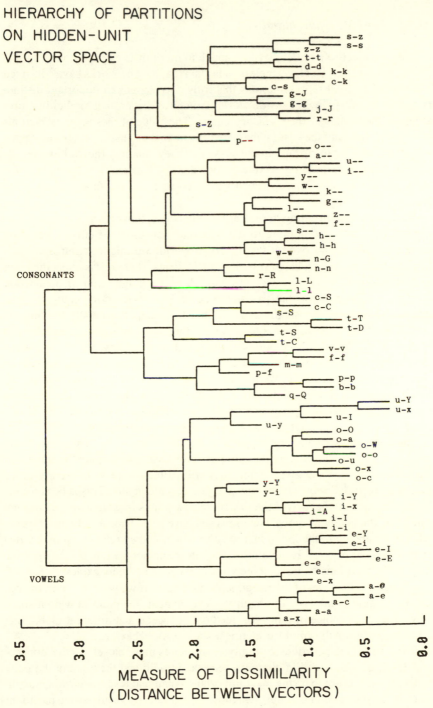

MEASURE OF DISSIMILARITY
(DISTANCE BETWEEN VECTORS)

Figure 8.

or computational role remotely rich enough to merit their assimilation to specifically human concepts. Nevertheless, it is plain that both networks have contrived a system of internal representations that truly corresponds to important distinctions and structures in the outside world, structures that are not explicitly represented in the corpus of their sensory inputs. The value of those representations is that they and only they allow the networks to 'make sense' of their variegated and often noisy input corpus, in the sense that they and only they allow the network to respond to those inputs in a fashion that systematically reduces the error messages to a trickle. These, I need hardly remind, are the functions typically ascribed to *theories*.

What we are confronting here is a possible conception of 'knowledge' or 'understanding' that owes nothing to the sentential categories of current common sense. An individual's overall theory-of-the-world, we might venture, is not a large collection or a long list of stored symbolic items. Rather, it is a specific point in that individual's synaptic weight space. It is a configuration of connection weights, a configuration that partitions the system's activation-vector space(s) into useful divisions and subdivisions relative to the inputs typically fed the system. 'Useful' here means 'tends to minimize the error messages'.

A possible objection here points to the fact that differently weighted systems can produce the same, or at least roughly the same, partitions on their activation-vector spaces. Accordingly, we might try to abstract from the idiosyncratic details of a system's connection weights, and identify its global theory directly with the set of partitions they produce within its activation-vector space. This would allow for differently weighted systems to have the same theory.

There is some virtue in this suggestion, but also some vice. While differently weighted systems can embody the same partitions and thus display the same output performance on any given input, they will still *learn* quite differently in the face of a protracted sequence of new and problematic inputs. This is because the learning algorithm that drives the system to new points in weight space does not care about the relatively global partitions that have been made in activation space. All it cares about are the individual *weights* and how they relate to apprehended error. The laws of cognitive evolution, therefore, do not operate primarily at the level of the partitions, at least on the view of things here being explored. Rather, they operate at the level of the weights. Accordingly, if we want our 'unit of cognition' to figure in the *laws* of cognitive development, the point in weight space seems the wiser choice of unit. We need only concede that different global theories can occasionally produce identical short-term behavior.

The level of the partitions certainly corresponds more closely to the 'conceptual' level, as understood in common sense and traditional theory, but the point is that this seems not to be the most important dynamical level, even when explicated in neurocomputational terms. Knowing a creature's vector-space partitions may suffice for the accurate short-term prediction of its behavior, but that knowledge is inadequate to predict or explain the evolution of those partitions over the course of time and cruel experience. Knowledge of the weights, by contrast, *is* sufficient for this task. This gives substance to the conviction, voiced back in section II, that to explain the phenomenon of *conceptual change*, we need to unearth a level of subconceptual combinatorial elements within which different concepts can be articulated, evaluated, and then modified according to their performance. The connection weights provide a level that meets all of these conditions.

This general view of how knowledge is embodied and accessed in the brain has some further appealing features. If we assume that the brains of the higher animals work in something like the fashion outlined, then we can explain a number of puzzling features of human and animal cognition. For one thing, the speed-of-relevant-access problem simply disappears. A network the size of a human brain—with 10^{11} neurons, 10^3 connections on each, 10^{14} total connections, and at least 10 distinct layers of 'hidden' units—can be expected, in the course of growing up, to partition its internal vector spaces into many billions of functionally relevant subdivisions, each responsive to a broad but proprietary range of highly complex stimuli. When the network receives a stimulus that falls into one of these classes, the network produces the appropriate activation vector in a matter of only tens or hundreds of milliseconds, because that is all the time it takes for the parallel-coded stimulus to make its way through only two or three or ten layers of the massively parallel network to the functionally relevant layer that drives the appropriate behavioral response. Since information is not stored in a long list that must somehow be searched, but rather in the myriad connection weights that configure the network, relevant aspects of the creature's total information are automatically accessed by the coded stimuli themselves.

A third advantage of this model is its explanation of the functional persistence of brains in the face of minor damage, disease, and the normal but steady loss of its cells with age. Human cognition degrades fairly gracefully as the physical plant deteriorates, in sharp contrast to the behavior of typical computers, which have a very low fault tolerance. The explanation of this persistence lies in the massively parallel character of the computations the brain performs, and in the very tiny contribution that each synapse or each cell makes to the overall computation. In a large network of 100,000 units, the loss or misbehavior of a single cell will not even be detectable. And in the more dramatic case of widespread cell loss, so long as the losses are more or less randomly distributed throughout the network, the gross character of the network's activity will remain unchanged: what happens is that the *quality* of its computations will be progressively degraded.

Turning now toward more specifically philosophical concerns, we may note an unexpected virtue of this approach concerning the matter of *simplicity*. This important notion has two problems. It is robustly resistant to attempts to define or measure it, and it is not clear why it should be counted an epistemic virtue in any case. There seems no obvious reason, either a priori or a posteriori, why the world should be simple rather than complex, and epistemic decisions based on the contrary assumption thus appear arbitrary and unjustified. Simplicity, conclude some (Van Fraassen 1980), is a merely pragmatic or aesthetic virtue, as opposed to a genuinely epistemic virtue. But consider the following story.

The rock/mine network of figure 5 displays a strong capacity for generalizing beyond the sample echoes in its training set: it can accurately discriminate entirely new samples of both kinds. But trained networks do not always generalize so well, and it is interesting what determines their success in this regard. How well the training generalizes is in part a function of *how many* hidden units the system possesses, or uses to solve the problem. There is, it turns out, an optimal number of units for any given problem. If the network to be trained is given more than the optimal number of hidden units, it will learn to respond appropriately to all of the various samples in its training set, but it will generalize to new samples only very poorly. On the other hand, with less than the optimal number, it never really learns to respond appropriately to all of the samples in its training set.

The reason is as follows. During the training period, the network gradually generates a set of internal representations at the level of the hidden units. One class of hidden-unit activation vectors is characteristic of rocklike input vectors; another class is characteristic of minelike input vectors. During this period, the system is *theorizing* at the level of the hidden units, exploring the space of possible activation vectors, in hopes of finding some partition or set of partitions on it that the output layer can then exploit in turn, so as to draw the needed distinctions and thus bring the process of error-induced synaptic adjustments to an end.

If there are far too many hidden units, then the learning process can be partially subverted in the following way. The lazy system cheats: it learns a set of *unrelated* representations at the level of the hidden units. It learns a distinct representation for each sample input (or for a small group of such inputs) drawn from the very finite training set, a representation that does indeed prompt the correct response at the output level. But since there is nothing common to all of the hidden-unit rock representations, or to all of the hidden-unit mine representations, an input vector from outside the training set produces a hidden-unit representation that bears no relation to the representations already formed. The system has not learned to see *what is common* within each of the two stimulus classes, which would allow it to generalize effortlessly to new cases that shared that common feature. It has just knocked together an *ad hoc* 'look-up table' that allows it to deal successfully with the limited samples in the training set, at which point the error messages cease, the weights stop evolving, and the system stops learning. (I am grateful to Terry Sejnowski for mentioning to me this wrinkle in the learning behavior of typical networks.)

There are two ways to avoid this *ad hoc*, unprojectible learning. One is to enlarge dramatically the size of the training set. This will overload the system's ability to just 'memorize' an adequate response for each of the training samples. But a more effective way is just to reduce the number of hidden units in the network, so that it lacks the resources to cobble together such wasteful and ungeneralizable internal representations. We must reduce them to the point where it has to find a *single* partition on the hidden-unit vector space, a partition that puts all of the sample rock representations on one side, and all of the sample mine representations on the other. A system constrained in this way will generalize far better, for the global partition it has been forced to find corresponds to something *common* to each member of the relevant stimulus class, even if it is only a unifying dimension of variation (or set of such dimensions) that unites them all by a similarity relation. It is the generation of that similarity relation that allows the system to respond appropriately to novel examples. They may be new to the system, but they fall on a spectrum for which the system now has an adequate representation.

Networks with only a few hidden units in excess of the optimal number will sometimes spontaneously achieve the maximally simple 'hypothesis' despite the excess units. The few unneeded units are slowly shut down by the learning algorithm during the course of training. They become zero-valued elements in all of the successful vectors. Networks will not always do this, however. The needed simplicity must generally be forced from the outside, by a progressive reduction in the available hidden units.

On the other hand, if the network has too few hidden units, then it lacks the resources even to express an activation vector that is adequate to characterize the underlying uniformity, and it will never master completely even the smallish corpus of samples in the training set. In other words, simplicity may be a virtue, but the system must command sufficient complexity at least to meet the task at hand.

We have just seen how forcing a neural network to generate a smaller number of distinct partitions on a hidden-unit vector space of fewer dimensions can produce a system whose learning achievements generalize more effectively to novel cases. *Ceteris paribus*, the simpler hypotheses generalize better. Getting by with fewer resources is of course a virtue in itself, though a pragmatic one, to be sure. But this is not the principal virtue here displayed. Superior generalization is a genuinely epistemic virtue, and it is regularly displayed by networks constrained, in the fashion described, to find the simplest hypothesis concerning whatever structures might be hidden in or behind their input vectors.

Of course, nothing guarantees successful generalization: a network is always hostage to the quality of its training set relative to the total population. And there may be equally simple alternative hypotheses that generalize differentially well. But from the perspective of the relevant microdynamics, we can see at least one clear reason why simplicity is more than a merely pragmatic virtue. It is an epistemic virtue, not principally because simple hypotheses avoid the vice of being complex, but because they avoid the vice of being *ad hoc*.

VI. How Faithfully Do These Networks Depict the Brain?

The functional properties so far observed in these model networks are an encouraging reward for the structural assumptions that went into them. But just how accurate are these models, as depictions of the brain's microstructure? A wholly appropriate answer here is uncertain, for we continue to be uncertain about what features of the brain's microstructure are and are not functionally relevant, and we are therefore uncertain about what is and is not a 'legitimate' simplifying assumption in the models we make. Even so, it is plain that the models are *inaccurate* in a variety of respects, and it is the point of the present section to summarize and evaluate these failings. Let me begin by underscoring the basic respects in which the models appear to be correct.

It is true that real nervous systems display, as their principal organizing feature, layers or populations of neurons that project their axons *en masse* to some distinct layer or population of neurons, where each arriving axon divides into multiple branches whose end bulbs make synaptic connections of various weights onto many cells at the target location. This description captures all of the sensory modalities and their primary relations to the brain; it captures the character of the various areas of the central brain stem; and it captures the structure of the cerebral cortex, which in humans contains at least six distinct layers of neurons, where each layer is the source and/or the target of an orderly projection of axons to and/or from elsewhere.

It captures the character of the cerebellum as well (figure 9a), a structure discussed in an earlier paper (Churchland 1986) in connection with the problem of motor control. I there described the cerebellum as having the structure of a very large 'matrix multiplier', as schematized in figure 9b. Following Pellionisz and Llinas (1982), I ascribed to this neural matrix the function of performing sophisticated transformations on incoming activation vectors. This is in fact the same function performed between any two layers of the three-layered networks described earlier, and the two cases are distinct only in the superficial details of their wiring diagrams. A three-layered network of the kind discussed earlier is equivalent to a pair of neural matrices connected in series, as is illustrated in figures 10a and 10b. The only substantive difference is that in figure 10a the end branches synapse directly onto the receiving cell body itself, while in 10b they synapse onto some dendritic filaments extending out from the receiving cell body. The actual connectivity within the two networks is identical. The cerebellum and the motor end of natural systems, accordingly, seem further instances of the gross pattern at issue.

But the details present all manner of difficulties. To begin with small ones, note

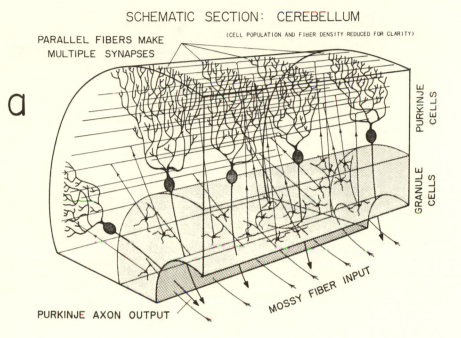

SCHEMATIC SECTION: CEREBELLUM

(CELL POPULATION AND FIBER DENSITY REDUCED FOR CLARITY)

PARALLEL FIBERS MAKE
MULTIPLE SYNAPSES

a

PURKINJE CELLS

GRANULE CELLS

PURKINJE AXON OUTPUT

MOSSY FIBER INPUT

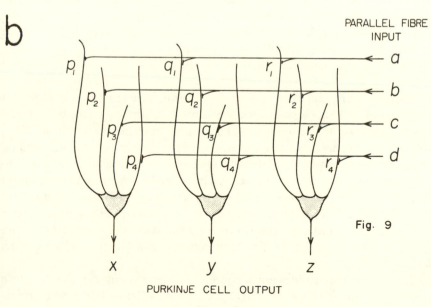

b

PARALLEL FIBRE
INPUT

p_1 q_1 r_1 $\leftarrow a$

p_2 q_2 r_2 $\leftarrow b$

p_3 q_3 r_3 $\leftarrow c$

p_4 q_4 r_4 $\leftarrow d$

Fig. 9

x y z

PURKINJE CELL OUTPUT

Figure 9.

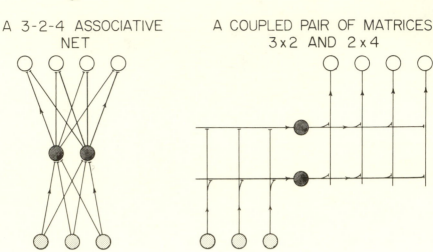

Figure 10.

that in real brains an arriving axon makes synaptic contact with only a relatively small percentage of the thousands or millions of cells in its target population, not with every last one of them as in the models. This is not a serious difficulty, since model networks with comparably pared connections still manage to learn the required transformations quite well, though perhaps not so well as a fully connected network.

More seriously, real axons, so far as is known, have terminal end bulbs that are uniformly inhibitory, or uniformly excitatory, depending on the type of neuron. We seem not to find a mixture of both kinds of connections radiating from the same neuron, nor do we find connections changing their sign during learning, as is the case in the models. Moreover, that mixture of positive and negative influences is essential to successful function in the models: the same input cell must be capable of inhibiting some cells down the line at the same time that it is busy exciting others. Further, cell populations in the brain typically show extensive 'horizontal' cell-to-cell connections *within* a given layer. In the models there are none at all (see, e.g., figure 4). Their connections join cells only to cells in distinct layers.

These last two difficulties might conceivably serve to cancel each other. One way in which an excitatory end bulb might serve to *inhibit* a cell in its target population is first to make an excitatory connection onto one of the many small *inter-*

neurons typically scattered throughout the target population of main neurons, which interneuron has made an inhibitory synaptic connection onto the target main neuron. Exciting the inhibitory interneuron would then have the effect of inhibiting the main neuron, as desired. And such a system would display a large number of short 'horizontal' intralayer connections, as is observed. This is just a suggestion, however, since it is far from clear that the elements mentioned are predominantly connected in the manner required.

More seriously still, there are several major problems with the idea that networks in the brain learn by means of the learning algorithm so effective in the models : the procedure of back-propagating apprehended errors according to the generalized delta rule. That procedure requires two things: 1) a computation of the partial correction needed for each unit in the output layer, and via these a computation of a partial correction for each unit in the earlier layers, and 2) a method of causally conveying these correction messages back through the network to the sites of the relevant synaptic connections in such a fashion that each weight gets nudged up or down accordingly. In a computer simulation of the networks at issue (which is currently the standard technique for exploring their properties), both the computation and the subsequent weight adjustments are easily done: the computation is done *outside* the network by the host computer, which has direct access to and control over every element of the network being simulated. But in the self-contained biological brain, we have to find some real source of adjustment signals, and some real pathways to convey them back to the relevant units. Unfortunately, the empirical brain displays little that answers to exactly these requirements.

Not that it contains nothing along these lines: the primary ascending pathways already described are typically matched by reciprocal or 'descending' pathways of comparable density. These allow higher layers to have an influence on affairs at lower layers. Yet the influence appears to be on the activity levels of the lower cells themselves, rather than on the myriad synaptic connections whose weights need adjusting during learning. There may be indirect effects on the synapses, of course, but it is far from clear that the brain's wiring diagram answers to the demands of the back-propagation algorithm.

The case is a little more promising in the cerebellum (figure 9a), which contains a second major input system in the aptly-named *climbing fibers* (not shown in the diagram for reasons of clarity). These fibers envelop each of the large Purkinje cells from below in the same fashion that a climbing ivy envelops a giant oak, with its filamentary tendrils reaching well up into the bushy dendritic tree of the Purkinje cell, which tree is the locus of all of the synaptic connections made by the incoming parallel fibers. The climbing fibers are thus at least roughly positioned to do the job that the back-propagation algorithm requires of them, and they are distributed one to each Purkinje cell, as consistent delivery of the error message requires. Equally, they might serve some other quite different learning algorithm, as advocated by Pellionisz and Llinas (1985). Unfortunately, there is as yet no compelling reason to believe that the modification of the weights of the parallel-fiber-to-Purkinje-dendrite synapses is even within the causal power of the climbing fibers. Nor is there any clear reason to see either the climbing fibers in the cerebellum, or the descending pathways elsewhere in the brain, as the bearers of any appropriately computed error-correction messages appropriate to needed synaptic change.

On the hardware side, therefore, the situation does not support the idea that the specific back-propagation procedure of Rumelhart et al. is the brain's central mechanism for learning. (Neither, it should be mentioned, did they claim that it is.) And it is implausible on some functional grounds as well. First, in the process of learning a recognition task, living brains typically show a progressive reduction in the reaction time required for the recognitional output response. With the delta rule, however, learning involves a progressive reduction in error, but reaction times are constant throughout. A second difficulty with the delta rule is as follows. A necessary element in its calculated apportionment of error is a representation of what would have been the *correct* vector in the output layer. That is why back-propagation is said to involve a global *teacher*, an information source that always knows the correct answers and can therefore provide a perfect measure of output error. Real creatures generally lack any such perfect information. They must struggle along in the absence of any sure compass toward the truth, and their synaptic adjustments must be based on much poorer information.

And yet their brains learn. Which means that somehow the configuration of their synaptic weights must undergo change, change steered in some way by error or related dissatisfaction, change that carves a path toward a regime of decreased error. Knowing this much, and knowing something about the microstructure and microdynamics of the brain, we can explore the space of possible learning procedures with some idea of what features to look for. If the generalized delta rule is not the brain's procedure, as it seems not to be, there remain other possible strategies for back-propagating sundry error measures, strategies that may find more detailed reflection in the brain. If these prove unrealizable, there are other procedures that do not require the organized distribution of any global error measures at all; they depend primarily on local constraints (Hinton and Sejnowski 1986; Hopfield and Tank 1985; Barto 1985; Bear et al. 1987).

One of these is worthy of mention, since something along these lines does appear to be displayed in biological brains. *Hebbian* learning (so-called after D. O. Hebb, who first proposed the mechanism) is a process of weight adjustment that exploits the temporal coincidence, on either side of a given synaptic junction, of a strong signal in the incoming axon and a high level of excitation in the receiving cell. When such conjunctions occur, Hebb proposed, some physical or chemical change is induced in the synapse, a change that increases its 'weight'. Of course, high activation in the receiving cell is typically caused by excitatory stimulation from many other incoming axons, and so the important temporal coincidence here is really between high activation among certain of the incoming axons. Those whose high activation coincides with the activation of many others have their subsequent influence on the cell increased. Crudely, those who vote with winners become winners.

A Hebbian weight-adjusting procedure can indeed produce learning in artificial networks (Linsker, 1986), although it does not seem to be as general in its effectiveness as is back-propagation. On the other hand, it has a major functional advantage over back-propagation. The latter has scaling problems, in that the process of calculating and distributing the relevant adjustments expands geometrically with the number of units in the network. But Hebbian adjustments are locally driven; they are independent of one another and of the overall size of the network. A large network will thus learn just as quickly as a small one. Indeed, a large network may even show a slight advantage over a smaller, since the temporal coincidence of incoming stimulations at a given cell will be better and better defined with increasing numbers of incoming axons.

We may also postulate 'anti-Hebbian' processes, as a means of reducing synaptic weights instead of increasing them. And we need to explore various possible flavors of each. We still have very little understanding of the functional properties of these alternative learning strategies. Nor are we at all sure that Hebbian learning, as described above, is really how the brain typically adjusts its weights. There does seem to be a good deal of activity-sensitive synaptic modification occurring in the brain, but whether its profile is specifically Hebbian is not yet established. Nor should we expect the brain to confine itself to only one learning strategy, for even at the behavioral level we can discern distinct types of learning. In sum, the problem of what mechanisms actually produce synaptic change during learning is an unsolved problem. But the functional success of the generalized delta rule assures us that the problem is solvable in principle, and other more plausible procedures are currently under active exploration.

While the matter of how real neural networks generate the right configuration of weights remains obscure, the matter of how they perform their various cognitive tasks once configured is a good deal clearer. If even small artifical networks can perform the sophisticated cognitive tasks illustrated earlier in this paper, there is no mystery that real networks should do the same or better. What the brain displays in the way of hardware is not radically different from what the models contain, and the differences invite exploration rather than disappointment. The brain is of course very much larger and denser than the models so far constructed. It has many layers rather than just two or three. It boasts perhaps a hundred distinct and highly specialized cell types, rather than just one. It is not a single n-layer network, but rather a large committee of distinct but parallel networks, interacting in sundry ways. It plainly commands many spaces of stunning complexity, and many skills in consequence. It stands as a glowing invitation to make our humble models yet more and more realistic, in hopes of unlocking the many secrets remaining.

VII. Computational Neuroscience: The Naturalization of Epistemology

One test of a new framework is its ability to throw a new and unifying light on a variety of old phenomena. I will close this essay with an exploration of several classic issues in the philosophy of science. The aim is to reconstruct them within the framework of the computational neuroscience outlined above. In section 5 we saw how this could be done for the case of theoretical simplicity. We there saw a new way of conceiving of this feature, and found a new perspective on why it is a genuine epistemic virtue. The hope in what follows is that we may do the same for other problematic notions and issues.

A good place to begin is with the issue of foundationalism. Here the central bone of contention is whether our observation judgments must always be theory laden. The traditional discussion endures largely for the good reason that a great deal hangs on the outcome, but also for the less momentous reason that there is ambiguity in what one might wish to count as an 'observation judgment' (an explicitly uttered sentence? a covert assertion? a propositional attitude? a conscious experience? a sensation?), and a slightly different issue emerges depending on where the debate is located.

But from the perspective of this essay, it makes no difference at what level the issue might be located. If our cognitive activities arise from a weave of networks of the kind discussed above, and if we construe a global theory as a global configuration of synaptic weights, as outlined in section 5, then it is clear that no cognitive activity whatever takes place in the absence of vectors being processed by some specific configuration of weights. That is, no cognitive activity whatever takes place in the absence of some theory or other.

This perspective bids us see even the simplest of animals and the youngest of infants as possessing theories, since they too process their activation vectors with some configuration of weights or other. The difference between us and them is not that they lack theories. Rather, their theories are just a good deal simpler than ours, in the case of animals. And their theories are much less coherent and organized and informed than ours, in the case of human infants. Which is to say, they have yet to achieve points in overall weight space that partition their activation-vector spaces into useful and well-structured subdivisions. But insofar as there is cognitive activity at all, it exploits whatever theory the creature embodies, however useless or incoherent it might be.

The only place in the network where the weights need play no role is at the absolute sensory periphery of the system, where the external stimulus is transduced into a coded input vector, for subsequent delivery to the transforming layers of weights. However, at the first occasion on which these preconceptual states have any effect at all on the downstream cognitive system, it is through a changeable configuration of synaptic weights, a configuration that produces one set of partitions on the activation-vector space of the relevant layer of neurons, one set out of millions of alternative possible sets. In other words, the very first thing that happens to the input signal is that it gets conceptualized in one of many different possible ways. At subsequent layers of processing, the same process is repeated, and the message that finally arrives at the linguistic centers, for example, has been shaped at least as much by the partitional constraints of the embedded conceptual system(s) through which it has passed as by the distant sensory input that started things off.

From the perspective of computational neuroscience, therefore, cognition is constitutionally theory laden. Presumptive processing is not a blight on what would otherwise be an unblemished activity; it is just the natural signature of a cognitive system doing what it is supposed to be doing. It is just possible that some theories are endogenously specified, of course, but this will change the present issue not at all. Innateness promises no escape from theory ladenness, for an endogenous theory is still a *theory*.

In any case, the idea is not in general a plausible one. The visual system, for example, consists of something in the neighborhood of 10^{10} neurons, each of which enjoys better than 10^3 synaptic connections, for a total of at least 10^{13} weights, each wanting specific genetic determination. That is an implausibly heavy load to place on the coding capacity of our DNA molecules. (The entire human genome contains only about 10^9 nucleotides.) It would be much more efficient to specify endogenously only the general structural principles of a type of learning network that is then likely to learn in certain standard directions, given the standard sorts of inputs and error messages that a typical human upbringing provides. This places the burden of steering our conceptual development where it belongs — on the external world, an information source far larger and more reliable than the genes.

It is a commonplace that we can construct endlessly different theories with which to explain the familiar facts of the observable world. But it is an immediate consequence of the perspective here adopted that that we can also apprehend the 'observable world' itself in a similarly endless variety of ways. For there is no 'preferred' set of partitions into which our sensory spaces must inevitably fall. It all depends on how the relevant networks are *taught*. If we systematically change the pattern of the error messages delivered to the developing network, then even the very same history of sensory stimulations will produce a quite differently weighted network, one that partitions the world into classes that cross-classify those of current 'common sense', one that finds perceptual similarities along dimensions quite alien to the ones we currently recognize, one that feeds its outputs into a very differently configured network at the higher cognitive levels as well.

In relatively small ways, this phenomenon is already familiar to us. Specialists in various fields, people required to spend years mastering the intricacies of some domain of perception and manipulation, regularly end up being able to perceive facts and to anticipate behaviors that are wholly opaque to the rest of us. But there is no reason why such variation should be confined to isolated skills and specialized understanding. In principle, the human cognitive system should be capable of sustaining any one of an enormous variety of decidedly global theories concerning the character of its commonsense *Lebenswelt* as a whole. (This possibility, defended in Feyerabend 1965, is explored at some length via examples in Churchland 1979. For extended criticism of this general suggestion see Fodor 1984. For a rebuttal and counterrebuttal see Churchland 1988 and Fodor 1988.)

To appreciate just how great is the conceptual variety that awaits us, consider the following numbers. With a total of perhaps 10^{11} neurons with an average of at least 10^3 connections each, the human brain has something like 10^{14} weights to play with. Supposing, conservatively, that each weight admits of only ten possible values, the total number of distinct possible configurations of synaptic weights (= distinct possible positions in weight space) is 10 for the first weight, times 10 for the second weight, times 10 for the third weight, etc., for a total of $10^{10^{14}}$, or $10^{100,000,000,000,000}$!! This is the total number of (just barely) distinguishable theories embraceable by humans, given the cognitive resources we currently command. To put this number into perspective, recall that the total number of elementary particles in the entire universe is only about 10^{87}.

In this way does a neurocomputational approach to perception allow us to reconstruct an old issue, and to provide novel reasons for the view that our perceptual knowledge is both theory laden and highly plastic. And it will do more. Notice that the activation-vector spaces that a matured brain has generated, and the prototypes they embody, can encompass far more than the simple sensory types such as phonemes, colors, smells, tastes, faces, and so forth. Given high-dimensional spaces, which the brain has in abundance, those spaces and the prototypes they embody can encompass categories of great complexity, generality, and abstraction, including those with a temporal dimension, such as harmonic oscillator, projectile, traveling wave, Samba, twelve-bar blues, democratic election, six-course dinner, courtship, elephant hunt, civil disobedience, and stellar collapse. It may be that the input dimensions that feed into such abstract spaces will themselves often have to be the expression of some earlier level of processing, but that is no problem. The networks under discussion are hierarchically arranged to do precisely this as a matter of course. In principle then, it is no harder for such a system to represent types of *processes*, *procedures*, and *techniques* than to represent the 'simple' sensory qualities. From the point of view of the brain, these are just more high-dimensional vectors.

This offers us a possible means for explicating the notion of a *paradigm*, as used by T. S. Kuhn in his arresting characterization of the nature of scientific understanding and development (Kuhn 1962). A paradigm, for Kuhn, is a prototypical *application* of some set of mathematical, conceptual, or instrumental resources; an application expected to have distinct but similar instances, which it is the job of normal science to discover or construct. Becoming a scientist is less a matter of learning a set of laws than it is a matter of mastering the details of the prototypical applications of the relevant resources in such a way that one can recognize and generate further applications of a relevantly similar kind.

Kuhn was criticized for the vagueness of the notion of a paradigm, and for the unexplicated criterion of similarity that clustered further applications around it. But from the perspective of the neurocomputational approach at issue, he can be vindicated on both counts. For a brain to command a paradigm is for it to have settled into a weight configuration that produces some well-structured similarity space whose central hypervolume locates the prototypical application(s). And it is only to be expected that even the most reflective subject will be incompletely articulate on what dimensions constitute this highly complex and abstract space, and even less articulate on what metric distributes examples along each dimension. A complete answer to these questions would require a microscopic examination of the subject's brain. That is one reason why exposure to a wealth of examples is so much more effective in teaching the techniques of any science than is exposure to any attempt at listing all the relevant factors. We are seldom able to articulate them all, and even if we were able, listing them is not the best way to help a brain construct the relevant internal similarity space.

Kuhn makes much of the resistance typically shown by scientific communities to change or displacement of the current paradigm. This stubbornness here emerges as a natural expression of the way in which networks learn, or occasionally fail to learn. The process of learning by gradient descent is always threatened by the prospect of a purely *local* minimum in the global error gradient. This is a position where the error messages are not yet zero, but where every *small* change in the system produces even larger errors than those currently encountered. With a very high-dimensional space, the probability of there being a simultaneous local minimum in every dimension of the weight space is small: there is usually some narrow cleft in the canyon out which the configuration point can eventually trickle, thence to continue its wandering slide down the error gradient and toward some truly global minimum. But genuine local minima do occur, and the only way to escape them once caught is to introduce some sort of random noise into the system in hopes of bouncing the system's configuration point out of such tempting cul-de-sacs. Furthermore, even if a local quasi-minimum does have an escape path along one or more dimensions, the error gradient along them may there be quite shallow, and the system may take a very long time to find its way out of the local impasse.

Finally, and just as importantly, the system can be victimized by a highly biased 'training set'. Suppose the system has reached a weight configuration that allows it to respond successfully to all of the examples in the (narrow and biased) set it has so far encountered. Subsequent exposure to the larger domain of more diverse examples will not necessarily result in the system's moving any significant distance away from its earlier configuration, unless the relative frequency with which it encounters those new and anomalous examples is quite high. For if the encounter frequency is low, the impact of those examples will be insufficient to overcome the gravity of the false minimum that captured the initial training set. The system may require 'blitzing' by new examples if their collective lesson is ever to 'sink in'.

Even if we do present an abundance of the new and diverse examples, it is quite likely that the delta rule discussed earlier will force the system through a sequence of new configurations that perform very poorly indeed when re-fed examples from the original training set. This temporary loss of performance on certain previously 'understood' cases is the price the system pays for the chance at achieving a broader payoff later, when the system finds a new and deeper error minimum. In the case of an artificial system chugging coolly away at the behest of the delta rule, such temporary losses need not impede the learning process, at least if their frequency is sufficiently high. But with humans the impact of such a loss is often more keenly felt. The new examples that confound the old configuration may simply be ignored or rejected in some fashion, or they may be quarantined and made the target of a distinct and disconnected learning process in some adjacent network. Recall the example of sublunary and superlunary physics.

This raises the issue of explanatory unity. A creature thrown unprepared into a complex and unforgiving world must take its understanding wherever it can find it, even if this means generating a disconnected set of distinct similarity spaces, each providing the creature with a roughly appropriate response to some of the more pressing types of situation it typically encounters. But far better if it then manages to generate a single similarity space that unifies and replaces the variation that used to reside in two entirely distinct and smaller spaces. This provides the creature with an effective grasp on the phenomena that lay *between* the two classes already dealt with, but which were successfully comprehended by neither of the two old spaces. These are phenomena that the creature had to ignore, or avoid, or simply endure. With a new and more comprehensive similarity space now generating systematic responses to a wider range of phenomena, the creature has succeeded in a small piece of conceptual unification.

The payoff here recalls the virtue earlier discovered for simplicity. Indeed, it is the same virtue, namely, superior generalization to cases beyond those already encountered. This result was achieved, in the case described in section 5, by reducing the number of hidden units, thus forcing the system to make more efficient use of the representational resources remaining. This more efficient use is realized when the system partitions its activation-vector space into the minimal number of distinct similarity subspaces consistent with reducing the error messages to a minimum. When completed, this process also produces the maximal *organization* within and among those subspaces, for the system has found those enduring dimensions of variation that successfully unite the diversity confronting it.

Tradition speaks of developing a single 'theory' to explain everything. Kuhn (1962) speaks of extending and articulating a 'paradigm' into novel domains. Kitcher (1981, 1989) speaks of expanding the range of application of a given 'pattern of argument'. It seems to me that we might unify and illuminate all of these notions by thinking in terms of the evolving structure of a hidden-unit activation-vector space, and its development in the direction of representing all input vectors somewhere within a single similarity space.

This might seem to offer some hope for a Convergent Realist position within the philosophy of science, but I fear that exactly the opposite is the case. For one thing, nothing guarantees that we humans will avoid getting permanently stuck in some very deep but relatively local error minimum. For another, nothing guarantees that there exists a possible configuration of weights that would reduce the error messages to *zero*. A unique global error minimum relative to the human neural network there may be, but for us and for any other finite system interacting with the real world, it may always be nonzero. And for a third thing, nothing guarantees that there is only *one* global minimum. Perhaps there will in general be many quite different minima, all of them equally low in error, all of them carving up the world in quite different ways. Which one a given thinker reaches may be a function of the idiosyncratic details of its learning history. These considerations seem to remove the goal itself—a unique truth—as well as any sure means of getting there. Which suggests that the proper course to pursue in epistemology lies in the direction of a highly naturalistic and pluralistic form of pragmatism. For a running start on precisely these themes, see Munevar 1981 and Stich 1989.

VIII. Concluding Remarks

This essay opened with a survey of the problems plaguing the classical or 'sentential' approach to epistemology and the philosophy of science. I have tried to sketch an alternative approach that is free of all or most of those problems, and has some novel virtues of its own. The following points are worth noting. Simple and relatively small networks of the sort described above have already demonstrated the capacity to learn a wide range of quite remarkable cognitive skills and capacities, some of which lie beyond the reach of the older approach to the nature of cognition (e.g., the instantaneous discrimination of subtle perceptual qualities, the effective recognition of similarities, and the real-time administration of complex motor activity). While the specific learning algorithm currently used to achieve these results is unlikely to be the brain's algorithm, it does provide an existence proof: by procedures of this general sort, networks can indeed learn with fierce efficiency. And there are many other procedures awaiting exploration.

The picture of learning and cognitive activity here painted encompasses the entire animal kingdom: cognition in human brains is fundamentally the same as cognition in brains generally. We are all of us processing activation vectors through artfully weighted networks. This broad conception of cognition puts cognitive theory firmly in contact with neurobiology, which adds a very strong set of constraints on the former, to its substantial long-term advantage.

Conceptual change is no longer a problem: it happens continuously in the normal course of all cognitive development. It is sustained by many small changes in the underlying hardware of synaptic weights, which changes gradually repartition the activation-vector spaces of the affected population of cells. Conceptual *simplicity* is also rather clearer when viewed from a neurocomputational perspective, both in its nature and in its epistemological significance.

The old problem of how to retrieve relevant information is transformed by the realization that it does not need to be 'retrieved'. Information is stored in brainlike networks in the global pattern of their synaptic weights. An incoming vector activates the relevant portions, dimensions, and subspaces of the trained network by virtue of its own vectorial makeup. Even an incomplete version of a given vector (i.e., one with several elements missing) will often provoke essentially the same response as the complete vector by reason of its relevant similarity. For example, the badly whistled first few bars of a familiar tune will generally evoke both its name and the rest of the entire piece. And it can do this in a matter of milliseconds, because even if the subject knows thousands of tunes, there are still no lists to be searched.

It remains for this approach to comprehend the highly discursive and linguistic dimensions of human cognition, those that motivated the classical view of cognition. We need not pretend that this will be easy, but we can see how to start. We can start by exploring the capacity of networks to manipulate the structure of existing language, its syntax, its semantics, its pragmatics, and so forth. But we might also try some novel approaches, such as allowing each of two distinct networks, whose principal concerns and activities are nonlinguistic, to try to learn from scratch some systematic means of manipulating, through a proprietary dimension of input, the cognitive activities of the other network. What system of mutual manipulation—what *language*—might they develop?

The preceding pages illustrate some of the systematic insights that await us if we adopt a more naturalistic approach to traditional issues in epistemology, an approach that is grounded in computational neuroscience. However, a recurring theme in contemporary philosophy is that normative epistemology *cannot* be 'naturalized' or reconstructed within the framework of any purely descriptive scientific theory. Notions such as 'justified belief' and 'rationality', it is said, cannot be adequately defined in terms of the nonnormative categories to which any natural science is restricted, since "oughts" cannot be derived from "ises". Conclusions are then drawn from this to the principled autonomy of epistemology from any natural science.

While it may be true that normative discourse cannot be replaced without remainder by descriptive discourse, it would be a distortion to represent this as the aim of those who would naturalize epistemology. The aim is rather to enlighten our normative endeavors by reconstructing them within a more adequate conception of what cognitive activity consists in, and thus to free ourselves from the burden of factual misconceptions and tunnel vision. It is only the *autonomy* of epistemology that must be denied.

Autonomy must be denied because normative issues are never independent of factual matters. This is easily seen for our judgments of instrumental value, as these always depend on factual premises about causal sufficiencies and dependencies. But it is also true of our most basic normative concepts and our judgments of intrinsic value, for these have factual presuppositions as well. We speak of *justification*, but we think of it as a feature of *belief*, and whether or not there are any beliefs and what properties they have is a robustly factual matter. We speak of *rationality*, but we think of it as a feature of *thinkers*, and it is a substantive factual matter what thinkers are and what cognitive kinematics they harbor. Normative concepts and normative convictions are thus always hostage to some background factual presuppositions, and these can always prove to be superficial, confused, or just plain wrong. If they are, then we may have to rethink whatever normative framework has been erected upon them. The lesson of the preceding pages is that the time for this has already come.

References

Barto, A. G. 1985. Learning by Statistical Cooperation of Self-Interested Neuronlike Computing Elements. *Human Neurobiology* 4:229–56.

Bear, M. F., Cooper, L. N., and Ebner, F. F. 1987. A Physiological Basis for a Theory of Synapse Modification. *Science* 237 (no. 4810).

Churchland, P. M. 1975. Karl Popper's Philosophy of Science. *Canadian Journal of Philosophy* 5 (no. 1).

—— 1979. *Scientific Realism and the Plasticity of Mind*. Cambridge: Cambridge University Press.

——. 1981. Eliminative Materialism and the Propositional Attitudes. *Journal of Philosophy. 78 (no. 2)*.

——. 1985. The Ontological Status of Observables: In Praise of the Superempirical Virtues. In *Images of Science*, ed. P. M. Churchland and C. A. Hooker. Chicago, University of Chicago Press.

——. 1986. Some Reductive Strategies in Cognitive Neurobiology. *Mind* 95 (no. 379).

—— 1988. Perceptual Plasticity and Theoretical Neutrality: A Reply to Jerry Fodor. *Philosophy of Science* 55 (no. 2).

Churchland, P. S. 1980. A Perspective on Mind-Brain Research. *Journal of Philosophy* 77 (no. 4).

——. 1986. *Neurophilosophy: Toward a Unified Understanding of the Mind-Brain*. Cambridge, MIT Press.

Feyerabend, P. K. 1965. Reply to Criticism: Comments on Smart, Sellars, and Putnam. In *Boston*

Studies in the Philosophy of Science, ed. M. Wartofsky. Dordrecht: Reidel. Reprinted in *Realism, Rationalism & Scientific Method, Philosophical Papers, vol. 1, Feyerabend, P. K. 1981*. Cambridge: Cambridge University Press, 1981.

———. *1980. Consolations for the Specialist. In Criticism and the Growth of Knowledge*, eds. I. Lakatos and A. Musgrave. Cambridge: Cambridge University Press.

Fodor, J. A. 1984. Observation Reconsidered. *Philosophy of Science* 51 (no. 1).

———. 1988. A Reply to Churchland's "Perceptual Plasticity and Theoretical Neutrality." *Philosophy of Science* 55 (no. 2).

Giere, R. 1988. *Explaining Science: A Cognitive Approach*. Chicago: University of Chicago Press.

Glymour, C. 1987. "Artificial Intelligence is Philosophy". In *Aspects of Artificial Intelligence*, ed. J. Fetzer. Dordrecht: Reidel.

Gorman, R. P., and Sejnowski, T. J. 1988. Learned Classification of Sonar Targets Using a Massively-Parallel Network. *IEEE Transactions: Acoustics, Speech, and Signal Processing*. Forthcoming.

Hempel, K. 1965. "Studies in the Logic of Confirmation". In *Aspects of Scientific Explanation*. New York: The Free Press.

Hinton, G. E., and Sejnowski, T. J. 1986. "Learning and Relearning in Boltzmann Machines". In *Parallel Distributed Processing: Explorations in the Microstructure of Cognition*, eds. D. E. Rumelhart and J. L. McClelland. Cambridge: MIT Press. 1986.

Hooker, C. A. 1975. The Philosophical Ramifications of the Information-Processing Approach to the Mind-Brain. *Philosophy and Phenomenological Research* 36.

———. 1987. *A Realistic Theory of Science*. Albany: State University of New York Press.

Hopfield, J. J., and Tank, D. 1985. "Neural" Computation of Decisions in Optimization Problems. *Biological Cybernetics* 52:141–52.

Hubel, D. H., and Wiesel, T. N. 1962. Receptive Fields, Binocular Interactions, and Functional Architecture in the Cat's Visual Cortex. *Journal of Physiology* 160.

Kitcher, P. 1981. Explanatory Unification. *Philosophy of Science* 48 (no. 4).

———. 1989. "Explanatory Unification and the Causal Structure of the World". In *Minnesota Studies in the Philosophy of Science, vol. 13, Scientific Explanation*, ed. P. Kitcher. Minneapolis: University of Minnesota Press.

Kuhn, T. S. 1962. *The Structure of Scientific Revolutions*. Chicago: University of Chicago Press.

Lakatos, I. 1970. "Falsification and the Methodology of Scientific Research Programmes. In *Criticism and the Growth of Knowledge*, I. Lakatos and A. Musgrave. Cambridge University Press.

Laudan, L. 1981. A Confutation of Convergent Realism. *Philosophy of Science* 48 (no. 1).

Lehky, S., and Sejnowski, T. J. 1988a. "Computing Shape from Shading with a Neural Network Model". In *Computational Neuroscience*, ed. E. Schwartz. Cambridge: MIT Press.

———. 1988b. Network Model of Shape-From-Shading: Neural Function Arises from Both Receptive and Projective Fields. *Nature* 333 (June 2).

Linsker, R. 1986. From Basic Network Principles to Neural Architecture: Emergence of Orientation Columns. *Proceedings of the National Academy of Sciences, USA*, 83:8779–83.

Minsky, M., and Papert, S. 1969. *Perceptrons*. Cambridge: MIT Press.

Munevar, G. 1981. *Radical Knowledge*. Indianapolis: Hackett.

Pellionisz, A., and Llinas, R. 1982. Space-Time Representation in the Brain: The Cerebellum as a Predictive Space-Time Metric Tensor. *Neuroscience* 7 (no. 12):2949–70.

———. 1985. Tensor Network Theory of the Metaorganization of Functional Geometries in the Central Nervous System. *Neuroscience*. 16 (no. 2):245–74.

Putnam, H. 1981. *Reason, Truth, and History*. Cambridge: Cambridge University Press.

Rosenberg, C. R., and Sejnowski, T. J. 1987. Parallel Networks That Learn To Pronounce English Text. *Complex Systems, 1:145–68*.

Rosenblatt, F. 1959. *Principles of Neurodynamics*. New York: Spartan Books.

Rumelhart, D. E., Hinton, G. E., and Williams, R. J. 1986a. Learning Representations by Back-Propagating Errors. *Nature*, 323.

——. 1986b. "Learning Internal Representations by Error Propagation". In *Parallel Distributed Processing: Explorations in the Microstructure of Cognition*, ed. D. E. Rumelhart and J. L. McClelland. Cambridge: MIT Press.

Salmon, W. 1966. *The Foundations of Scientific Inference*. Pittsburgh: University of Pittsburgh Press.

Scheffler, I. 1963. *The Anatomy of Inquiry*. New York: Knopf.

Sejnowski, T. J., Kienker, P. K., and Hinton, G. E. 1986. Learning Symmetry Groups with Hidden Units: Beyond the Perceptron. *Physica D*: 22.

Stich, S. P. 1990. *The Fragmentation of Reason*. Cambridge: MIT Press.

Suppe, F. 1974. *The Structure of Scientific Theories*. Chicago: University of Illinois Press.

Van Fraassen, Bas 1980. *The Scientific Image*. Oxford: Oxford University Press.

Zipser, D., and Elman, J. D. 1988. Learning the Hidden Structure of Speech. *Journal of the Acoustical Society of America* 83(4):1615–25.

Are Economic Kinds Natural?

I

This essay is primarily about some foundational problems in economic theory. I shall approach these problems by focusing on the natural kinds postulated in economic theory and physical theory. I hope to develop some understanding of how we can think of natural kinds in scientific theories, but my main goal is to try to answer some more practical questions about the foundations of economics. In what follows, 'natural kinds' means natural kinds in the world—the *kinds* of *things* that there are. Furthermore, I do not mean the kinds of things that fall into the extensions of the so-called natural-kind *terms* of the vernacular. Instead, this essay is about the kinds of things that are picked out by our best scientific theories. Perhaps I am not justified in appropriating the term 'natural kind' in this way, but for the present it can be regarded as a stipulative definition.[1]

How do scientific theories pick out natural kinds? Let's assume that a scientific theory is a sort of linguistic object containing some items, natural-kind terms, that are semantically related to the natural kinds in the world. This assumption needs plenty of defending, but none will be provided here. Which items in the theory are the natural-kind terms? If we could identify within theories scientific *laws* of the form, "All Ps are lawfully associated with Qs," then it would be sensible to think of 'P' and 'Q' as natural-kind terms. There are too many problems with this approach for us to count it a general analysis, but in this essay I shall concentrate on some real examples from physics and especially from economics for which the analysis has some plausibility.

The particular question about economics dealt with in this essay is this: Why

I received many helpful comments when presenting some of the ideas in this paper to audiences at the following institutions: University of Minnesota; Caltech; Northwestern University; University of California, Irvine; University of Arizona; Indiana University; and the University of California, Berkeley. Special thanks to Philip Kitcher, Wade Savage, and the participants at the NEH seminar at Minnesota. I have also specially profited from discussions and correspondence with Arthur Fine, Scott Gordon, Daniel Hausman, Alexander Rosenberg, and Paul Teller.

does economics do so much worse than the physical sciences in applications when its formal structure often rivals the others with respect to mathematical sophistication, rigor, and aesthetic appeal?[2] Many answers have been proposed. Pessimists say it is because social science is inherently impossible, or because contemporary economics surreptitiously incorporates bad ideology, or because it is simply a failure like phlogiston theory. Optimists say it is because it is a young science, or because the phenomena it attempts to deal with are so complex, or because we cannot run carefully controlled, repeatable experiments.

I want to defend another answer that has both pessimistic and optimistic elements. I shall provide new arguments for the old pessimistic position that says there are very good reasons for supposing that economics does not apply to the world successfully because its central concepts do not represent natural kinds. The optimistic part comes from a characterization of the sorts of empirical results we would need to get in the future to show that the natural-kind terms of economics do refer to natural kinds after all. I shall also have something to say about procedures that might be successful for obtaining these results. To begin the argument we require an account of what the natural-kind terms in economics are.

II

The economic theory I shall concentrate on is General Equilibrium Theory (GET). GET is about economies of individual, independent agents, both consumers and producers of commodities. It analyzes economies with attention to the interdependence of the economic decisions they make. It is assumed for the sake of simplicity that decisions invariably eventuate in the intended economic actions (although these actions may not have all the intended results). The basic goal of the theory is to determine what configurations of relative prices of goods will yield outcomes consistent with rational behavior on the parts of all agents. The properties of these so-called equilibrium outcomes are then studied. It is called *general* equilibrium when the prices of all commodities are more or less constant. GET is especially appropriate for the purposes of this essay because of its great generality. Since almost all branches of contemporary capitalistically inspired economics are either forms of GET or special cases of it,[3] understanding it is close to understanding virtually all of modern academic economics. Moreover, GET is characterized by a fruitful and well-developed formalism, so there is little doubt that the empirical problems of economics are the result of a logically or mathematically defective formalism. It must be stressed, however, that the empirical problems of economics do not result simply from clumsy attempts to apply the general and highly abstract mathematical formalism of GET to particular actual cases. Good empirical work is supposed to be, in principle, conformable to the GET framework, but almost everyone realizes that empirically useful results require careful extensions and applications of the basic GET framework.

The theory's structure can be briefly summarized as follows.[4] It contains a small number of foundational elements. Each of these elements, or theoretical concepts, can be thought of as representing a natural kind. To understand them, it is necessary to study them together, because the concepts are closely intertwined.

Consumers. These represent people – the principal agents in the economy. It is their actions, their *exchanges* of commodities, that determine the nature of most economic phenomena, including production. From the standpoint of GET and its special cases, individual consumers' identities consist solely of their *preferences* for and *endowments* of commodities, so one can think of them simply as repositories of preferences and endowments – they are not quite actual people. Thus their economically interesting actions consist of only exchanges of commodities that maximize utility (or perhaps tend to maximize utility, or perhaps exchanges that the consumer thinks maximize utility, and so on). It is also true of every consumer that he has the capacity to annihilate and create commodities by consuming and laboring, but the mechanisms that are involved in these activities are not part of the subject matter of economics.

Preferences. Each consumer has preferences that put all of the bundles of commodities available to him in an *order*.[5,6] They are represented in GET by an ordering of vectors having an element for each kind of commodity. Preferences are often conveniently represented by a *utility function* that assigns a scalar measure of utility to the consumer for every commodity vector. In virtually all cases of interest, use of utility functions is equivalent to use of preferences, and I shall follow economic practice in exploiting this fact. In particular, I shall freely shift back and forth between utility functions and preferences, depending on which make the point in question more perspicuously.[7]

Endowments. Each consumer has an endowment, a bundle of commodities that she *owns*.[8] The endowment is represented by a vector of the same commodities as the preference vector. (Since few have everything that they like, some elements will be equal to zero.) How much utility a consumer "gets," how well her preferences are satisfied, depends on how much and on what commodities are in the endowment. These commodities include not only goods in the sense of "dry goods," but also claims to "services" of all kinds (TV repair, foot massages, musical performances, etc.). Perhaps the most important services owned by the consumer are the ones that she can provide herself, in short, her labor. The endowment, or parts of it, can be either exchanged on the market for other commodities, or consumed.

Firms. Firms are organizations of people (i.e., consumers) and commodities (i.e., capital) that transform bundles of commodities into other, different bundles of commodities. The idea is that the firm employs "factors of production," labor and relatively raw materials, and *produces* "finished" commodities. In the same way that consumers are fully characterized by their preferences, endowments, and propensity to maximize utility, firms are fully characterized by their technological possibilities for converting commodity inputs to commodity outputs— their "production possibilities set." The stream of commodities produced by the firm is owned by a group of consumers, the stockholders—it becomes part of their endowments. Since consumers are utility maximizers, the firm's owners will direct it to maximize output in a way that benefits the owners in terms of utility.

Netput Vectors. The firm's production possibilities set can be represented by a set of vectors of quantities of commodities. The elements represent inputs and outputs of each commodity, hence the name. Each netput vector represents a production process that is technologically possible for the firm. If an element of one of these vectors is positive, that commodity is an output of the process (a product); if an element is negative, that commodity is an input (a factor of production). It is often convenient to represent the set of netput vectors by a function that takes inputs of factors of production to outputs of product, a "production function."

Prices, the Market, and Equilibrium. It is assumed that commodities exchange at prices that are standardized for the entire economy. Some elaborations of GET attempt to model the institution of fiat money (intrinsically worthless money like bank notes, as opposed to intrinsically valuable bullion), but in general a price is simply a relative rate of exchange—five avocados for three bananas. In the absence of fiat money we can, for the sake of simplicity, calculate all prices in terms of a "numeraire," for example "cost in number of avocados." A market is constituted by the existence of a price system. It is usually, but not always, assumed that the price of finding exchange partners and then transacting exchanges is zero. Equilibrium is then said to obtain when the prices of commodities are such that quantities demanded equal quantities produced for every commodity because all consumers (including stockholders) achieve utility maxima by exchanging parts of their endowments at those prices.

III

Two concepts clearly emerge as the most fundamental from this way of laying out the structure of GET. One is the *commodity*. I did not give it its own paragraph because it is difficult to fully explicate in terms of the others (though one might say that commodities simply are the things that give consumers utility), but it was prominently featured in the explication of each of the other concepts. Consumers are exchangers of commodities. (And producers and annihilators of them, though these processes are not objects of economic scrutiny.) Preferences are for commodities. Utility functions are from quantities of commodities to an index of utility. Endowments are of commodities. Firms transform some commodities into others. Netput vectors are of quantities of commodities, and production functions are from quantities of commodities to quantities of commodities. Equilibrium obtains when all agents' plans for doing things with commodities are mutually consistent. And prices are relative exchange rates of commodities.

The other fundamental concept is *preferences* (or utility functions). Even commodities can almost be characterized in terms of them. Karl Marx, for example, said, "A commodity [*Ware* in German] is, in the first place, an object outside us, a thing that by its properties satisfies human wants of some sort or other" (1967, 35). Marx did not have a GET, but we might try to adapt his idea. If we take preferences as the most fundamental concept and then say that commodities are those things over which consumers have preferences, we can continue with the above explication of all the other key concepts in terms of *commodity*.

It would be very interesting to establish a complete hierarchy of fundamentality for economic concepts, but the main argument of this essay does not require it. I do, however, want to concentrate attention on *commodity* and *preference* and leave aside the others, which are all arguably less fundamental than these two. I think it is plausible to take *commodity* as fundamental in the *description* of economic phenomena. What we observe and wish to explain are exchanges of commodities and the properties of these exchanges. *Preferences*, then, are fundamental in the *explanation* of the phenomena. We understand the commodity exchanges as having the properties they do primarily because of facts about preferences.

First, however, a potential objection needs to be considered. It is sometimes claimed that economics is primarily about relative price levels and the fact that equilibrium exists, instead of about facts about consumers. But relative price levels are nothing more than the rates at which commodities exchange; prices are properties of what is really basic, commodities. The fact of equilibrium (if it is indeed a fact) is also, at root, a fact about preferences and commodities. Equilibrium obtains when all relative prices are constant and all agents are maximizing their utility given that price level and their endowments.

This encapsulation of the GET framework is, I hope, an aid to understanding the structure of the theory, how the parts of the model of actual economic phenomena fit together. I now want to proceed to the matter of interpreting the model. What does GET tell us about actually observed economic events? It is convenient to begin with an analysis of the concept of *utility*.

IV

Theoretically coherent economics that is clearly ancestral to contemporary GET can be regarded as beginning around the 1870s, with what is called the "Marginalist Revolution." Utility was at this time incorporated into economic theory as a psychological or psychophysical quantity that was measurable in principle. It came to be called "cardinal utility" because it was believed that a person's utility level could be assigned a numerical index that was objective in the sense that it served for making interpersonal comparisons of utility levels.[9] Economists became disaffected with the commitment to cardinal utility in the face of growing evidence that most kinds of human behavior are not primarily motivated by cardinal utility maximization and that there probably is no such psychophysical quantity in the first place.[10] It was, however, not possible for this disaffection to take root because there was no competing theory with as much potential for explanation and for guiding economic political policy as the utility theory.

This changed around the beginning of the 1930s with the speedy development of a well-developed and mathematically consistent theory based on "ordinal" instead of "cardinal" utility. J. Hicks and R. Allen, in their seminal paper (1934), showed how demand functions for individuals, theoretically and empirically important functions from prices of commodities to the quantities of that commodity exchanged for, could be derived from something less than psychophysically based rankings of available commodities. All that is mathematically required is utility functions that preserve only the consumer's ordering of options and not the "cardinality" or magnitudes of the utility differences between options.[11] In other words, ordinal utility functions come in classes that are equivalent up to linear transformation. Therefore, they demonstrated that cardinal utility functions contain more information than is needed to derive demand curves from first principles about individual economic agents. (The derivation of demand curves was, and still is, considered very important because some of our best economic data is for market demand curves.)

It was natural to hope that the excess content in cardinal utility functions was exactly the part of the theory that was in conflict with the psychological facts that had been emerging. In 1938, P. Samuelson showed how this hope could be realized and, in a way, exceeded. He mathematically derived the most characteristic and important results of consumer theory from a formalism that did not require the postulation of even ordinal utility functions. Instead of deriving the properties of individual demand curves from ordinal utility functions and budget constraints, points on the demand curves were taken as given. Samuelson was then able to prove that these two hypotheses are equivalent:

i) The commodity bundles chosen by the consumer (i.e., the points on the demand curve) conform to the results of maximizing a well-behaved ordinal utility function.

ii) Points on a demand curve that satisfy the Strong Axiom of Revealed Preference (SARP) fully characterize a well-behaved preference map.[12]

The interpretive significance of Samuelson's formalism is considerable. Since SARP is a constraint on the *choices* that consumers make and not on their utility functions, the new formalism suggested[13] that the old interpretation, the one in which the theoretical concept of utility (or preference) plays a role in explaining how economic behavior is partly caused by psychological facts about consumers, be abandoned. In short, it suggested that economists instead use overtly observable choices and SARP to *construct* utility functions when they might prove *convenient* as instruments for prediction. The machinery of consumer theory was thereby almost completely severed from the psychological thinking that had begotten it. It was easy to see this as emancipating economics from constraints imposed by another science whose relevance to purely economic concerns was no longer entirely clear.[14] As Samuelson put it,

> The discrediting of *utility* as a psychological concept robbed it of its only possible virtue as an *explanation* of human behavior in other than a circular sense, revealing its emptiness as even a construction. . . . The introduction and meaning of [any fact about indifference curves][15] independent of any psychological, introspective implications would be, to say the least, ambiguous, and would seem an artificial convention in the explanation of price behavior. . . . I propose, therefore that we start anew in direct attack upon the problem, dropping off the last vestiges of the utility analysis. This does not preclude the introduction of utility by any who may care to do so . . . (1938a, 61–62)

This provides a partial characterization of an interpretation of consumer theory based on the Revealed Preference formalism. I shall call it the "Utility-as-Revealed interpretation."

Utility-as-Revealed Interpretation

Consumers' psychological states are not part of the economic explanation of their behavior. Any reference to concepts such as utility and preference is for the sake of convenience and is fully eliminable from the theory.

Something like this is often called the "Revealed Preference Interpretation." It is good to avoid this name because Revealed Preference is not an interpretation or even a theory—it is an axiomatic formalization of a part of consumer theory.

According to this interpretation, we are not to think of observed choices as revealing what the *preferences* are at all. If a consumer makes choices that are consistent with SARP, then he acts *as if* he were maximizing utility or getting onto the highest indifference curve, but there is no ontological commitment to these mentalistic things. And since the theory is not ontologically committed to them, then any facts that other scientists (e.g., psychologists) discover about them can safely be ignored.[16]

So the revealed preference interpretation provides a rationale for disregarding the severe problems that empirical psychologists were discovering in cardinal utility theory. The ordinal utility theory, however, does not *require* the revealed preference interpretation.

There is a second important interpretation in which preferences (represented by ordinal utility functions) are understood to occupy a much more important position in the theory. They are regarded as providing part of a causal *explanation* of why economic agents behave as they do, instead of simply describing their behavior; I shall call it the "Utility-as-Explanatory Interpretation."

Utility-as-Explanatory Interpretation

Consumer theory explains economic behavior as well as describing and predicting it. The explanation comes from a causal story: a) An agent is confronted with a choice set of commodity bundles that is partly determined by his endowment and income and partly by availability; b) he compares this choice set to his indifference map; and c) he chooses the bundle that lies on the highest indifference curve.

Thus, indifference maps are taken to be a partial *cause* of the observed behavior, qua *economic* behavior. Alternatively, since well-behaved preference maps are fully represented by ordinal utility functions, we might substitute for b) and c): b') He plugs this information into his utility function, and c') chooses the bundle that is calculated to give the greatest utility. Hence, the name. It is quite certain that little of the "comparing" or "calculating" referred to is done consciously, but that is to be expected. Many complicated psychological tasks involve substantial amounts of cognition that does not take place at the conscious level. The understanding and production of language is an example. In what follows, I shall argue that the Utility-as-Explanatory interpretation is superior to Utility-as-Revealed.

The crucial differences between Utility-as-Revealed and Utility-as-Explanatory can be brought out by familiar episodes from the history of other sciences in which analogous distinctions have been important. Consider the mathematical expression for the Balmer series of the hydrogen atom. Before scientists discovered how to derive this expression from the Bohr model of the atom, it appeared to be a fortuitous piece of numerology concocted to conform to existing data. The fact that it gave excellent empirical results and was eventually shown to predict even the frequencies of previously unobserved spectral lines contributed to scientists' confidence in the relationship, but Balmer's formula did not *explain* the phenomena that it described. In this case, an explanation of the frequencies required some account of what *causes* the lines to come out the way that they do. The explanation was provided only after the appearance of spectral lines was understood as resulting from differences in the energies of various excitation states of the atom's electrons.

According to Utility-as-Revealed, utility functions must have the status that Balmer's formula had before physicists knew how to derive it from more fundamental laws. Utility functions are concocted to fit the available data, and we hope that they are able to correctly predict previously unobserved phenomena. It is not, according to Utility-as-Revealed, in order to inquire as to *why* a certain utility function accounts for the data. Economists who believe in Utility-as-

Revealed are not interested in deriving the utility functions from anything more fundamental or more general, nor are they interested in trying to understand how a utility function might be connected with the causes of the behavior that it describes. But according to Utility-as-Explanatory, theoretical economic entities like utility functions ought to explain phenomena by providing a story about their causation. Accepting Utility-as-Explanatory commits one to attributing a causal role to the facts about human economic agents described by utility functions. Analogously, we now understand Balmer's formula (or Balmer's law, as it appears to us now in light of its derivation) to describe causally significant properties of entities such as atoms, electrons, and photons.

If a causal role cannot be found for utility functions, then there is some danger that they are not like Balmer's law as much as they are like Bode's law. Bode's law (which is really not a *law* at all) was a mathematical expression concocted to give the distance of the planets, Mercury through Saturn, from the Sun. It turned out, like Balmer's law, to be an astonishingly accurate predictor. The mean orbital radii of the asteroid belt and Uranus fit Bode's law almost perfectly. Even its predictions for Neptune and Pluto were within reason.[17] But Bode's law is obviously only a historical curiosity. Unlike Balmer's law, it is not recognized as a law, even though it has never been convincingly disconfirmed.[18]

According to Utility-as-Revealed, economic descriptions of choice behavior are more like the description given by Bode's law than the one given by Balmer's law. It says that utility functions are supposed to do no more than describe actual choices, thus their variables and parameters are specified post facto so that they necessarily yield the correct results. There is no presumption about the nature of any underlying causal mechanism. It may then be hoped that they are also useful for prediction. Many commentators have pointed out that this practice can reduce utility functions to tautologies. It is possible to concoct a utility function that can give an empty account of almost any behavior. This means, moreover, that if at time T we assign utility function U_i to agent i and her behavior at $T + t$ disconfirms U_i, it is a similarly trivial matter to modify U_i to account for the new behavior as well. The result of this uninteresting process is clearly a mere description and not a scientific theory at all. The situation is directly analogous to, say, providing neo-Ptolemaic derivations of all ephemeridical data through today's date. With enough circles and epicycles, there is little question that it is possible to do this, but no one would care much about the results. Providing Ptolemaic derivations today would be little more than an exercise, because the results could not afford us with an explanation of any astronomical phenomena. The Ptolemaic setup has no chance of being even approximately true, or true *ceteris paribus*, given our current understanding of astronomy.

The problem with unsatisfactory theories like Ptolemaic positional astronomy or Bode's law is not only that they can be manipulated to account very accurately for a large number of observations. Reasonable theories like Newtonian mechanics can likewise account for any collection of celestial motions by simply postulating sufficient ad hoc forces. Instead, theories like Ptolemy's fail primarily because of the severely limited range of phenomena that they can account for; Ptolemaic theory and Aristotelian spheres apply only to the motion of the Sun, Moon, and six planets, while Newtonian mechanics applies to countless kinds of motions. What makes contemporary Ptolemaic theory trivial is the a priori restriction on the range of phenomena that have evidential bearing on it. It is not possible to use unrelated, well-confirmed theories to get access to any other information that could independently confirm the problematic theory. In contrast, the Newtonian account of celestial motions can be independently corroborated by terrestrial motions. The Newtonian formulas fitted to celestial motions are constrained by phenomena other than the ones they are devised to describe. There is an analogous difference between Utility-as-Revealed and Utility-as-Explanatory. In the former there is, by stipulation, no observational or experimental access to utility functions independent of the instances of behavior that they describe.

In Utility-as-Explanatory, however, the whole point of mentioning utility is to give an insightful characterization of how and why the observed behavior comes about. Moreover, since the maximization of utility is given a part in a causal story, it has real ontological standing. This means that there are, in principle, ways of finding out about it aside from observing its effects on overt economic exchanging behavior. This is important. Seventeenth-century natural philosophers would have been much more suspicious of universal gravitation if it were introduced only as a means of accounting for celestial motions. It was a great advantage that the same force postulated to account for those phenomena had already proven its worth in (what seemed to be) an entirely different domain—terrestrial motions. In another article (Nelson 1986), there is a fairly detailed treatment of how this might work in economics. I can only briefly summarize the results here.

The Utility-as-Explanatory interpretation eventually requires experimental investigation of individual economic behavior. One basic, but powerful consideration in favor of this position is that once preferences are assigned a causal role, they must be treated as things or entities in the strongest sense. And when one is scientifically investigating *things*, one does not deliberately ignore any available information about these things without compelling reasons to do so. An excellent source of information about individuals' preferences, perhaps the best source, is the actual economic behavior of the individual in question. Attending primarily to individuals does not have the terrible result that interesting generalizations cannot be made across individuals. Economics would not be of much interest if it produced a separate theory for every agent in an economy. Even if consumers differ in their relative appreciation of avocados and bananas, however, their utility functions may be represented by similar mathematical forms. One consumer's preferences might, for example, be of the form (3 × number of avocados + 2 × number of bananas) while another's might be of the form (number of avocados + 5 × number of bananas), but both of these consumers have utilities that are linear in avocados and bananas. What needs to be done to determine what particular individuals' utility functions look like is some experimental investigation in either the laboratory or the marketplace.[19]

V

So far, the argument for preferring Utility-as-Explanatory to Utility-as-Revealed has relied on considerations familiar to philosophers of science. While these are, I think, quite convincing in their own right, it is initially worrisome that many, perhaps most, economists who have considered the matter say that they prefer Utility-as-Revealed. If there were reasons deeply rooted in the necessities of economic practice for their stated preference, we would have a very perplexing problem about the foundations of economics. Therefore, let us continue by examining not what economists say about how the theory should be interpreted, but instead the interpretation that emerges from their own discussions of economic, and not philosophical problems.

In a very revealing passage in an important textbook, J. Henderson and R. Quandt write,

> At the beginning of this chapter, the cardinal approach to utility theory was rejected on the grounds that there is no reason to assume that the consumer possesses a cardinal measure of utility. By the same token one could question whether she even possesses an indifference map. It can fortunately be proved that a consumer who always conforms to the axioms of revealed preference must possess an indifference map. . . . If the consumer does not conform to the axioms, she is said to be "irrational". Her inconsistent actions mean that she does not possess an indifference map, *and the shape of her utility function cannot be determined by observing her behavior*. (1980, 46, emphasis added)

Is this the correct way to handle inconsistent actions (e.g., actions that seem to reflect intransitive preferences) according to Utility-as-Revealed? It seems strange to say that in these circumstances the consumer "does not possess an indifference map"; Utility-as-Revealed says that descriptions of the indifference maps are constructed by simply recording the choices that the consumer would make in given circumstances (or, according to Revealed Preference, recording the actual choices and smoothly interpolating the missing points on the map). This can be done no matter what the choices turn out to be. The strangeness can be resolved by remembering that 'indifference map' is shorthand for 'indifference map conforming to some basic assumptions'. Thus, in standard economic usage, what constitutes an indifference map cannot be just any batch of points in commodity space.

So we are supposed to say that the irrationally behaving consumer does not possess an indifference map. This is in accord with the teaching of the Revealed Preference Interpretation. But Henderson and Quandt go on to conclude that, "the shape of her utility function cannot be determined by observing her behavior." That clearly implies that a utility function does exist and that it does conform to the standard restrictive assumptions, but in this case the consumer does not "pay attention" to it while making some economic choices. This implication is inconsistent with Utility-as-Revealed. It is a mathematical fact (and true on any interpretation of the theory) that ordinal utility functions are equivalent to well-behaved indifference maps. Therefore, if we are going to say that the errant consumer possesses no indifference maps, then we must also say that she possesses no utility function. The only alternative is to maintain that she does have a utility function or the equivalent indifference map, but (sometimes) does not pay attention to it, or does not use it, when choosing. But Utility-as-Revealed says that indifference maps must be understood as being mere constructs of choices. The internal inconsistency is unavoidable.

I think that this confusion finds its way into the quoted passage because Utility-as-Explanatory is so reasonable that one must be extremely vigilant if one wishes to avoid believing it at all costs. A very neat account of the errant consumer's behavior has already been hinted at. She has a utility function. How is it then that she goes astray while making an economic choice? It is because something has interfered with her accurately maximizing her utility function. Perhaps her knowledge of the choice that she was making was deficient, or perhaps she was nervous or short of time so that she made an error, or perhaps the choice was so complicated that it was beyond her mathematical abilities to make the maximizing choice, and so forth. If we are to come up with *economic* explanations, it is probably wise to avoid trying to account for every conceivable type of interference with purely economic phenomena.

This is probably what Henderson and Quandt meant, but if one is committed to Utility-as-Revealed, it is quite difficult to say what one means. If a utility function is merely a description of an agent's behavior, and we do not want to count behavior that has been interfered with in ways that obscure the economically relevant features of the behavior, then how can the function be characterized? We must say something like this: a utility function describes the choices an agent *would* make *if* the choices that *were* to be made were not interfered with in *certain* ways. The counterfactuals in this characterization must evidently be cashed out psychologically, but Utility-as-Revealed proposes to ignore this resource. So we are returned to the fact that Utility-as-Revealed precludes actually obtaining an indifference map for an actual individual economic agent. That task would involve either dabbling in psychology or employing aggregate data.[20]

Another context in which economists seem to be committed to Utility-as-Explanatory in spite of themselves is in discussions of what commodities ought to appear as the independent variables (or "arguments") of utility functions. This is illustrated by microeconomic treatments of money (and some macroeconomic treatments of wealth that have microfoundations in GET). It is overwhelmingly obvious that one thing actual economic agents enjoy getting their hands on is money. This is true even when the stuff that serves as money in a particular economy has no other important practical use. Rational agents are just as anxious to acquire bits of paper printed by governments or banks whose only value lies in that fact that they are, for whatever reasons, considered to be money. They will prefer a ten-dollar bill to nine dollars' worth of gold, even though the gold can be useful and valuable for purposes other than exchange. This kind of intrinsically useless money is called "fiat money."

The point emerges when we consider a revealed preference experiment. A rational agent is asked to choose between a bucket of dollar bills and an avocado. She is then asked to choose between a dime and a bucket of avocados, and so forth for a very large number of combinations of avocados and U.S. currency. After completing the experiment, it is an easy matter to construct an indifference map for this agent between avocados and fiat money, and therefore partially to construct an ordinal utility function that includes avocados and fiat money as dependent variables. It is significant that economists are extremely reluctant to do this. Neil Wallace writes,

> The principal way of abandoning intrinsic uselessness [the thesis that fiat money is never wanted for its own sake] is to make money an argument of utility functions or engineering production functions. But this begs too many questions. Is it fiat money or commodity money that appears in these functions? What if there are several fiat moneys, those of different countries? Do all appear, and if so, how? Does Robinson Crusoe have fiat money as an argument of his utility function? And what about other pieces of paper? . . . All of this is to say that theories that abandon intrinsic uselessness will be almost devoid of implications. (1980, 49)

James Tobin is more blunt,

> Clearly enough, the value of paper money does not derive from the beauty of the engravings; the practice of putting money stocks in utility functions is reprehensible. (1980, 86)

Why is it "reprehensible" to include certain dependent variables in utility functions if the functions that result are the best at saving the phenomena? Why does including some extra dependent variables render a utility function "devoid of implications"? It is especially surprising to see a greatly respected economist worried about begging empirical questions. Is it fiat money or commodity money? Why not obtain the answer by repeating the revealed preference experiment I suggested, substituting a commodity money for avocados? Is it U.S. money or U.K. money? Why not find out by offering combinations of buckets of each variety? Which, if any, does Robinson Crusoe have in his utility function? If we can find him, we can experiment on him. These utility functions are empirically very rich and not at all "almost devoid of implications."

I conclude that the powerful sentiments against including money in utility functions stem from the idea that utility is a measure of something real. Perhaps it is not very plain what the "something" is supposed to be, whether it is a measure of psychic satisfaction, or literally something that has utility as a property, or something else. Anyway it appears that according to these and many other extremely influential economists, a piece of paper cannot give one this something (except in small quantities when it is used as wallpaper or to light a cigarette); fiat money can only be exchanged for another commodity, another *real* commodity, that does confer the mysterious something on a agent. But any theoretical entity that is separable in this way from the observation of actual choices of economic agents cannot be given a meaningful interpretation by Utility-as-Revealed. If utility is going to be considered an objective psychological construct with causal efficacy, then any economic theory that is going to be a theory of this construct requires an interpretation something like Utility-as-Explanatory. Therefore, the popular view of money expressed in the above quotations is committed to a denial of Utility-as-Revealed.[21]

A third example of the discomfiture that can be caused by adherence to Utility-as-Revealed is provided by Milton Friedman's analysis of demand (1976). He writes,

> . . . [we] shall suppose that the individual in making these decisions acts *as if* he were pursuing and attempting to maximize a single end. This implies that different goods have some common characteristic that makes comparisons among them possible. This common characteristic is usually called *utility*. . . . We observe that people choose; if this is to be regarded as a deliberative act, it must be supposed that the various things among which choice is made can be compared; to be compared, they must have something in common.
>
> Let X, Y, Z, etc., stand for the quantities of various commodities. Then the notion that these commodities have some element in common and that the magnitude of this common element, utility, depends on the amounts of the various commodities can be expressed by writing utility as a function of X, Y, Z. . . . (35–36)

The most striking thing about this treatment is that it makes utility a property of the commodities themselves. It is more common to treat utility as a representation or index of the consumer's preferences. The commoner treatment, then, makes the amount of utility associated with any commodity object essentially relative to the consumer who consumes it. Each consumer has his own, perhaps unique, utility function that determines how much utility he "gets" from a given bundle of commodities. But according to Friedman, the total amount of utility from a commodity bundle is obtained by combining the amounts of utility inherent in each commodity in the bundle. Facts about the consumer, therefore, are irrelevant to determining the utility level.

Friedman's motivation for this nonstandard approach is clear. The already sparse ontological requirements of Utility-as-Revealed appear to have been further purified. Utility-as-Revealed allows a specially defined property of utility to be *attributed to* individual consumers, or at least to an idealized average individual. Friedman does not allow even this hint of economic cognition on the part of consumers; utility is a completely objective property of physical objects. Consumers use information about this utility to make otherwise inscrutable deliberations resulting in observable choice.

The proposal is difficult to interpret. It apparently entails that there is only one utility function, because utility is a "common element" in commodities, and a utility function simply gives the total magnitude of the common element that results from combining commodities (and hence, utility) in bundles. If follows that consumers' deliberations are all based on the same central economic fact—the amount of utility in the bundles under consideration. This means that if two consumers are weighing the choice between a bowl of beans and a bowl of rice, the only economically relevant difference in their deliberations will be their incomes. So if they have identical incomes, any difference in observed choice must result from economically inscrutable aspects of their deliberations. To most economists, this result will seem daft. If one consumer chooses the beans and the other the rice, microeconomics ought to say that the first *prefers* rice to beans, and the second prefers beans to rice. Or, equivalently, one *gets* more utility from rice than from beans, and the other more utility from beans than from rice. According to Friedman, microeconomics must say that one deliberated differently from the other, and the whole explanation of this must be found in another science, like psychology.

Another way of expressing this position is to say that, from an economic point of view, consumers are identical up to their budget constraints. Instead of identifying consumers by *their* utility functions, all economically explainable behavioral differences are to be attributed to diverse budget constraints resulting from differences in incomes or prices. This consequence of Friedman's view has seemed attractive to a few other notable economists (though for different reasons), but it has some grave flaws. Perhaps the worst of these is that it is easy to construct counterfactual economic situations that ought to be explainable by any account, but are not by this one. Consider again the two consumers with identical incomes who, when faced with the same choice between a bowl of beans and a bowl of rice, make different choices. There is a natural and satisfying explanation of this phenomenon in the Utility-as-Explanatory interpretation of microeconomics, and there may be at least a pretense of explanation in Utility-as-Revealed. The Utility-as-Element-of-Commodity interpretation however, like the hapless anti-Copernican who denied the responsibility of explaining planetary phases, arbitrarily and unnecessarily restricts the scope of the theory. Since there are no mitigating circumstances, the rejection of the interpretation is indicated.

Both an investigation of economist's actual practice and independent methodological considerations lead us to the conclusion that the best interpretation of utility theory requires us to think of concepts such as utility as robustly real. The reality of these concepts does not derive its significance by extending economic theory to additional results about economic markets; the theory's usefulness for dealing with markets is unaffected by (sensible) methodological interpretation. The Utility-as-Explanatory interpretation does have some important empirical implications for psychology (utility is to be psychologically real; see Nelson, 1986) and also for the economic behavior of individuals. In the next section, I shall argue that there are further implications of considerable philosophical interest. One might, however, object at this point that these conclusions are too strong. From an examination of economic theorizing and some basic lessons from the history of science, I seem to have concluded that some kind of scientific realism is true, and instrumentalism and other anti-realisms are false. Surely such a conclusion would require more purely philosophical reasoning.

But my conclusions do not involve any controversial philosophical doctrines about realism and instrumentalism. I have argued that Utility-as-Explanatory opens the possibility that important facts about utility functions can be inferred from observations of individual economic behavior and, perhaps, even from some kinds of noneconomic behavior of interest to psychologists. There is no sensible scientific instrumentalism of the general sort that is inconsistent with this. Sophisticated instrumentalists say that we have no warrant to regard unobservable theoretical constructs as ordinary entities.[22] This is a purely *philosophical* stance to adopt towards a scientific theory; it does not have any implications for how the science is to be conducted. In particular, it obviously does not proscribe any means of investigating the properties of the theoretical constructs. Proper scientific method, according to the sophisticated instrumentalist, requires us to behave *as if* the objects apparently referred to by the theory exist. It may even be psychologically helpful for the scientist to actively pretend that they exist.

But this is all that is required by the arguments I have given for Utility-as-Explanatory. This interpretation does not require any particular philosophical stance on the precise ontological standing of theoretical constructs that are, in some sense, unobservable. So, for example, if we are to pretend that utility functions exist, the pretense commits us to recognizing the potential relevance and importance of experiments done in psychology laboratories that (pretend to) investigate them. Neither Utility-as-Explanatory nor sophisticated instrumentalism permits us to ignore psychological effects of utility functions on the grounds that they do not really exist. Similarly, an instrumentalist does not argue against the construction and use of electron microscopes on the grounds that electrons do not really exist. It is enough that electrons seem to exist. Thus, the argument of this paper has no direct bearing on realism/anti-realism issues. In what follows, I shall take this for granted and speak of utility functions, commodities, and the like as "existing."

VI

I have argued at some length for the Utility-as-Explanatory interpretation. This interpretation reinforces the plausible idea that the concepts of utility and preference as they appear in GET are more than convenient fictions. They seem instead to be closely related to corresponding putative natural kinds in the world. This has important implications for our initial problem about the degree of empirical success of economic theory. I have already described how Utility-as-Explanatory places potential empirical constraints on economics, constraints that come from psychology, neurology, or some other science.

One obvious explanation of the empirical difficulties of economics would be that it is a false theory precisely because these constraints are not met. It may be that experimental studies of the psychology of economic behavior, or of neurology, etc., will show that nothing that can be sensibly represented by preferences or maximizing goes on when humans behave economically. It is very unfortunate that very little of this kind of experimental study of economic behavior is being done. We are beginning to see a good deal of data on overt economic behavior under laboratory conditions (there is the kind of work initiated by Vernon Smith and his associates, see e.g., Smith 1982) and some data on animals like pigeons and rats. We will need to wait quite a long time before we have enough results from investigations of the right kinds of "nonovert," "internal" economic phenomena to be able to see whether economics actually meets the relevant constraints (again, see Nelson 1986 for more discussion).

If I am right, then interpreters of economics are faced with a dilemma. They must either adopt the mistaken Utility-as-Revealed interpretation or sit and wait for the verdict of currently unpracticed experimental procedure. In what follows, I shall be examining a potential way out of this dilemma. Those who fear that economics might be unfairly convicted by constraints imposed by arguments concerning a difficult theoretical notion such as utility will welcome the possibility of this way out.

The position is developed by shifting our focus from the concepts of utility and preference back to the other central concept, the commodity. Because of the close connection between *utility* and *commodity*, the adoption of Utility-as-Explanatory yields a parallel interpretation of *commodity*. Commodities, like utility functions, are more than a picturesque aid to describing actual choices. They are the *things* that real agents choose to exchange and produce; hence, in GET they are an essential part of the explanation of economic behavior. This follows from Utility-as-Explanatory and the fact that utility functions can be *defined* as mappings from quantities of commodities to levels of utility.

The case for a robust interpretation of commodities may be even stronger than the case for so interpreting utility. Utility is an indispensible link in the explanatory chain, and there is potentially a variety of means by which its effects might be observed. Still, we never directly observe a utility function; we infer their existence from direct observations of their various effects. Commodities seem different in this respect. Although the theoretical notion of a commodity is central to the very abstract GET, we are inclined to think that modern people have a clear pretheoretical notion of what a commodity is. Utility functions, in contrast, would probably not occur to a layperson trying to grapple with interesting economic phenomena—a "folk economist." Some things that turn out to count as commodities in the theory are admittedly somewhat surprising: shares of stock in financial institutions and currency futures, for example. But the paradigm cases of actual commodities are such familiar items as avocados, shoes, and theater tickets. Even sophisticated treatments take it for granted that GET's *commodities* are to be closely related to, or even identified with, what we pretheoretically think of as commodities:[23] "A general equilibrium theory is a theory *about* both the quantities and the prices of all commodities" (Arrow and Hahn 1971, 2, emphasis added). Most economists, if asked to describe what phenomena they are theorizing about, would reply with facts about ordinary objects and their aggregates. The price of bananas at the supermarket, the output of the auto industry, the wages of steelworkers, and the glut of avocados. These seem to be the natural economic kinds that we want to have a theory of; we do not need a theory to tell us what they are. Similarly, it is good to have a theoretical understanding of water, but we are inclined to think that we know much about water before we learn any science. We want to say it is manifestly a natural kind.

So it seems extremely plausible to think that the central concepts of economics, the natural-kind *terms* of the theory, do pick out natural economic kinds in the world. And, one might think, we do not need to worry about the interpretive dilemma posed by the analysis of utility. Even if we became convinced that GET and related theories were simply bad and ought to be rejected, it would be plausible to think that we had bad theories *of* the kinds of things these theories are about. We would seek better theories of the same things: the purchases of consumers, fluctuations in banana prices, the relative wage rates of Wall Street analysts and professors, and so on. These are the kinds of things that are salient in our economic life, and we want a scientific theory to make events involving them explainable, predictable, and intelligible.

Here we can draw an interesting analogy with the philosophy of psychology. Many think that there is something called *folk psychology*, an informal theory that we all learn without formal study.[24] This theory, or prototheory, is relied upon in our everyday lives to explain intentional behavior as resulting from our acting on our beliefs to fulfill our desires. Folk psychology may be further articulated to deal with other kinds of mental states like worries, suspicions, imaginings, and rememberings. Some philosophers say that folk psychology is almost useless to the modern scientific psychologist. They think that, as a matter of empirical fact, the vocabulary of folk psychology as it is presently understood does not refer to natural kinds. They expect that the advance of cognitive psychology and neuroscience will force us to either eliminate concepts like *belief* and *desire* from a true scientific theory of behavior, or to heavily revise our folk-theoretic understanding of them. There is, however, a second view of the status of folk psychology. The proponents of the second view argue that pretheoretic concepts like *belief* are not subject to revision by science. If some scientific theory tells us that there is no such thing as beliefs, then this is ipso facto not a theory *about* intentional behavior. It might be a fine theory for what it is about, but it would be quite irrelevant to a range of phenomena we cannot help but regard as centrally important. Similarly, if some biological theory were to tell us there is nothing scientific to be said about *human beings* because the important generalizations are over some other kinds, then we would stubbornly look for another theory to tell us what we want to know about humans — humans are inescapably interesting.

If this second view is correct, then there are some domains for which we know what some of the natural kinds are before we do any real science. Before we do any psychology, for example, we are supposed to know what we want psychology to do for us: explain how our beliefs interact with our desires to contribute to the production of our behavior. Of course, the second view is not correct for all sciences. People used to mistakenly think that Earth was essentially different from other planets. A science whose development was constrained by the necessity of respecting pretheoretical beliefs about the Earth would not have turned out (did not turn out) very well.[25] It might be true, nevertheless, that we are not always prepared to let science have the first cuts when we carve up the world into intelligible slices.

Perhaps an analogue of this second view about folk-psychological concepts is appropriate for economic concepts as well. *Commodities*, *preferences* for commodities, *prices* of commodities, and so forth, may be so deeply ingrained in our pretheoretical thinking about economic life, in our folk economics, that we shall insist that a fully adequate scientific theory of economic phenomena be *about* them. Another theory that dealt with some aspects of economic life, but did not employ these concepts simply wouldn't be *economics*. If this is right, then it is no surprise that a highly coherent theory employing exactly these concepts should have developed. Perhaps I need not have produced a long argument to establish what the natural economic kinds are; instead, the argument might be viewed as a checkup on the health of the theory. Had the theory been built around other concepts, it would be unhealthy economics. Is this optimistic, conservative outlook justified?

To answer this question it is helpful to begin by reconsidering the nonconservative view of the scientific viability of folk psychology. On what ground might nonconservativism be justified? The most obvious would be the clear prospect of a successful empirical theory of human behavior that made no use of folk-psychological terms as they are ordinarily conceived. But psychologists have not yet found anything like this; the best we have are some arguments that research programs that are not committed to folk-psychological concepts are more promising. Another convincing though less powerful ground for nonconservativism would be a demonstration that scientific psychology based on folk concepts was quite hopelessly stagnant. In this case, we might abandon conservative psychology even in the absence of a competitor; we might search for a replacement *in vacuo* or even conclude that scientific psychology is impossible. Again, as a matter of empirical fact, this ground does not obtain.

Do either of these considerations, that is, the prospect of a better nonconservative theory or a demonstration of conservative stagnation, come to bear in the case of economics? We might be persuaded that there are serious competitors to economics in the image of GET, but relatively few economists will claim that we are in possession of a *better* theory. There is, however, some consensus that GET is in some danger of stagnation. Whether the theory itself is stagnating is controversial. The theory is constantly being beautifully articulated, but some would argue that the details of existing theory are being highly polished without any fundamental advances taking place. But it is uncontroversial that our ability to *apply* the theory to predict, control, or even explain actual phenomena is stagnating. As noted at the start of this essay, many prominent economists have argued that economics is almost a paradigm case of a stagnating science.

The question whether we should revise or abandon our folk-psychological concepts will not have much urgency until we see some decisive empirical developments. But in economics, the argument from stagnation is strong enough — I shall assume this in this essay — to make us take seriously the possibility that the root of the problem with GET-based economics is its conservative nature. The argument from stagnation is apparently reinforced by some examples of progress in the physical sciences. In physical science, the starting point for the development of a theory is often a pretheoretical notion — a part of folk science. For example, the untutored human intellect tends to see mechanical phenomena in a rather Aristotelian impetus framework. Bodies suffering impacts gain something, impetus, that propels them into motion and sustains motion until it is depleted.[26]

Other examples: it used to be thought that the Earth was uniquely distinguished from the celestial bodies and that it, therefore, was governed by a wholly different set of physical principles. Also, it was very widely thought that the heart was the center of most bodily functions and of life itself, and that electricity was a kind of fluid—it is easy to go on.

In each of these cases, beginning with scientific concepts picked out by terms in the vernacular, that is with folk-scientific concepts, led to empirically bad scientific theory. One might even think that it is partly *characteristic* of scientific inquiry (as opposed to other ways of finding out about the world) that our inherently uninformed folk concepts be rejected in favor of those that allow us to increase the precision and intelligibility of our predictions and explanations. In light of the considerations raised by historical cases, we might judge it an error to fixate on folk concepts when doing science.

Could this be the root of the persistent empirical difficulties facing economics? Is it the case that economics has uncritically taken folk notions and built a vast theoretical edifice upon such unsubstantial foundations? If *commodity* is inescapably a vernacular kind, a folk concept, and hence not a natural kind, and if, moreover, we require that economics be about actual commodities, then the beginning of an apparent explanation of the theory's shortcomings presents itself.

It can be shown, however, that this is not a good explanation. There is, I shall argue, at least one imposing example of a very successful scientific theory that was founded on a folk concept. I have in mind Newtonian mechanics—the theory as it appears in Newton's *Principia* and not later elaborations and improvements of the theory. Just as modern economics is founded on the *commodity*, Newtonian mechanics is founded on the *body*.

I first need to produce some evidence for this idea. There appear to be competitors for the status of foundational concept in Newtonian mechanics, or it appears, at least, that there is no single central-kind term. There are *motions* (and this even seems to be a folk concept), both uniform and accelerated; there are *forces*; and there are *interactions*. I shall first argue that the concept of *body* is the folk concept. I shall then show how the other important kinds in the theory are derivative in the same way that important economic kinds can be derived from *commodity*.

The word *body* (*corporis* in Latin)[27] first appears in the *Principia* in Newton's own preface to the first edition. There he says, "But since the manual arts are chiefly employed in the moving of *bodies*, it happens that geometry is commonly referred to their magnitude, and mechanics to their motion" (C xvii, emphasis added). Manual arts, of course, deal with what we prescientifically take to be bodies, or "physical objects." So mechanics is to be *about* the motions of ordinary bodies. But after reading this prefatory remark, one might still wonder whether the *theory* deals with these motions in virtue of dealing with some idealized kind of body that only resembles or represents actual, ordinary bodies. Such is not the case. The famous first law of motion says, "Every body continues in its state of rest . . . ", and the elucidation goes on to say:

> Projectiles continue in their motions, so far as they are not retarded by the re-
> sistance of air. . . . A top . . . does not cease its rotation, otherwise than
> it is retarded by the air. The greater bodies of the planets and comets . . .
> preserve their motions both progressive and circular for a much longer time.
> (C 13)

There is clearly nothing of an ideal or representational character about "projec-
tiles retarded by air," "tops retarded by air," and "planets and comets." All of these
bodies are precisely the kinds of things we are familiar with before any investigations in the science of mechanics (though *recognizing* the celestial bodies as bodies takes some science).

Futhermore, there is never any analysis of *body* in the *Principia* to suggest that
in the theory the concept has some complexity not present in the common concept. When, in the Scholium to the Definitions (C 6–12), we are given clarifying
definitions of some fundamental terms that are "well known to all" such as 'time',
'space', and 'motion', the term 'body' is conspicuously absent. We do learn that
there are "particles of bodies" (C xviii) that compose them, but since at least some
particles are themselves divisible and not essentially atomic, and since some bodies are so small as to be insensible (C 399), it seems that the things Newton calls
particles are themselves bodies. So bodies are made up of smaller bodies, quite
in keeping with the common, pretheoretical concept.

Finally, most other apparently fundamental and apparently theoretical terms
are straightforwardly definable in terms of *body*. *Motion*, both relative and absolute, is simply "the translation of a body from one absolute/relative place to another" (C 7). *Place* is "part of a space which a body takes up" (C 6). All that remain are *space*, *time*, and *forces*. These, like body, are unanalyzable, though
Newton's space and time are probably not pretheoretical concepts, and forces certainly are not. But unlike space and time, forces, both inertial and impressed, cannot be understood apart from bodies. "The innate force of matter [subsequently
called *vis inertiae* or inertia] is a power of resisting by which every body . . .
continues in its present state. . . . " and "An impressed force is an action exerted upon a body. . . . " (C 2). So inertial forces are essentially connected to
bodies, and impressed forces are necessarily forces on bodies. Moreover, any
force impressed on a body somehow originates with some other body or bodies;
consider gravitational force, magnetic force, tensions, etc. Similarly, a body's inertial force will not be exerted unless there is a force impressed on it. Therefore,
a perfectly isolated body would neither exert nor feel any forces, so forces do not
exist except in the presence of more than one body. Moreover, once we have bodies and the forces that act on them, we have everything. Newton says the whole
burden of philosophy is, " . . . from the phenomena of motions to investigate

the forces of nature, and then from these forces to demonstrate the other phenomena . . ." (C xvii–xviii).

Thus the only things that are, in a sense, ontologically independent in Newtonian mechanics are space, time, and body. But Newton did not think that his theory was in any way *about* space and time (though today we may be inclined to disagree with him about this). It is about bodies *in* space and time. It did not seem strange to Newton to accord space and time this peculiar status; for him they are charged with theological significance (C 545). I conclude that the whole of Newtonian mechanics, as it is expressed in the *Principia*, rests on the concept of *body*; bodies are the *kinds* of things that the theory is about. And these bodies are understood prescientifically – the extension of the vernacular term 'body', one is tempted to say.[28]

The theory of Newton's *Principia* has to its credit many empirical successes, many of these described in the same pages by Newton himself. Moreover, the folk-theoretical kind *body* plays a ubiquitous role in the development of the science. Thus, it cannot be concluded that it is generally true that successful science cannot be based on kinds that are really accessible to us prior to theoretical articulation. Let us return to the example of economics. It now seems most unlikely that its difficulties in satisfactorily connecting up with observed phenomena can be blamed on the centrality of the kind *commodity*. The example from physics shows that there is nothing wrong in principle with having a science *of* some things or phenomena that are individuated prescientifically. We still seem to be left with a puzzle about why the impressive theoretical framework of economics is so difficult to apply to the world successfully.

It might be objected at this point that my treatment of *body* in the *Principia* was too hasty and that the conclusions drawn are too strong. Perhaps Newton does *begin* with a folk-theoretic kind, but in the course of developing his theory, it might be that the concept *body* also develops. Perhaps, by the end, the characteristic scientific enterprises of generalizing and systematizing leave behind a concept that is at some remove from the common notion of body. Perhaps when we look at the entire *Principia* in context, and not just at isolated passages as I have done above, we can discover ways in which *body* evolves from a folk-scientific concept into a concept that is almost as plainly theoretical as *quark* is.

If so, we should consider the possibility that the source of the empirical differences between Newtonian mechanics and economics can be located here. Newtonian mechanics may owe its predictive and explanatory successes with bodies ordinarily conceived to its employment of a more highly developed, idealized version of the folk concept to model or represent the actual objects. I shall call such an idealized concept a *refinement* of a folk-theoretic or pretheoretic concept.[29] the refined theoretical concepts often turn out to be surprising or counterintuitive to those not sufficiently familiar with the theories and the problems they were meant to solve. The objects and properties that these new concepts involve are also usually further removed from straightforward observability than those involved by folk concepts. For instance, *impetus* and *heaviness*, almost sensually determined kinds, are refined into *momentum* and *mass*. On the face of it, scientific progress is characterized by improvements in prediction, explanation, and intelligibility that are all made possible by the refinement first of folk-scientific concepts, and then by the successive refinement of more sophisticated concepts.

If this is right, then the symmetry between *body* and *commodity* would be broken if Newtonian mechanics provided a refinement of *body*, but economics failed to refine the folk theoretic *commodity*. I can briefly indicate to what extent *body* gets refined in the *Principia*; a complete treatment is difficult because Newton himself seems to have been confused about this aspect of the foundation of his theory. What is important for this essay, however, is that it is quite clear that, contrary to what I have been heretofore supposing, *commodity* also undergoes some surprising refinement in economic theory. The relevant "structural" symmetry between physics and economics that I have been exploiting is not, therefore, broken by considerations of refinement.

The most significant refinement of *body* first explicitly occurs in Corollary IV to the famous third law of motion, with the introduction of the concept of the *center of gravity* common to two or more bodies (C 19). When applying the theory to determine actual trajectories, tensions, and so on, almost all calculations are performed using the device of centers of gravity (today we quite properly prefer to speak of centers of mass) and not with the surfaces of the bodies themselves. When calculating we often treat systems of bodies as single bodies, all of whose mass is concentrated in a single point of space, the center of gravity of the system. The "body" so obtained is plainly not a real body at all in the ordinary sense. Even the center of gravity of a single body (what modern texts call a point mass) is not an ordinary body—it occupies only one dimensionless point of space.[30] Without at least this refinement of *body*, it is hard to see how Newtonian mechanics could yield any empirically useful results at all.

If economics were unable to offer an analogous refinement of its central-kind term, we might naturally suspect that this was the root of the difficulties with obtaining good quantitative results. But there does seem to be a strong analogy between economics and Newtonian mechanics with respect to refinements. Just as powerful mathematical techniques cannot be applied directly to unrefined features of bodies, the technical apparatus of economics (the most powerful being associated with GET) cannot cope with the complexity of the unrefined concept of commodity. One complexity that can be made to appear obvious has to do with the overwhelming variety of commodities in any modern economy. Part of our economic behavior as consumers involves making choices among avocados, bananas, cucumbers, dates, eggplants, figs, etc., and that is only for produce. Then there are economic decisions among airline tickets, boat cruise tickets, canoe rentals, dog sled rentals, elevated train tokens, etc., for transportation. These lists can be extended almost indefinitely, and there are almost indefinitely many more lists to be constructed. Even if a consumer did have a well-defined utility function for this vast array of commodities, there would be no practical hope of discovering what it was. If we experimented on a particular person for his entire lifetime, we would not have enough data to infer even approximately how his choices among bundles containing indefinitely many commodities were structured. But there are methods for coping with this situation.

In the first place, economists almost never attempt to predict or explain the behavior of particular consumers. Instead, the unit of analysis is a "representative" or "average" consumer, whose properties are constructed from data about entire markets that is often plentiful and reliable. Since such a representative individual (call him R) is a fiction, we must think of what he consumes as also something less than fully tangible. This is already beginning the process of refinement. A second sort of refinement takes place when we begin to study the hypothetical behavior of R. If R's utility function is to be subject to standard mathematical maximization techniques, its potentially indefinite list of variables (commodities) need to be *aggregated* into a manageably small list. We might, for example, take R's utility to be a function of three variables: food, shelter, and entertainment. Each of the three categories is understood to comprise the appropriate specific commodities that R is, or might be, confronted with, like avocados, apartment rental, and computers.

Commodities can be aggregated in various ways into variously constructed aggregates. The skillful economist will choose aggregates that make problems of interest mathematically tractable. For example, if we are interested in market behavior concerning facsimiles of the manuscript of Brahms's Fourth Symphony, we may wish to suppose that R's utility is a function of two variables: 1) copies of the facsimiles, and 2) everything else that R might choose. If we are interested in R's choices of investments, we may aggregate all commodities into these two: 1) things consumed in the first half of R's life, and 2) things consumed in the second half of R's life. There is some well-developed theory regulating the conditions under which it is permissible to aggregate commodities. There is, for example, Hick's composite commodity theorem that states, " . . . if a group of prices move in parallel, then the corresponding group of commodities can be treated as a single good [commodity]" (Deaton and Muellbauer 1980, 121). In practice, of course, we can only hope that our aggregates do not violate the restrictions imposed by Hick's theorem, because we do not have dependable data on the pairwise price movements for the whole indefinitely long list of actual commodities. But I am presently interested in theoretical refinement.

The examples I have presented from physics and economics are much more complex and subtle than I can indicate here, but I think that these very brief sketches suffice to bring out a significant point. In Newton's theory, the ordinary concept of body was refined into something unrecognizable to the untrained intellect—the center of gravity. In GET, the ordinary concept of commodity is similarly refined, for the "commodities" arising from aggregation are surprisingly unfamiliar objects. What manner of *thing* is *everything consumed in the first half of a lifetime*, or *everything except facsimiles of the manuscript of Brahms's Fourth Symphony*? And it is notoriously difficult to think of a person's generalized capacity to labor as a commodity, something to be bought and sold as though it were in principle completely alienable from the laborer. No one would normally have occasion to think of such things as these except in the context of a sophisticated scientific theory.[31] I mentioned a related refinement earlier, when characterizing GET. In GET, a *consumer* is a thing with a utility function that exchanges, annihilates (i.e., consumes), and helps create (i.e., by supplying labor) *commodities*. We can think of GET's *consumer* as an extreme refinement of the pretheoretic concept of a person or the folk-economic concept of a consumer. It is indeed extreme; GET's consumers are paradigms of what we would ordinarily apply the vernacular term 'inhuman' to! Similarly, centers of gravity that are at single points of space are good candidates for the vernacular terms 'incorporeal' and 'immaterial'.

VII

In the absence of further arguments, it is reasonable to conclude that Newtonian mechanics and GET both begin with folk-theoretic kinds and refine these in strikingly analogous ways. The difference in empirical applicability cannot be accounted for by differences in the type of refinements the important kinds undergo. In this section I want to examine the radical idea that the kind terms of economics do not pick out natural kinds in the world. This idea is radical because it suggests that economics is essentially different from natural sciences. Successful natural sciences produce reliable laws and formulas that can be thought of as expressing relationships among natural kinds or relationships among the properties of a single natural kind. If there are no economic kinds that are natural, economics cannot do this. This way of formulating the problem is not new. A. Rosenberg (1983) has written,

> Philosophers have shown that the terms in which ordinary thought and the behavioral sciences describe the causes and effects of human action do not describe "natural kinds," they do not divide nature at the joints. (301–2)
> The predictive weakness of theories couched in intentional vocabulary reflects the fact that the terms of this vocabulary do not correlate in a manageable way the vocabulary of other successful scientific theories; they don't divide nature at the joints. . . . [Some] insist that we must jettison "folk psychology" and its intentional idiom if we are to hit upon an improvable theory in the science of psychology. This choice extends, of course, beyond psychol-

ogy to all the other intentions sciences, of which economics, with its reliance on expectation and preference, is certainly one. (303)

In this essay, I want to remain neutral as to what philosophers have or have not shown about the existence of intentional natural kinds and offer another sort of diagnosis of the failure of *commodity* to pick out a natural kind. Let us return to the fact that there is a pretheoretical concept of a body or object and to the assumption (I shall question this assumption in what follows) that there is a pretheoretical concept of a commodity.

Why is there the pretheoretical concept of a body? It is tempting to answer by saying it is because there *are* bodies, and human beings have evolved to be extremely good detectors of bodies that are roughly their own size. It seems safe to conjecture that any mature normal human being will have a concept of body. We need to be good at recognizing things that move, can be grasped and thrown, can be eaten and mated with, and can forcibly collide with us. One could also approach the question from the other direction and say with Kant that it is the fact of appearances of bodies as spatiotemporal that yields the transcendental ideality of space and time as founding the possibility of these appearances. Either way, the result is that humans inescapably perceive the world as containing bodies, and that fact itself is some kind of evidence for the veracity of these perceptions. Newtonian mechanics, therefore, has helped itself to a pretheoretical concept that is very well suited, perhaps uniquely well suited, to scientific refinement. Now let us ask the same question for economics.

Why is there the pretheoretical concept of a commodity? In one sense, there simply isn't one, because it is very unlikely that this concept is universally possessed in the same way that the concept of *body* is. We know that there have been some societies in which the practice of private ownership is extremely attenuated for most items that we would consider commodities. In such a society, these items would not be conceived as objects of production or even of exchange. It is even unlikely that anything except food would be thought of as being *consumed*, since the practice of leasing whereby time-slices of durable goods are consumed requires a specialized system of strong property rights. Anthropologists could probably identify the concept of a 'potentially useful portable object', but that is simply a *body*—commodities are not kinds of things for such a society and, a fortiori, not natural kinds. It is, I think, questionable whether the concept is *pre*theoretical even in advanced capitalistic economies. By the time one has reached the age at which things like paper money, salaries, rents, banking and investment, and so on are somewhat comprehensible, one is surely already in possession of a fairly substantial economic folk theory. One understands that higher prices mean fewer sales, inflation is good for debtors and bad for those with fixed incomes, depressed stock prices are bad for pension funds, and so forth. It is not important for present purposes to draw a sharp distinction between what is pretheoretical and what is folk theoretical; I merely want to stress the point that whether people can recognize commodities as kinds of things, or even individuate them at all, is determined by the society and culture that they live in. This does distinguish economics from Newtonian mechanics. Economics did not begin, as physics did, by helping itself to a ready-made pretheoretical central concept that lent itself to successful refinement.

Now it does not *follow* from the fact that *commodity* is not a universally recognized pretheoretic kind that it is not a natural kind after all. In general, it is entirely possible that people not have the right kind of mental equipment to recognize real kinds. In like manner, it does not follow from the simple fact that there is a pretheoretic concept of body, that *body* is a natural kind. But it is. We can see this from the fact that lawlike relationships obtain among bodies. Furthermore, the universality of the experience of bodies seems closely related to their constituting a natural kind. This property of universality might be a key to understanding the status of *commodity*.

Suppose that the reason commodities are not universally recognized is that there just aren't any of them in some nonpathological human societies. Suppose, that is, that insofar as commodities do exist, they are brought into existence by virtue of some distinctive properties of some societies. (I shall consider some points that may count against these suppositions below.) Now consider that an intuitively extremely plausible condition on something's being a scientific law is that it be universal, that it apply to all the things in its domain.[32] Newtonian mechanics applies to all bodies, electrochemistry applies to all solutions, and so on. Does GET-based economics have this kind of universality? This question cannot be answered until it is determined what the domain of the theory is supposed to be. If we think of economics as a *social* science, a science dealing with the economic aspects of life in society, then economics is not universal, for we are supposing that there are societies to which it does not apply in virtue of their not having any commodities. This is close to the conclusion that economics is not a science in the same sense that natural sciences are.

If economics is taken to have a more modest domain, the economies of modern capitalistic societies, for example, then the level of empirical success that the theory enjoys becomes important to the investigation of its foundations. If the theory were an extremely good predictor and explainer for the limited domain, then we would be inclined to say that modern capitalistic societies are essentially different from any others to which the theory does not apply, and that this difference is part of what makes them appropriate objects of scientific economic inquiry. If, however, we are prepared to admit that the theory cannot be applied with great success even to a specially chosen domain, then I think we should be less inclined to see a deep divide between modern capitalistic societies and other human societies. If there is no such divide, and if all the other suppositions I have been entertaining are well founded, then economics is not universal and does not express relations among natural kinds and their properties. This is stronger than claiming that economics sometimes gets it more or less right (when applied to modern capitalistic societies) and sometimes gets it wrong (when applied to other societies). It is claiming that economics *never* gets it really right because *commodity* is not a natural kind. If commodities are not natural kinds in *any* society, there cannot be an empirical science about them. If I am right, we should not think of economics as a false theory about things that are in the world; its lack of success is, instead, inevitable because the things that it is supposed to be dealing with are not there.

This powerful conclusion is arrived at with the help of a few assumptions. I shall conclude by considering two kinds of interesting objections to the assumptions I have made. First, however, I want to point out an objection that won't be discussed. It is often asserted (more commonly in conversation than in print) that economics is terrifically good at prediction and explanation after all. I think that most scholars, even most economists, believe that this is wrong. Applications passed off as great successes usually turn out to be either post facto reconstructions of aggregate data from properties of imaginary representative individuals (see Nelson, 1989), or else things readily available to folk economists such as, "people buy smaller automobiles when gasoline prices are higher." It is not clear how this disagreement can be resolved.

A more interesting objection comes from combining the assertions that economics does tolerably well for modern capitalistic societies, that in principle it can do extremely well for these, *and* that it can, in principle, do extremely well for other kinds of societies as well. Some economists, following the prominent example of G. Becker, think that virtually all individual, and hence all societal, human behavior can be understood by means of GET-based economics. They presumably think that this is the case even when the members of the society do not conceive their actions within this framework. Empirical work on precapitalistic societies has been initiated, and some of it is very interesting even though it is far from convincing. (C. Dahlman's 1981 treatment of the feudal English open field system is a good example). As with economic studies of contemporary societies, we cannot determine a priori that surprisingly good empirical results are not forthcoming at some future date. We seem to be in store for a long wait.

Another related, but stronger objection could be mounted if we could discover some naturalistic foundation for *commodity*. Such a foundation might be provided by psychology in the way indicated above on pp. 108–9. (See Nelson 1986 for some details.) It might be true of the cognitive faculties of human beings that they include something representable by a utility function and that economic behavior (perhaps narrowly, perhaps broadly construed) is partially caused by a process representable as the maximization of this function. Another way to demonstrate that commodities are natural kinds, despite the argument of this essay, would be to bypass psychology and go straight to biology. It is undeniable that human beings have in common certain biological needs, like nutrition and shelter, and that there are various choices among strategies to be implemented in attempting to meet these needs. In response to the selectional pressures these needs place upon the species, humans might have evolved specialized neurological structures that compute what the efficient courses of need-satisfying action are. These neurological structures might turn out to have accurate descriptions in the language of economics; they might realize utility functions, for example. In this way, economic theory might turn out to be in our genes. One striking thing about both of these proposals is that the *utility function* or the *preferences* replace *commodity* as the fundamental concept in the theory. If there is a sense in which human minds or brains actually unconsciously compute utility maxima, then this phenomenon can be directly studied with the help of psychology or neurology. What commodities there actually are becomes a practical, not a theoretical issue. Of course, there may be a further "transcendental" argument from our having utility functions to there being *commodities*, like the argument from the appearances we have to the existence of bodies. This would give *commodities* a sort of ontological priority to *utility*, but the investigation of this connection between *commodity* and *utility* would not be part of economics, it would be part of biology.

It must be conceded that these possibilities that *commodity* would turn out to be a natural kind are remote. And few economists show any interest in pursuing them. Still, careful scientific investigation might reveal one of these possibilities to be the truth. It is, however, sometimes open to philosophers to argue about what is and is not possible given what we presently know.

In view of the extreme claims made on all sides about economics, I think that the position I have argued for in this essay can be considered moderate. I have tried to show that there is nothing incoherent or wrongheaded about thinking that the central-kind terms of economics pick out natural kinds. But if they are, then we are committed to an interpretation of the concepts connected with the kind terms as having explanatory force. We are committed to what I called the Utility-as-Explanatory interpretation. When conjoined with the fact that economics has a poor empirical record, however, this interpretation has the consequence that *commodity*, and along with it the other key concepts in the theory, is not a natural kind. Since physics, chemistry, biology, and closely related disciplines contain lawlike relationships among natural kinds and their properties, this fact about economics distinguishes it from these "natural" sciences. After studying other social sciences and other approaches to economics in the way that GET-based economics has been studied in this essay, we might arrive at the same conclusion about them. If so, we would be closer to understanding the peculiar scientific status of the social sciences.

Notes

1. My own, apparently unpopular, view is that insofar as vernacular "natural"-kind terms have a class of referents tidy enough to merit the title "extension" (quite unlikely, I think), this extension is very different from the extension of the corresponding scientific term. For example, I think that the semantic properties of the vernacular term 'water' differ considerably from the chemical term 'water'. These positions are strongly argued for in Donnellan (1983, 85ff.):

> . . . the terms obviously are not borrowed from the vocabulary of science and were part of English long before the advent of modern science. . . . [a]lthough one might suppose that if terms for natural kinds are to be found anywhere the language of science would be replete with them, it is not obvious that the Kripke-Putnam theory is applicable to kind terms in science. Nor is it obvious that it will apply to terms which the vernacular has borrowed from the language of science. . . .

I shall just assume in this essay that there are natural kinds of things in the world, but I really only need the notion as a tool for analyzing particular scientific theories. Little depends on the metaphysical stance adopted toward the kinds. In particular, the argument does not require that natural kinds are ontologically prior to scientific laws.

2. I shall not argue here that economics is not as empirically successful as physics is. The claim seems to require enumeration of cases more than argumentation anyway. For the view of an important economist see Leontief (1971). A philosophical treatment of the relative success of economics and the physical sciences is in Rosenberg (1989).

3. Anything that cannot be put into the GET framework is usually regarded as suspect economics. (Economists disagree among themselves about this to some extent.) The situation is similar to that in classical mechanics in the nineteenth century. Any physical theory that was not somehow a part of, or reducible to, classical mechanics did not clearly count as physics. Even a "nonmechanical" domain like electromagnetism is subsumed by virtue of the forces it describes figuring as components in F=ma. A good example of the generality of GET is it applicability to parts of Marxian economics in Roemer (1981). For a discussion of the problem of fitting macroeconomics into GET see Weintraub (1979).

4. My exposition of this material is unusual, but its economic content is based on Arrow and Hahn (1971), the book that H. Varian calls one of the two "definitive modern treatments" and "the most up to date treatment" of systematic GET (1984, 210, 211).

5. The intuitive idea is that different bundles of commodities are ordered according to preferences the consumer has for one bundle over another: Arrow and Hahn write of "levels of satisfaction" (1971, 80), and Varian uses the phrase "the consumer thinks that the bundle x is at least as good as the bundle y" (1984, 80). But economists often wish to dispense with such psychological terminology as I explain in what follows.

6. It is often convenient (and will be in this essay) to think of the order given by the consumer's preferences as inducing an *indifference map*. An indifference map is a set of *indifference curves* in R^n (where n is the number of commodities in the economy), each of which contains points in the space of commodity bundles that "occupy the same place in the preference order" — the consumer is "indifferent" to any two points on the same indifference curve.

7. Any set of consistent preferences, preferences that are transitive, can be represented by some utility function — by brute force if necessary. Not every consistent preference map will be representable by a *well-behaved* function. Lexicographic preferences, for example, cannot be represented by a continuous utility function.

8. This is the place where most of the ideological, noneconomic content of the theory is introduced. The concept of an endowment entails a far-reaching theory of property rights. I shall have more to say about this later.

9. See Stigler (1965, 84–144) for a summary of the relevant history.

10. See Coats (1976, 50–59), Schumpeter (1954, 1056–61), and Stigler (1965, 144–48, 151–55) for a summary of the relevant history.

11. As is usual in the history of science, the first fully satisfactory treatment was prefigured in various ways; Allen and Hicks did not invent ordinal utility theory, but they found a statement of it that made a clean break with the cardinal theory possible. For some historical details, see Schumpeter (1954, 1062–66) and Samuelson (1974, especially the appendix).

12. SARP requires, roughly, that the consumer's choices not violate a transitivity condition. For details, see Samuelson (1938b and 1950).

13. It is only a suggestion. It would not be inconsistent to adopt the formalism of revealed preference for its formal virtues and still interpret the theory as being about cardinal utilities.

14. The import of this historical episode has various interpretations. Rosenberg (1981) argues that psychological facts disconfirmed the economic theory and that subsequent developments were unjustifiable, ad hoc attempts to save the theory. Cooter and Rappoport (1984) think that economists' interests shifted away from the welfare economics that required interpersonal comparisons towards more theoretical concerns.

15. Samuelson here refers to marginal rates of substitution.

16. In particular, this interpretation tells us to ignore the (disastrous) possibility that preferences change over time. A consumer who behaves as if his preferences change is considered irrational and, therefore, not subject to the theory.

17. Given the additional, not terribly ad hoc, assumption that Pluto was once a satellite of Neptune's. For a brief account of the history of Bode's law see Holton and Roller (1958, 198–201). For Balmer's formula and the Bohr model, see Holton and Roller (607–33).

18. Bode's law does not fail to be a real law because of a conceptual flaw; it seems that we can imagine a universe constructed along Keplerian lines in which it falls out of other cosmological facts. Its success is simply accidental. Accidental successes of this magnitude demand searches for underlying mechanisms; in this case the search was soon shown to be futile.

19. For more details and a defense of this kind of procedure against methodologically inspired objections, Nelson (1986) must be consulted.

20. Both alternatives involve surprising complications. For a discussion of the former, see Nelson (1986); for the latter, Nelson (1984, and 1989).

21. An analogous situation arises in the theory of the firm. Production functions contain factors of production as independent variables. Should money be included as a factor of production? One might think not because the physical money, the bank notes and coins, do not help produce output. One might shovel notes into the furnace along with a genuine factor of production like coal, but this would be similar to a consumer's using the notes as wallpaper to get utility. If, however, one thinks that production functions, like utility functions, are descriptive and not explanatory, then it seems that economists should use dependent variables resulting in the functions that best fit observed phenomena—and these probably will include money. See Sinai and Stokes (1981) and the references listed there for a sample of the literature.

22. See, for example, Van Fraassen (1980).

23. Later, I shall question whether there really is a good pretheoretical notion of a commodity.

24. For discussion, see Stich (1983).

25. Experimental studies show that modern, well-educated people have generally poor physical intuitions about basic mechanical processes (McCloskey 1983). We do need a theory to tell us what are natural kinds of motions.

26. Again, see McCloskey (1983) for a review of the old physics and for some psychological experiments supporting the claim that this is indeed the folk physics of mechanics.

27. *Corpore* is used in two ways by Newton. It usually means some particular body, but it can also mean matter or what makes up bodies. It appears in both senses in Definition I, one line apart!

134   Alan Nelson

"It is this quantity that I mean hereafter everywhere under the name of body [*corporis*] or mass. And the same is known by the weight of each body [*corporis*] . . . " (Cajori 1934 [1726], 1 [I shall refer to this book as "C"]; Koyre and Cohen 1972 [1726], 1:40). The first sense predominates and I won't discuss the second here. M. Jammer seems to suggest that Galileo struggled with the distinction (1961, 52).

28. The primary vernacular sense is, of course, as in the bodies of animate creatures (*corpore* in Latin). The *Oxford English Dictionary*, however, suggests that this sense is very close to the sense of "material object." Samuel Clarke, writing to Leibniz more or less on behalf of Newton, suggests that we understand bodies and their interactions on the model of the interaction between our own bodies and our minds (or God). (Alexander, ed., 1956 [1717], 116–17).

29. It is not necessary here to give an analysis of what I am calling refinement. It is, perhaps, related to Toulmin's account of the evolution of concepts (1972, 200–36).

30. So the folk concept *body* is implicitly refined in the very first corollary, "A body, acted on by two forces simultaneously, will describe the diagonal of a parallelogram . . . " (C 14) because a parallelogram and its diagonal are mathematical and not physical figures. Only a *center* of gravity could describe a line; an actual body would describe a solid wormlike figure. I think that this lack of expository precision on Newton's part indicates that he was not clear in his own mind about the distinction between the folk notion *body* and its refinement. It is interesting to note that the same refinement is employed when Newton discusses the *arcs* that planets describe.

31. There are, of course examples of more familiar aggregates, such as *food*, but if a utility function has many arguments at this familiar level of aggregation, calculation again becomes impossible.

32. The intuitive idea seems reasonably clear, although it is notoriously difficult to state clearly. "Applying to everything in its domain" is different from being "exceptionless." A law has an exception if it applies to something, but gets at least one thing wrong about *it*. *Really wanting ice cream* might be lawfully related to *trying to get ice cream*, although there are many exceptions—some external phenomenon may intervene before the wanter manages to try. But we might agree that *everyone* who is a wanter will be a tryer *unless* there is intervention. So the relation might be universal without being exceptionless.

References

Alexander, H., ed. 1956 (1717). *The Leibniz-Clarke Correspondence*. Manchester: Manchester University Press.

Arrow, K., and Hahn, F. 1971. *General Competitive Analysis*. San Francisco: Holden-Day.

Coats, A. 1976. "Economics and Psychology: The Death and Resurrection of a Research Programme." In *Method and Appraisal in Economics*, ed. S. Latsis. Cambridge: Cambridge University Press.

Cooter, R., and Rappoport, P. 1984. Were the Ordinalists Wrong about Welfare Economics? *Journal of Economic Literature* 22: 507–30.

Dahlman, C. 1980. *The Open Field System and Beyond*. Cambridge: Cambridge University Press.

Deaton, A., and Muellbauer, J. 1980. *Economics and Consumer Behavior*. Cambridge: Cambridge University Press.

Donnellan, K. 1983. "Kripke and Putnam on Natural Kind Terms." In *Knowledge and Mind*, eds. C. Ginet and S. Shoemaker. New York: Oxford University Press.

Friedman, M. 1953. "The Methodology of Positive Economics." In *Essays in Positive Economics*. Chicago: University of Chicago Press.

Friedman, M. 1976. *Price Theory*. Chicago: Aldine.

Hahn, F. 1984. *Equilibrium and Macroeconomics*. Cambridge: MIT Press.

Hicks, J., and Allen R. 1934. A Reconsideration of the Theory of Value. *Economica* 1 (new series): 52–76 and 196–219.

Henderson, J., and Quandt, R. 1980. *Microeconomic Theory*. New York: McGraw-Hill.

Holton, G., and Roller, D. 1958. *Foundations of Modern Physical Science*. Reading, Mass.: Addison-Wesley.

Jammer, M. 1961. *Concepts of Mass*. New York: Harper and Row.

Karenken, J., and Wallace, N., eds. 1980. *Models of Monetary Economics*. Minneapolis: Federal Reserve Bank of Minnesota.

Leontief, W. 1971. Theoretical Assumptions and Nonobserved Facts. *American Economic Review* 61: 1-7.

Marshall, A. 1961 (1890). *Principles of Economics* (9th ed.). London: Macmillan.

Marx, K. 1967 (1887). *Capital*, vol. 1. Trans. from the third German ed. by S. Moore and E. Aveling. New York: International Publishers.

McCloskey, M. 1983. Intuitive Physics. *Scientific American* 248: 122-30.

Nelson, A. 1984. Some Issues Surrounding the Reduction of Macroeconomics to Microeconomics. *Philosophy of Science* 51: 573-94.

——. 1986. New Individualistic Foundations for Economics. *Nous* 20: 469-90.

——. 1989. Average Explanations. *Erkenntnis* 30: 23-42.

Newton, I. 1934 (1726). *Newton's Principia*. Ed. and trans. F. Cajori. Berkeley: University of California Press.

——. 1972 (1726). *Isaac Newton's Principia*. 3rd ed. of 1726 assembled and edited in two vols. by A. Koyre and I.B. Cohen. Cambridge: Harvard University Press.

Quirk, J., and Saposnik, R. 1968. *Introduction to General Equilibrium Analysis and Welfare Economics*. New York: McGraw-Hill.

Roemer, J. 1981. *Analytical Foundations of Marxian Economic Theory*. Cambridge: Cambridge University Press.

Rosenberg, A. 1981. A Skeptical History of Microeconomics. *Theory and Decision* 12: 79-93.

——. 1983. If Economics Isn't Science, What Is It? *The Philosophical Forum* 14: 296-314.

——. 1989. Why Generic Prediction is Not Enough. *Erkenntnis* 30: 43-68.

Samuelson, P. 1938a. A Note on the Pure Theory of Consumer Behavior. *Economica* n.s. 5: 61-71.

——. 1938b. The Empirical Implications of Utility Analysis. *Econometrica* 6: 344-56.

——. 1950. The Problem of Integrability in Utility Theory. *Economica* 17: 355-85.

——. 1974. Complementarity: An Essay on the 40th Anniversary of the Hicks-Allen Revolution in Demand Theory. *Journal of Economic Literature* 12: 1255-89.

Schumpeter, J. A. 1954. *History of Economic Analysis*. New York: Oxford University Press.

Sinai, A., and Stokes, H. 1981. Money and the Production Function. *The Review of Economics and Statistics* 63: 313-18.

Smith, V. 1982. Microeconomic Systems as an Experimental Science. *American Economic Review* 72: 923-55.

Stich, S. 1983. *From Folk Psychology to Cognitive Science*. Cambridge: MIT Press.

Stigler, G. J. 1965. "The Development of Utility Theory." In *Essays in the History of Economics*. Chicago: University of Chicago Press.

Tobin, J. 1980. "Discussion." In (Karenken and Wallace, eds., 83-90).

Toulmin, S. 1972. *Human Understanding*. Princeton: Princeton University Press.

Van Fraassen, B. 1980. *The Scientific Image*. Oxford: Clarendon Press.

Varian, H. 1984. *Microeconomic Analysis*. 2d ed. New York: Norton.

Wallace, N. 1980. "The Overlapping Generations Model of Fiat Money." In (Karenken and Wallace, eds, 49-82).

Weintraub, E. R. 1979. *Microfoundations*. Cambridge: Cambridge University Press.

Foundational Physics and Empiricist Critique

1

Empiricism is not a very popular methodological or epistemological stance nowadays, and we all know why. It presupposes the existence of sensory contents both as uncontaminated by theory and as present to us in a way so as to serve as the foundation of our corpus of rational belief. But Gestalt considerations tell us that the presuppositionless empirical content is a myth, and considerations familiar since the idealists tell us that only the propositional can play a role in the inferential structure of justification, and not "objects" as sense contents were supposed to be (if "objects" of a peculiar sort). To go beyond rational belief in particular facts, empiricism requires, also, rules of inductive inference supported on a priori grounds. But both traditional Humean skepticism about the rationality of induction and more recent objections based on the absence of any a priori basis for a selection of "natural kinds" relative to which inductive generalizations are to be formed, cast grave doubts on the possibility of a coherently formulated and rationalizable empiricist account of the confirmation of generalities.

Even if we could rationally found our general beliefs about observables in the manner that the empiricist suggests, his doctrine would still be inadequate as a reconstruction of scientific knowledge. For despite all the effort made to overcome the well-known long-standing problems, we still don't understand how, beginning with empiricist preconceptions, we can explicate our understanding of theoretical concepts, realistically construed, nor rationalize our inference to theory that posits what, to the empiricist, are unobservable entities and structures in the world. Indeed, if the perceptual basis is confined to the familiar empiricist realm of the private contents of sensory awareness, it is hard to understand how we could, on empiricist precepts, develop an intersubjective account of the world at all.

Worse yet are the difficulties the empiricist faces in giving us an adequate reconstruction of how we come to grasp the meaning of the terms essential to frame our scientific hypotheses. The empiricist semantic foundation of basic words acquiring meaning in the presuppositionless way by means of the intentional association of the word with an "idea in the mind" ostensively presented is, for a variety of reasons (some of which overlap the skeptical arguments against the empiricist notion of the "given" as an epistemic foundation), a dubious ground on which to base a theory of meaning or the grasp of meaning. And—in a manner similar to the epistemic doubts that, even given an empiricist foundation for knowledge, we could on the basis of that foundation reconstruct the body of objective, scientific knowledge—the empiricist semantic foundation, even if it existed, provides too subjective and too flimsy a ground on which to base our comprehension of the meanings of the terms in a public, scientific language fully possessed of the resources to refer to and describe even the unobservable.

In the face of these difficulties with traditional empiricism, we have been presented with alternative models of scientific meaning accrual and scientific epistemic justification. Various varieties of "realism" and "pragmatism" share common objections to empiricist presuppositions and common alternatives to some empiricist claims, even though they at least seem, in places, to differ from one another in their own reconstructive approaches.

For the realist the analysis of any such notions as the delimitation of the domain of the observable, the reference of language to the world, or the distinction between justified and unjustified rules of epistemic inference is a matter for our best available natural science to explore, not any matter of a priori armchair philosophizing. We can, indeed, explore our place as observers in the realm of nature by means of neurophysiology and empirical psychology, but cannot hope to found our already existing best-available science on some myth of the prescientific given. We can also naturalistically study our place as language users in the natural world, but can't hope for some philosophical semantics that will offer us an analysis of what meaning in general must be, again in advance of any naturalistic scientific theory being accepted by us. And we can, from the standpoint of an accepted scientific world view, ask which rules of epistemic inference will reliably lead us to the truth; but we can have no hope of justifying, in advance of accepting a scientific picture of the world, some general principles of rational inference that would ground the scientific enterprise in the first place.

The pragmatist shares with the realist skepticism toward any element grounding epistemology or semantics outside our currently accepted scientific framework. But he shows rather more sensitiveness to the frequent objection to realist naturalism that its purely "externalist" stance fails to do justice to the questions the empiricist tried to answer. For the pragmatist the question of the origin of the "normative" aspect of meaning and justified belief is to be found in the practices we actually engage in. Observations ought to play a special role in our accepting and rejecting theories, because they are the statements we do, in fact, accept and come to consensus about. Rules of inference from data to theory, even rules relying on apparently arbitrary classifications of phenomena into chosen "natural kinds" for inductive projection, or rules as seemingly unconnected with "truth" as rules for accepting simpler hypotheses in preference to the less simple, are again reasonable rules because they are but the normative idealizations of the rules for inference we are inclined to accept. Here, of course, the familiar objections about the arbitrariness of our conceptual scheme, the possibility of equally "rational" alternative schemes, and so on, are deflected either by arguments concerning translation (denying the real possibility of alternative conceptual schemes) or by deflationary views about truth and correspondence combined with arguments about the incoherence of even stating (from within a chosen framework) the relativistic thesis.

And of course there are those who think they can have the virtues of both positions (by being "internal realists") or, rather, that properly understood the positions coincide. All, in any case, are agreed about a number of things. While a "soft" distinction between what is observable and what is not may play some role in our epistemology and in our semantics, no hard-and-fast, once-and-for-all, theory-independent distinction of this kind can be drawn. While the data of observation may indeed play a role of some special importance, again in both our understanding of how meaning accrues to our terms and our understanding of how believability accrues to our theories, no special importance of the kind the empiricist attached to the "data of immediate awareness" can be attributed to the role of observation. Observation may be part of the web or network of meaning accrual and belief accrual, and a part with some distinctive virtues, but a *foundation* for meaning and rational belief it is not. For there is no such foundation, and the whole "hierarchical" model of semantics and epistemology that words like *foundation* suggest is a misconstrual of how we get on in science.

2

But one still finds many philosophers of science enamored of the empiricist approach to theories. It is something of a paradox, I suppose, that the empiricist approach to theory seems to be favored most by those philosophers of science who spend their time dealing with the structure of fundamental physical theory. But, on reflection, this is perhaps not so surprising. It is all very well from a general methodological standpoint to tell us that our epistemology and semantical theories, insofar as there can be such things at all, ought to rest upon the "accepted best available scientific theory to date," having in mind, usually, the most contemporary version of fundamental theoretical physics. But it is another matter to look closely at these foundational physical theories and discover how little naive confidence one ought to have in them, and how puzzling the understanding of their fundamental concepts can be. At this point, the idea that the epistemology and semantics of theories itself rests upon the naive acceptance of these fundamental theories becomes disorienting indeed, and one looks for some other access into questions of justification and meaning accrual.

It can be argued, with some plausibility, that empiricist preconceptions are, in fact, "built-in" to many aspects of our most fundamental physical theories. Of course, if this is true it might be true merely because the theories as usually presented are infected with misguided philosophy of science. Could it not be that while the theories, properly understood, do indeed serve as our "best available physical theories to date"—for who in the philosophical community will these days have the courage (or foolhardiness) to refute the conclusions of the physicists by theorizing from his armchair—their empiricist elements are inessential to them? If this were so we could disabuse the philosophers of physics impressed by the apparent empiricist preconceptions of foundational physical theory of the misunderstanding that if they accept these physical theories they must accept elements of empiricism as well.

But to begin exploring this, let me first rehearse some of the foundational physics that does at least appear to many to rest upon empiricist (or, at least, empiricistlike) preconceptions, beginning with theories of space and time.

3

Faced with the null results of the round-trip light experiments designed to determine which inertial reference frame was the ether frame, physics constructed many "compensatory" theories, designed to save the ether frame by "explaining away" the null results as the joint interaction of two compensating changes, the changed velocity of light due to the observer's motion compensated by the effect of motion with respect to the ether on the observer's rods and clocks.

Einstein's revolutionary reinterpretation of the facts that these theories were trying to deal with is justly famous. A new theory is proposed that makes no predictions that would not follow from the "compensatory" theory if its consequences had been fully followed out. But the new theory is, rightly, considered by all a vast improvement over its predictively equivalent but inferior predecessors. And, at least as surface appearances go, the novelty of the new theory rests almost entirely on its empiricistically motivated critique of some of our most familiar concepts. The fundamental move is to focus on our understanding of what it is for two events to be "simultaneous" when they are spatially separated from one another. Since whether or not two such events are simultaneous is not, it is argued, a matter open to our direct observational knowledge, some *inference*, mediated by some causal process, must ground our beliefs about distant simultaneity.

Next Einstein criticizes several approaches to establishing distant simultaneity relations that might have worked but that, in the face of new observational results, cannot. Limitations on the velocity of transmission of causal signals makes it impossible to establish simultaneity by using causal signals of arbitrarily high velocity and so ever narrowing down the transmission time from the emission to the reception of the signal. We might think of using transported clocks to establish distant simultaneity, but the very necessity of the compensatory theory introduced to explain away the null results of the round-trip experiments makes it clear that transported clocks will not even provide a unique specification of the event at a distance simultaneous with a given event. Finally Einstein suggests the famous "radar" method, using reflected light beams and local clocks, to establish simultaneity; and he clearly demonstrates how such a specification of the simultaneity relation will lead to one that gives different pairs of events as simultaneous relative to the state of motion of the agent carrying out the radar stipulation of simultaneity.

The point to be made here is just how strong one empiricist assumption is that is being built-in to this whole critical program. Simultaneity for events at a point and continuity along lightlike paths are taken for granted as "observables" in the Einstein argument. But the empiricism doesn't really rest in any assumption that these relations are themselves immune to a critique that shows them not to be in some sense theoretically untainted "direct observational" features of the world. Rather it is in the assumption that, whatever counts as observational, distant simultaneity is *not* a legitimate observational feature of the world that is the characteristically empiricist move. Without the assumption that simultaneity for spatially separated events is "in principle" unavailable to our direct inspectional access, now and forever, independently of which theoretical framework we pick as the correct one to describe the world, it is hard to see how Einstein's argument ever gets going.

Once this general assumption is made, there are, to be sure, many alternative reconstructions of the relativistic argument that can be formulated. Einstein himself, of course, at least at this stage in his thinking, takes the method for the determination of simultaneity he has proposed as "definitional." It is, according to him here, a *stipulation* on our part that it is events so related by reflected light signals that we will take to be simultaneous. This approach leads to a long development of the school of thought that takes this aspect of the space-time structure (and other aspects as well) to be a matter for "conventional choice" on our part, and to a general line that the underdetermination of full theoretical content by all possible observational data can only lead to skepticism, permissivism, or conventionalism with regard to theoretical truth.

Certainly that is not the only direction in which we could move to reconstruct the epistemology of the situation. A major alternative to such a conventionalist reading of the situation would be any one of the realist approaches to theory that attempts in one way or another to rationalize the selection of the special relativistic space-time against the number of observationally equivalent theories as the "most rationally believable" theory that accounts for the data. Whether the approach be one through the construction of a confirmation theory that assigns differing degrees of "probability" to theories that are observationally indistinguishable, or one that proposes some defensible notion of "inference to the best explanation," or one or another of those approaches that relies upon considerations of simplicity or methodological conservatism to motivate the choice of theory, one is (unless these lead too quickly to a pragmatist rereading of the whole situation) still within the realm of a basically empiricist approach to theoretical belief, so long as the very special role of the observable as "ground" of epistemic access to reasonable belief is held onto. Only if one were to begin to deny the Einsteinian presupposition that a feature of the world like the simultaneity of separated events was truly distinguishable in principle from local simultaneity and continuity along a lightlike path, in being, unlike them, "in principal immune to observational determination," would one begin to reconstruct the selection of special relativity in a fundamentally anti-empiricist way.

A similar characterization of the situation holds when we look at the Einsteinian semantical analysis of distant simultaneity as holds when we explore the epistemological aspects of his critique. Einstein, taking distant simultaneity to be outside the range of observable features, argues that the concept 'simultaneous' must be explicity defined, using only concepts referring to "observables" in order to have meaning within a physical theory. Those skeptical of any hard-and-fast analytic/synthetic distinction among the propositions of physical theory will likely allow for a looser connection between the terms purporting to refer to the unobservables and those referring to the observables. A favorite account is the familiar one in which terms in theories get their meaning from the place they hold in the entire theory so that the meaning to be attributed to a term referring to an unobservable becomes dependent upon the role played by the term in the theory as a whole and upon the global structure of the theory which, as a whole, gives rise to its body of observational consequences.

Here, once again, there will be many options and many consequences of them. If one's theory of meaning accrual for the nonobservational part of the vocabulary is one that results in all theories having the same observational consequences being declared "equivalent" to one another on the basis of the way meaning is acquired by the theoretical vocabulary, one will have an easy time undercutting the threat of skepticism raised by a plethora of observationally equivalent theoretical alternatives. But one will find one's "realism" with regard to theoretical reference slipping out of one's grasp, and the pressure toward some "eliminationist" account of theoretical reference hard to resist. On the other hand a finer-grained notion of theoretical equivalence, perhaps demanding that theories be taken to "say the same thing" only when they both "save the same phenomena" and bear to one another some appropriate structural interrelation at the theoretical level (such as interdefinability of their theoretical terms), will make it easier to remain a realist but will make it harder to solve the epistemic problems raised by underdetermination. Is the special theory of relativity "saying the same thing" as the compensatory ether theories, but in a "descriptively simpler" way; or is it the case, rather, that the theories are quite distinct in "what they say about the world," with, perhaps, special relativity to be preferred as the *true* alternative on the basis of some notion of simplicity as a mark of believability, where simplicity is not merely simplicity of expression?

The main point here is that all of these familiar alternatives are basically in the empiricist mold. A theory of meaning accrual that dismisses the semantic import of the distinction between observables and nonobservables, or one that bases the very notion of meaningfulness on features not related to observation and ostensive definition, is anti-empiricist. But one that places observation terms in a special, "grounding" category, and that tries to understand the accrual of meaning by the other terms in the vocabulary by their connection through theory to the observables (however weakened that connection is taken to be from the rigid demand of explicit definition) is basically an empiricist theory of meaning. And the standard variants on the role of terms like *simultaneous* for separated events in the foundation of special relativity almost all gravitate toward some variant or other of such an empiricist semantic account.

But it is not only in the foundations of the special theory of relativity that we find empiricist epistemological and empiricist semantic assumptions "built-in to" the physical theories themselves. The general theory of relativity, Einstein's second great contribution to modern space-time theories and second great revolution in our conception of space and time, rests in a structure of argumentation that almost exactly parallels his founding of the special theory in a basically empiricist critique of the compensatory ether theories. The problem with the compensatory ether theories is that there are too many of them all equally compatible with the same class of observational data. Special relativity, with its denial of a preferred inertial frame, obviates the need to pick one inertial frame as special. But the same problem infects the theory of gravitation, even Newtonian gravitation. As Maxwell[1] argued in the late nineteenth century, within the Newtonian gravitational theory we could not observationally distinguish a world with no uniform gravitational force from one with a uniform gravitational force everywhere of whatever magnitude you liked. Later researchers exploring the possibility of cosmological models in the Newtonian framework realized that the most common models, of a uniformly filled cosmos in slowing expansion or accelerating contraction, would, in the Newtonian framework, have as comoving observers one distinguished by being inertial while all the others were accelerated. Yet each of the accelerated observers would be unaware of their acceleration since, being due to gravity, it would not be indicated on their acceleration measuring devices attached to their reference frames. Both of these "paradoxical" elements of Newtonian theory of gravitation are due, of course, to the special "universal" nature of gravitational force, which accelerates everything on which it acts to the same degree independently of the constitution of the test object and its size. The result of this special nature of the gravitational force is that within Newtonian theory there will be many possible worlds indistinguishable from one another by any observational test.

It is just this otioseness of theory that Einstein, once again, obviates by the theory of general relativity. Here gravitational force is replaced by the curvature of space-time, with the paths of "free" particles taken to be timelike geodesics in the space-time and the paths of light rays to be null geodesics, where 'free' now means free of all forces other than gravity. Of course this new theory doesn't differ from the Newtonian in just this way. The greatest modification necessary to the Newtonian picture is one that is almost forced on the theorist by relativistic considerations. The net result of the relativistic considerations is a series of plausibility arguments to the effect that in a relativistic context we may expect gravity to have metric effects, revealed by the standard measuring instruments of measuring tapes and clocks, as well as its familiar dynamical effects. These additional observationally determinable results of the presence of gravity are neatly encompassed also in the picture of gravity as curvature of the space-time manifold.

What needs to be emphasized here is the importance for Einstein's arguments of aspects of the empiricist account of our access to the world and to meaningful assertion about it. Again a standard repetoire of features is presupposed open to observational determination: the paths of free particles and of light rays; coincidences among measuring tapes; and the readings of the standardized clocks, which also tick in coincidence when brought to the same place. Much more importantly, there is, in this physical construction, a presupposed standard repetoire of quantities that are taken without question to be in principle, forever, *immune* to observational determination. Were the structure of space-time itself — rather than that which is revealed to us by moving particles, light rays, measuring tapes, and clocks — available to our direct inspection, the whole ground of the plausibility of the Einstein arguments for expecting gravity to reveal itself in nonflat metric aspects of space-time would be severely weakened. And were "the gravitational field strength" itself available to our inspection, rather than the effect of gravity as revealed in motion and in metric measuring instruments, the virtue of general relativity (as opposed to the theories that posit gravity as a field superimposed over an underlying space-time structure), that is, its virtue of replacing a manifold of observationally indistinguishable possibilities by a single space-time model, would vanish.

The pattern of thought here, which consists in characterizing a portion of the consequences of the existing theories as immune in principle from observational determination, of then locating the features of the theory that result in the theory having parameters immune in perpetuity from observational determination, and of then replacing these theories with one with a "thinned-down" ontology less subject to the underdetermination by observation difficulties; exactly parallels the way in which special relativity is introduced as superior to the compensatory ether theories. In both cases an epistemological and semantic critique of already existing theory, founded on presuppositions that at least appear to be empiricist, is at the core of the scientific revolution.

At this point, of course, many alternative options for understanding the epistemology and semantics of the situation can be imagined. Reductionist approaches take the meaning of the theories to be encompassed in their observational consequences, and vitiate the skeptical threat by declaring all of the observationally equivalent alternatives mere alternative expressions of one and the same theory. Anti-reductionists take the theories (general relativity vs. any of the flat-space-time-plus-gravitational-field alternatives) to be inequivalent to each other, relying upon some account of meaning in theory that, one way or another, attributes meaningful content over and above the sum of observational content to the theories. Then they try, again one way or another, to rationalize our theory choice among alternatives observationally indiscriminable from one another—be this by simplicity, methodological conservatism, a priori plausibility of theory, or whatever. But the common empiricist aspects to all these accounts are clear. Some parts of theory are in principle immune from observational determination. If we are to understand the meaning of the parts of theory dealing with the in principle unobservable, it must be by some manner in which understood meaning of the observational part of the theory works upward into the part dealing with the unobservable. And if we are to believe the assertions of the theory dealing with the in principle unobservable, it can only be by an upward motion of confirmation from the observable confirmed by its connection with empirically available experimental result, however such observational results are utilized in some confirmatory scheme.

4

We have looked at two cases, both from space-time theories, in which theories have been "thinned" of some of their concepts and their ontologies "thinned" of posited features of the world on the basis of a semantic and epistemological critique that is fundamentally empiricist in nature. It might be worthwhile looking at at least one case, again from space-time theories, that illustrates the claim that such conceptual and ontological pruning of theories is not a completed task, but one that will undoubtedly result in further critical assaults on current concepts and current ontology as our theory develops.

There is a long-standing program in the philosophy of space-time that advocates the claim that some or all of our space-time features can be "reduced" to causal features of the world. Sometimes it is said that the direction of time is "causally definable." Sometimes the claim is that the space-time metric or the space-time topology is "reducible to causal features of the world." There is not just one program called the "casual theory of space-time," but, rather, a plethora of such programs. And the programs differ from one another in their claims, in their motivations, and in the empirical and philosophical arguments they use to defend their theses. While some of the claims of "reducibility" would like to argue that space-time features of the world "reduce" to causal features in a manner analogous to the way light reduces to electromagnetic radiation or tables reduce to arrays of molecules, other causal reductionist theories bear a closer analogy to the phenomenalist's claim that material objects "reduce" to sense data. The former style of programs relies upon some alleged scientific discovery (or possibility of scientific discovery) that one class of entities of features in the world "is identical to" some other class of entities of features, an identity allegedly established by empirical discovery in science. The latter style of program relies, rather, on a semantic-epistemic critique of theories, in the empiricist vein, similar to the critiques we have discussed above. It is to the latter style of "causal theory of space-time" that I will direct my attention.

We can begin with Robb's attempt,[2] early in this century, to formulate the space-time of special relativity entirely in terms of the single primitive 'after', taken as the relation between events when one is, in relativistic terms, absolutely after the other, i.e., after it and causally connectible to it. Refinements of Robb's approach show us that we can, indeed, define all of the metric features of the space-time of special relativity solely in terms of one event being causally connectible to another. This suggests the possibility of at least some version of a "causal theory of the space-time metric" in the theoretical framework of special relativity.

For reasons I will not go into here I think it implausible to claim that the results of Robb, and of those who have refined his methods, really do show us that within the context of special relativity we really can (or, rather, ought to) define the space-time metric in terms of causal connectibility among events. In any case the program falls apart when one moves from the world of special relativity to the many different space-time worlds allowed by the general theory of relativity. In this broader context, it turns out that there are many space-time worlds whose metric structures differ greatly from one another, but that are exactly alike as far as the causal connectibility relations among the events in the worlds are concerned. These are all the worlds that, although metrically unalike, are related to one another by a so-called conformal isomorphism.

For this reason, in the general relativistic context it is usually the weaker structure of the topology of the space-time, rather than its metric structure, that is alleged to be "causally definable." Here the claim is that the full specification of the causal structure of the space-time is already a full specification of its topological structure. But, again, the issue is not a simple one.

If we take causal connectibility among events as the basic causal structure of the world to which all topological structure of space-time is to be reduced, then we run into a problem blocking this program if we allow as a possible general relativistic world any space-time world consistent with the basic equations of the theory. There are possible space-time worlds that have "pathological" causal structure. An example is a world with a closed timelike loop in it, which is a one-dimensional collection of events, all timelike related to one another, but whose topology is that of the circle rather than the open line. Events in this loop cause later events (in the local sense), but the casual chain ultimately arrives back at the initiating event. Even if we bar such closed timelike lines, weaker causal pathologies are still possible in the form of "almost closed" causal (i.e., timelike or lightlike) one-dimensional paths. A variety of restrictions on such causal pathologies can be imposed, the strongest being the demand for "stable causality," which is the demand that there not be any closed causal paths obtainable by any infinitesimal distortion of the actual space-time structure. For present purposes what is interesting is that if causal pathologies are permitted, then there can be, in violation of our first version of a causal account of space-time topology, worlds that are alike in causal connectibility structure, but that are unalike topologically.

One solution to this dilemma is to insist that causal pathologies not obtain, although it isn't fully clear why one ought to believe that this is necessarily so. A more interesting move is to try to restore the possibility of casual specification for the topology by moving to a richer notion of the causal structure of the space-time. An important result (due to D. Malament[3]) tells us that, if we confine our attention only to the standard manifold topologies usually taken to be the topologies of space-time, then in any world compatible with general relativity the totality of continuous causal paths fully determines the topology of the space-time. That is where the totality of facts as to whether or not two events are connected by some continuous causal path of events or other will not fully fix the topoplogy; the specification of what does or does not constitute a continuous segment of causally connectible events will do the job.

Why is such a result relevant to our pursuit of empiricist presuppositions in physical theories? The relevance is clear when one considers the importance granted to "causal" paths in these results. Why should we *care* whether or not the set of continuous causal (or even merely timelike) paths determines the topology? The set of continuous spacelike paths, for example, also fully fixes the manifold topology. I think exploration into the motivation behind this version of a "causal" theory of space-time topology reveals to us the presupposition that it is continuity along timelike paths that is the revelation to us by observational means of the topological structure of the space-time. Continuity along timelike paths is available to us because timelike paths are the kinds of paths we as observers, or idealized pointlike versions of us, can traverse, and they are, therefore, the kinds of paths whose continuity we can "observationally" determine. The presupposition here is similar to Einstein's presupposition in his critique of distant simultaneity, that we are entitled to assume that we can determine the selfsame identity through time of the light signal sent, reflected, and received, to determine which event at a distance is simultaneous with a given local event. From this perspective the so-called causal theory of space-time topology seems misnamed. It is rather an epistemologically motivated critical examination, just like those that founded special and general relativity, but now into the components of the newer theories, once again looking for the "hard observational facts" underlying the meaning accrual of the terms in the theory and the warrant accrual of the propositions. But in this case any relevant "thinning down" of theoretical concepts and theoretical ontology remains in the future. For we do not yet have the theory that would replace the full space-time theory in the event our critique showed it to be still otiose from the point of this empiricist-critical examination.

That the theory still is otiose in this way seems to be indicated by some additional facts about topology. The results we have discussed above tell us that, so long as we stick to one of the usual manifold topologies presupposed by general relativity, the totality of information about what constitutes continuous timelike (or timelike and lightlike, i.e., causal) paths in the space-time fully fixes the topology of the space-time. But there are many other possible topologies than these usual manifold topologies. Typical alternative topologies of mathematical interest are the "finest topologies compatible with the continuity along causal curves" generated by taking as a basis of open sets in the space-time all those sets of points (i.e., event locations) whose openness is compatible with the continuity specifications along the causal world lines. Since the specification of the basis of open sets in a space fully determines its topology, this specifies a new kind of topology for the general relativistic space-times. Such topologies, in certain interesting ways, "code" the causal structure of the space-time more naturally than the usual manifold topologies.

But such topologies also lead us to talk of the topology of the space-time in a radically different way from our usual description of it in terms of the standard manifold topologies. What are we to make of these novel topologies? The situation seems familiar on reflection. Once again it seems as though we have a variety of theoretical accounts of the space-time of the world from which we must select an accepted theoretical account. Once again we have the intuition that our full body of possible observational data, now taken to be all the facts about continuity along causal (or, perhaps, epistemically traversable) paths, is insufficient to do the selection for us. Once again the usual thoughts arise. Shouldn't we say that the alternative topologies, all of which agree on continuity along causal paths, are all "equivalent" to one another, presenting merely alternative *descriptions* of one and the same world? Or, rather, should we maintain that these topologies present genuinely alternative pictures of the world, and that we must seek for some methodological rule (simplicity, conservatism, a priori plausibility, or what have you) to tell us which of these empirically equivalent, but not fully equivalent, accounts of the structure of the world is the most reasonable to believe? Or should we just despair as skeptics of ever knowing the true structure of the world? Or advocate one or another version of permissive conventionality?

It will be interesting to await further developments in the physical theory of space-time. It seems clear that we cannot assure ourselves that the kind of critical "thinning" of ontology that Einstein gave us in the discovery of the special and general theories of relativity has reached its climax. One cannot foretell what future physics will look like, but it seems clear that we can expect this critically motivated endeavor to continue.

5

One further case might be useful to illustrate the point being made here, since it, unlike the others, is not drawn from the physics of space and time.

The early history of quantum mechanics is a familiar story. Working from quite different perspectives Heisenberg (attempting to turn traditional Fourier analysis into a kind of "two-dimensional Fourier analysis" in order to predict intensities of spectral lines, known to be coded by two parameters instead of the traditional one parameter specifying harmonic level) and Schrödinger (looking for an equation whose solution would be the de Broglie wave posited by the latter as being dual to all particle motions in analogy with the Planck-Einstein wave-particle duality of light) both invented theories that were able to predict the same quantities, the energy levels of electrons in atoms undisturbed and disturbed, frequencies and intensities of spectral lines, etc. Astonishingly the predictions were numerically identical despite the apparently radically different nature of the two theories.

The situation was finally clarified when Schrödinger, in a justly famous paper, showed the "equivalence" of the two theories, by demonstrating a structural relation between them that guaranteed identity for their "observational predictions." Later theorists, in the so familiar manner, found ways of expressing the common features of the two theories in a more abstract format, which relinquished some of the aspects of the two approaches where their apparent differences lay. And physicists from that time on spoke of the "Heisenberg representation" and the "Schrödinger representation" of quantum mechanics. Both "representations" assign mathematical structures to systems "prepared" by appropriate experimental procedures. Both "representations" assign distinct mathematical structures to quantities we wish to observe by some measuring experiment. Both compute the theories' "outcomes," now taken to be "possible values" of the observable to be measured and assignments of probability distributions over the set of such possible outcomes, by combining their mathematical structures for prepared "states" and selected "observables" in appropriate ways. And both predict the same possible outcomes for any observable; and, for a given preparation, a given time interval, and a given dynamic intervention into the system in question over that time interval, predict the same probability distribution over those observable values. Where they differ, fundamentally, is that the Schrödinger picture puts the changes over time into a mathematical evolution of the representative for the prepared state, and the Heisenberg picture puts it into a mathematical evolution over time of the representative of the measured observable.

For our purposes the structural similarities with the space-time cases are clear. In order for this understanding of quantum mechanics to go through, there must be firm agreement that certain quantities, the mathematical representatives of prepared states and of observables, be taken to be, in principle, not quantities open to observational determination. Otherwise we could not be assured that mere commonality of probability distributions over observable values, or such commonality plus the appropriate isomorphism at the theoretical level, would be enough for us to declare the two accounts mere "representational variants" of one another. It is our assurance that "states themselves" cannot ever be an object of "direct observational determination," which makes us reject the claim someone might make that Schrödinger and Heisenberg don't "say the same thing," because in the Schrödinger picture states change and in the Heisenberg picture they don't; and that we might "someday be able to 'look and see' which really occurs." And beginning with this assurance that some aspects of the theoretical structure are, in principle, immune from observational determination, we continue with the familiar program of "thinning out" the theoretical structure to get rid of the otiose elements that make alternative representations give the false appearance of offering alternative theories of the world.

6

All of these cases have, then, a series of common elements. First there is an intuitively accepted distinction between what is observable and what is nonobservable, which is taken to be a distinction "in principle" and one relative to which a critical semantic and epistemological attitude toward the existing theoretical structure is necessary. This requires not an intuitive agreement of what are to count, now and forever, as the true "observables," relative to which the theory can be tested, but, rather, an agreement that some features spoken of by the theories are, now and forever, in principle *unobservable*. It requires, that is, some assurance that we will never be able to bring "direct observation," however we understand that phrase, to bear on determining what goes on with regard to some specific quantities putatively posited of the world by the existing theoretical structure.

Next there is the awareness that a multiplicity of theories talking about such putative unobservable structures of the world, and saying incompatible things about it, will agree on the remaining facts derivable from them. That is, that there are many apparently distinct theories, all of whose distinctness is "trapped" in the realm taken as being that of in principle unobservability.

Finally there is the program of reconstructing the existing theory in a novel way that will, in response to this problem of the underdetermination of theory by evidential data, look for some new alternative account that is at least less infected with undeterminable features, features posited but characterized by some class of parameters whose values can't be fixed by observational determination, than were the existing theories that are now realized to be so infected.

It is interesting to reflect a bit on some of the reasons why it is fundamental physics that seems to attract this continuing program of semantic-epistemic critique. One reason, surely, is that the striking and unexpected facts revealed to us by the ever-increasing power of experimental observation are frequently so startling in their nature as to require some radical revision or other of our concepts and our hypotheses. Faced with the null results of the round-trip experiments that experimentally ground the special theory of relativity, or the manifold of puzzling and anomalous results that led to quantum mechanics, the scientific community is faced with the realization that something quite radical must be done to make our physical theory accord with the observational facts. In the face of such a need for radical revision of one sort or another, it is not too surprising that at least some insightful thinkers are impelled to reexamine the presupposed fundamental concepts and hypotheses of the most familiarly accepted background theories to see if something in them must be challenged. In the face of extraordinary puzzles with the data, the comforting admonition to rely upon "general accepted background science" in a conservative way seems less than persuasive as the most fruitful way to go about things. It is in circumstances like these that the empiricistically motivated doctrine of looking at the earlier theories, to see to what degree the "hard facts" really impelled us to accept them, becomes a promising scientific strategy.

The fact that the novel hypotheses introduced to account for the novel data are themselves so radical in nature also cries out for a semantic-epistemic appraisal of them. Special relativity introduces space-time, with its denial of an absolute simultaneity relation, to replace our traditional notion of space through time. General relativity introduces pseudo-Riemannian curved space-time to replace traditional flat space-time and traditional gravitation as a field within space-time. Quantum theory introduces the novel notions of the state function and the observables to replace the traditional notion of a state as a point in classical phase space.

But how are we to understand these novel concepts and hypotheses? And why ought we to believe in the hypotheses that utilize them? Here the very novelty of the conceptual structure seems to demand of those introducing them that something substantial be done to explain to the scientific community how the concepts and the hypotheses in which they appear are to be understood (i.e., what meaning is to be attributed to them) and how they are related to possible observational data in ways appropriate to allow the data available to us to provide observational tests of the truth of the new hypotheses. We cannot take such questions for granted, as understood, or as built-in to previous scientific practice in well-understood ways, when such radically new conceptual structures are being proposed. Even "analogy" with older concepts and hypotheses, the sort of thing that works when we explain molecules, say, as "like tiny billiard balls too small to see," fails us in contexts like these. Instead, a radical appraisal of how these new concepts and the hypotheses in which they appear are related to possibilities of evidential experience seems to be what is in order; and this is just the sort of appraisal the importance of which is the core of the empiricist program in philosophy.

Another feature that tends to give rise to such semantic-epistemic critical appraisals in these foundational physical contexts is the ability of these problem situations to realize in concrete form a familiar philosophical puzzle. Since Descartes and Hume we have been concerned, as philosophers, with alternative hypotheses to our familiar ones that "save the same phenomena." Descartes introduces perpetual dreams and malevolent demons. Hume tells us to consider the fact that our belief in the continuity of phenomena when we are not perceptually aware of them is grounded more on imaginative projection than any counsel of reason. But the alternatives to our usual world scheme proposed by such philosophers are usually taken to be puzzles, rather than seriously alternative hypotheses about the nature of the world. We take it, as philosophers, that our job is to refute the skeptical threat such imaginative alternative world views suggest, but that we ought not to seriously consider that the world might really be so radically different from the way we usually take it to be. Witness the pragmatist distinction between real doubts and the "mere" skeptical doubts of the philosophers, or Hume's admission that, no matter how seriously we will think with the learned in our philosophical studies, we will always act with the vulgar in the world.

But in the context of foundational physics we seem to have presented for our epistemic appraisal radically different alternative world views. They are fully filled out in all detail, unlike the mere sketches offered by the philosophers. And nothing initially seems to tell us that only one of these alternative world hypotheses is worth serious consideration, the others to be taken even as hypotheses for consideration only with a grain of salt. All of the underdetermined alternatives seem, at least initially, on a par as genuine scientific alternatives, and it is no surprise that, in this context of *genuine* scientific indecision, a reappraisal of how observational data is to relate to scientific belief seems in order.

Finally we ought to note that there are special features of the particular foundational theories we have been dealing with that also play a role in suggesting that, in our scientific search for new hypotheses, semantic-epistemic critical appraisal is in order. The postulation of space itself, or space-time itself, as a theoretically inferred structure of the world, has always been one that has been treated with skeptical doubt. Witness the skepticism toward Newton's "space itself" from Leibniz through Berkeley to Mach. Both the peculiarity of "space" as an "entity" over and above the ordinary material inhabitants of the world, and such special problems as those suggested by Leibniz in his use of the principle of the identity of indiscernibles to criticize the postulation of a space, all of whose infinity of points were indistinguishable by anything in their own nature from one another, have led many to be all along skeptical of the legitimacy of positing space itself as one more inferred theoretical structure in the world. For this reason scientists and philosophers are "primed" in the kinds of situations of scientific revolution described above to be receptive to empiricistically minded semantic-epistemic critiques that throw doubts on the way in which such theoretical concepts have been used without sufficient questioning in the past.

A similar situation holds in quantum mechanics. The theory itself arises, at least in part and from the Heisenbergian point of view, out of a despair with the prospects of reconciling some traditional theoretical notions (the usual notions of definite trajectories for the subatomic charged particles and the usual rules associating such motions with detectable emitted radiation) with the rules governing the observables discovered in experiment. From the very beginnings of the theory, there is a tendency to avoid premature commitment to some definitive theoretical structure and to look for some calculational system that will predict the right correlations among observables without saying too much about explanatory theoretical structures that might account for these correlations. Heisenberg, at least, is quite self-consciously "positivistic" in his motivations behind the discovery of matrix mechanics.

Later the manifest peculiarities of the theoretical structures introduced in the developed quantum mechanical formalism, peculiarities such as the instantaneous change of the state vector upon measurement in a manner totally unaccountable in the rules for its dynamic evolution in ordinary processes, or the peculiar nature of quantum states as introducing probabilistic correlations for events at spacelike separation that cannot be accounted for in any straightforward way in terms of previously fixed local parametric hidden variables, again lead to a skepticism, based on the special nature of this particular physical theory, with regard to taking its apparent theoretical posits in an uncritical manner. Once again an empiricist injunction to "see what the hard data actually forces upon us" will be welcomed in the context in which an already critical and skeptical attitude to the apparently posited theoretical structure has been forced upon us by its intrinsic peculiarities.

7

The considerations we have gone through indicate, then, that at the level of foundational physical theories, and especially when these theories are subject to the severe difficulties encountered when they are radically in conflict with the observational data, a skeptical attitude with regard to the meaning of some of the fundamental concepts and a skeptical attitude with regard to the warrant with which fundamental hypotheses had previously been supported, become not the idle speculations of the philosopher but integral parts of the ongoing process by which our most important and most general theories of the world are refined and, if necessary, replaced.

That doubts about the intelligibility of concepts or about the warrant that really accrues to hypotheses will cause one to reflect on "what the hard data of observation really provides us" seems inevitable. It is in precisely such conditions of instability of fundamental theory that are incipiently revolutionary that the consoling advice of the pragmatist to rely upon "what we have previously practiced" or of the realist to rely on "what our best available science tells us about the world" seem out of place. It is just such notions as "what we have practiced" and "what our theories have told us" that are in doubt in these contexts. It is when a scientific "form of life" has proven to be a failed form, by that most important of test of success in scientific life — the ability to correctly predict the data of observation, that the sense of a need for a radical exploration into the grounds by which we have taken meaning to accrue to our theoretical concepts and warrant to accrue to our theoretical hypotheses is intuitively felt to be called for. And it is in just these circumstances that the resort to "what we really have available to us from observational experience" naturally arises as the touchstone on which such a critique of existing concepts and hypotheses is to be grounded.

Can the notion of "what the data provides us" be understood here in a pragmatist or realist vein? I think not. For the pragmatist, the distinction between what is observable and what is not is a fluid one. Most pragmatist accounts will rely on some notion of spontaneous communal agreement to pick out some body of sentences quickly assented to by the population in general in a stimulus situation to fix on what, if anything, can be called observational assertions. From this point of view, what would be to the empiricist inferred propositions (but inferred on the basis of inference licenses so deeply rooted as to result in unconscious and spontaneous inference) would, to the pragmatist, count as assertions just as fully observational as those the empricist takes to be truly based on observation alone. But examination of such features of the physical critique we have been exploring as Einstein's critique of our universally held idea that we knew what we were talking about when we spoke of events "being at the same time," and that we meant the same thing by this for spatially separated as for coincident events, shows us that the notion of observability (or, again, more importantly, nonobservability) that the physicist has in mind in these critiques is not the merely psychological feature of spontaneity of judgment that the pragmatist has in mind.

Nor, I think, does what the "realist" tells us about the observational/nonobservational distinction do full justice to the critical context we have been exploring. From the realist perspective, "what is observable" is a matter for our best available science, including psychology, neurophysiology, etc., to determine for us. But in the critical contexts of physics, where distant simultaneity is rejected as an observable simply because it is not a local notion; where the gravitational field magnitude is rejected as an observable because it is not by itself expressed in the motions of bodies or light rays or in the readings of metrical measuring instruments; where the spacelike continuity of lines is rejected as observable because of its "in principle nonsurveyability;" and where the state vector itself of quantum mechanics is rejected as observable because its values are not the values of what is taken, in the theory, as a primitive, undefined notion of the results of measurement processes; there seems to be an at least quasi a priori element presupposed in the characterization of where to draw the line between that which at least might be characterized as directly observable and that which certainly can not so be characterized.

Once this characterization is drawn, of course, then "what the theories themselves tell us about the world" does, indeed, become essential to the critical program. If we take local relations among events and continuity along paths traversable by an "observer" as the full body of observational data relevant to a determination of the topology of space-time, for example, then it is indeed our "best available theory," relativity, that tells us that some paths, the spacelike ones, posited in our theory are "out of bounds" when it comes to critically examining the semantic and epistemological basis on which our account of the world's topology rests. For given this scientific theory as true, paths of that kind are not so traversable. But the restriction to local data as the only data available to an observer in the first place, the core of what gets the critical program going, is not a consequence of our "best available physical theory" in any reasonable sense, but is presupposed by us on some other basis.

The immediate inspiration for the semantic-epistemological critique of existing theory is invariably the discovery that, in the light of one's presuppositions, certain facts are in principle immune from direct observational determination, in the light of novel observational facts, and in the light of the realization that novel theoretical options are available to us, it turns out that our theoretical structure of the world is, in its current form, apparently radically underdetermined by the data. Too many theories we take to be inequivalent to one another all seem equally good in the face of all possible observational results: too many ether theories, once the null results of the Michelson-Morley experiment are in; too many gravitational theories, once some version of the principle of equivalence is established; too many space-time topologies, once it is realized that a variety can be made compatible with all topological facts along traversable paths; too many versions of the dynamics of quantum states compatible with the probability distributions among the values of observables that are the outcomes of the theoretical manipulation.

It is in the face of this newly realized problem of underdetermination (or, in some cases, in a reawakened interest in an underdetermination problem that had existed for a long time but had been swept under the rug) that the critical examination of our access to theoretical meaning and our warrant for theoretical assertion is undertaken. And it is in these contexts that the suggestion is inevitably made that the most reasonable response is to thin out our theoretical conceptual structure, eliminating some apparent theoretical reference and some putative theoretical assertion, in order to arrive at a "cleaner" version of theory less subject to the infection of undeterminable theoretical parameters and underdetermination of the theory by data in general.

Need such empiricistically motivated semantic-epistemic criticism of existing theory always lead us down a slippery slope to an ultimately positivist view of the world, where all theoretical reference is taken as at best instrumental in nature, with the only real reference being reference to some fixed basis of "directly, immediately observational sensory contents"? It just isn't clear. For we don't yet really understand enough about how such empiricist critique from within an ongoing context of realistic science is supposed to work. Much depends upon how we ultimately resolve crucial questions about meaning accrual and warrant accrual.

Consider, for example, the "realist" view with regard to theories that, while starting from the empiricistic critical position we have been exploring, holds that the end of such criticism is not some phenomenalistic eliminativist program of taking all observationally equivalent theories to "say the same thing;" but is, instead, the selection of a preferred theory from the set of all the phenomenally equivalent alternatives, or, rather, from all those alternatives presently under consideration whose disagreements with one another are, we are sure, trapped at the nonobservational level.

Special relativity, the argument frequently goes, is preferred to the alternative ether theories since it, unlike them, does not introduce absolute velocity as an undeterminable parameter. It is preferred on the basis of ontological and conceptual parsimony or simplicity. But it is still a realistic theory positing "space-time itself" as an explanatory structure. The "inertial frames" of space-time remain as "not directly observable," but as the reference frames relative to which acceleration shows up in mechanical effects (inertial forces) and optical effects (non-null results for round-trip experiments, for example). But special relativity is to be preferred also, it is alleged, to the mere set of its observational consequences taken as a theory. For that latter "theory," unlike the special theory with its theoretical space-time structure, fails to offer genuine *explanations* of the observable phenomena.

An almost identical stance can be taken with regard to general relativity as opposed to any of the flat-space-time-plus-gravitational-field alternatives to it with the same observational consequences. Here again it can be argued that general relativity is preferable to its theoretical alternatives on grounds of conceptual and ontological parsimony and in not having, as the other theories do, an intrinsically undeterminable parameter at the theoretical level. And, again, it can be argued that it is superior to its purely phenomenal alternative in that it still offers an explanatory account of the observational facts.

Such a realist approach is still basically empiricist in its structure. It is now a variant, not of the kind of eliminative empricism that results in phenomenalism, but of the realist empiricisms that, while taking observation as the starting point for meaning and warrant accrual, allow for the possibility of semantic principles that allow for the legitimation of meaning attribution (of a not merely instrumentalist or representationalist sort) to terms going beyond reference to the observable, and for the possibility of epistemic principles legitimizing warrant accrual to hypotheses whose content outruns the observable and correlations among its particular contents.

What seems clear, however, is that realist and pragmatist accounts of meaning and warrant will not do full justice to the semantic-epistemological situation in these cases where it is the most general and fundamental background theories of our scientific world view, and the most entrenched of our scientific communal linguistic and inferential practices, that are up for questioning. At least some of the ingredients of an empiricist approach are called for here: a rather rigid and theory-independent notion of what is in principle immune from direct observational determination, a scrutiny of meaningfulness for theoretical concepts rooted in an examination of how they relate to what is available to us in direct experience, and a critical scrutiny of the warrant by which we have held our theoretical hypotheses, again rooted in an exploration of the extent to which they can be held to account by the observational data or by that data in combination with some inferential rules that are themselves "above" the standards of justification provided by resort to "what our present science tell us" or "what is entrenched in our present scientific practice."

This kind of empiricism-as-critique, practiced in the context of an ongoing *refinement* of the existing theories in our background science, would seem to require that we always be in a state of readiness to rethink the grounds of meaningfulness for our theoretical concepts, the grounds for positing our theoretical ontology, and the grounds for accepting our most basic and pervasive theoretical hypothesis from what is, essentially, an empiricist perspective. This is not an empiricism that tells us that we can find some theoryless observational language to which all of our theoretical assertions can be reduced by translation. For, at least on its surface, it neither requires for its critical stance belief in the accessibility of some once-and-for-all-time "detheorized" observational language as basis, nor a belief in the reducibility or translatability, once and for all time, of all the theoretical discourse we have or will encounter into such a pretheoretical observation language.

The kind of empiricism that seems to be called for by the essential role played by semantic and epistemic critique in foundational physical theory does require that we take a "hierarchical" attitude toward our concepts and hypotheses, giving those "closer to the pure observable" primary status in the work of meaning accrual and the accrual of warrant for belief. The reader will, perhaps, be reminded in this context of some of C.I. Lewis's attitudes toward the "given" as a kind of inexpressible limit ideal and of Popper's remarks in *The Logic of Scientific Discovery* on the "conventionality" of what at any time is taken as the level that counts for theoretical purposes as the level of "observation." What would be called for to make this sense of empiricism more plausible as a viable account of our scientific method would be to accompany this attempted demonstration of the way in which such a critical empiricist attitude is presupposed, and, indeed, "built-in" to our contemporary foundational physics, with some extended understanding of how such a critical empiricism differs from more traditional "bottom-up" foundational empiricism and how, when all of the consequences of the critical empiricism are thought through, it can avoid the familiar difficulties with that more traditional empiricism that drove away so many of its sympathizers to the alternative pragmatist or realist perspectives.

Notes

1. J. Maxwell, *Matter and Motion* (New York: Dover, 1954), 85.

2. A. Robb, *A Theory of Time and Space* (Cambridge: Cambridge University Press, 1914); and *The Absolute Relations of Time and Space* (Cambridge: Cambridge University Press, 1921).

3. D. Malament, "The Class of Continuous Timelike Curves Determines the Topology of Spacetime," *Journal of Mathematical Physics* 18(1977): 1399–1404, esp. p. 1401, Theorem 2.

Theories as Mere Conventions

1. Expunging Conventional Elements.

Conventionalism, as an approach to understanding scientific theory, is hardly new. Mach[1] construed the Newtonian relation $f = ma$ as a conventional characterization of the unobservable quantity "force." Poincaré[2] argued that the geometrical structure we attribute to space is a matter of convention. These two examples are interestingly different. Poincaré writes as a mathematician: whatever the true laws of dynamics may be (provided they satisfy very general and uncontroversial conditions) they can be expressed in terms of either Euclidean or non-Euclidean space. Mach writes as an experimentalist: the only fundamental way you can tell about forces is to measure masses, times, and distances—i.e., masses and accelerations. Yet these widely different points of view lead Mach and Poincaré to the same general view regarding certain theoretical claims.

To Duhem[3] is attributed the thesis that *theoretical* as opposed to *experimental* laws are essentially conventional and irrefutable. The link between irrefutability and conventionality is firmly established in Duhem. And the modern representative of this view, though in somewhat attenuated form, is Quine.[4] Quine's thesis, though bracketed with Duhem's, is somewhat different, for Quine sees the poles, "conventional" and "substantive," as unrealized ideals. That is, every statement in our scientific corpus partakes of both analyticity and empiricalness. The distinction is one of degree: some statements are more resistant to counterevidence than others.

Quine's view will be our starting point. As metaphored by Quine and Ullian,[5] our body of scientific knowledge should be viewed as a web. At the center of the web are the truths of logic and mathematics—the statements most resistant to modification in the face of recalcitrant experience. At the periphery of the web are statements most directly related to experience, to what happens to us. These are the occasion sentences on which our whole body of scientific knowledge is to be based. At the center, then, according to this picture, we have sentences that are highly structured and regimented, but evidentially remote from our ordinary

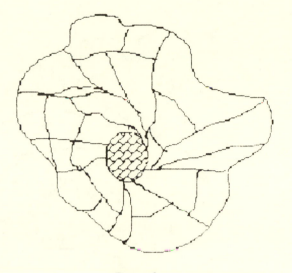

Figure 1.

experiences. At the edges, we have sentences that are more or less directly warranted by ordinary experience, but that have little or no structural connection to the rest of our body of beliefs. The general picture is reflected in figure 1.

Let us now consider how this view of things would be changed if we were to drop from our representation the arbitrary elements. That is, if logic is thought

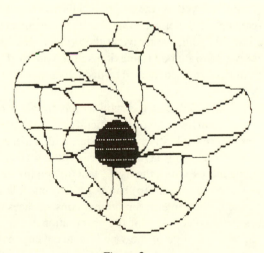

Figure 2.

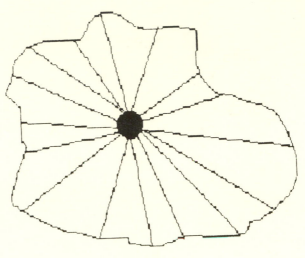

Figure 3.

of as being arbitrary or conventional, let us delete from our picture the elements that correspond to logic. The view we have then is illustrated in figure 2. There is still a web, embodying many interconnections, on which experience impinges only at the periphery. We may think of the internal nodes of this web as representing "theoretical" terms, while the nodes at the edges represent "observation" terms.

But then, as Craig[6] showed us, the theoretical terms are arbitrary and inessential additions – conventional elements that, but for convenience, could be eliminated. If we follow this path, we arrive at the prettier web illustrated in figure 3, in which there is a central node (the axioms of the physical theory, expressed in purely observational terms) connected by deductive spokes to the observational consequences on the rim.

If this picture of knowledge strikes you as *too* simple, you can, as I once pointed out,[7] impose your own favorite structure, involving your own favorite theoretical entities. More explicitly: you can start with a theory T_1, choose your favorite theoretical entities, your favorite (consistent) laws concerning these entities, and be sure that there exists a theory T_2 that (a) employs as its only theoretical entities those you have chosen, (b) has as its theoretical laws those you have chosen, (c) has exactly the same observational consequences as T_1, and (d) requires those laws for the deduction of the observational consequences.

It appears that only the edge of the web is nonconventional. If we eliminate all the conventional elements we are left with figure 4.

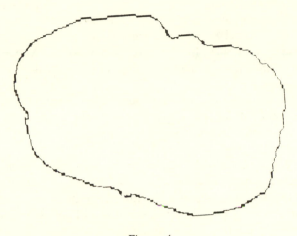

Figure 4.

All this has been predicated on a very shakey assumption—namely that the edge of the web, the observation sentences, can be sharply distinguished from the interior of the web, sentences involving theoretical terms. But to reject that distinction leads to an even worse state. If we associate irrefutability with convention, as we have been doing, then even the observation sentences begin to seem conventional. As Quine pointed out long ago,[8] even observation sentences are not uniquely determined by what happens to us. Alternatively, we may reflect that *any* observation sentence, if it has content at all, may turn out to be in error.

If we give up the distinction between theoretical and observational vocabulary, and disregard the conventional elements, we are left with a web of knowledge that has neither interior nor periphery.

This is clearly unproductive; we must somehow escape this consequence. The way to do so is to take a closer look at conventionality and the way in which it functions in scientific knowledge. Up to this point I have been using the term "convention" in what I take to be an ordinary philosophical sense. This is the sense discussed by David Lewis,[9] in which arbitrariness is the essence of convention. Indeed, Lewis writes "It is redundant to speak of an *arbitrary* convention. Any convention is arbitrary because there is an alternative regularity that could have been our convention instead" (p. 70).

To say this immediately sets lights flashing. Poincaré certainly did not regard Euclidean geometry as conventional in *this* sense! In fact, he (mistakenly) argued that Euclidean geometry would always be the geometry of choice for physical theory, on the ground of its simplicity. This is exactly to claim (a) that the choice of a geometry is conventional, and (b) that there are reasons for choosing one convention over another. If there are reasons for choosing one convention over another, these conventions are not "arbitrary." It is hard to believe that any conventions are entirely arbitrary; "arbitrary convention" seems to me more like an oxymoron than a redundancy. But be that as it may, we may certainly single out a class of conventions that may be adopted for *epistemic* as opposed to *nonepistemic* reasons.

Conventions that are adopted for nonepistemic reasons, we shall call *arbitrary*. Conventions that are adopted for epistemic reasons, we will not call arbitrary, even though "there is an alternative regularity that could have been our convention instead."

And what is a "nonepistemic" reason? Familiarity, elegance, and the like. I shall also construe simplicity as nonepistemic, though, as is well known, some philosophers take the simplicity of a scientific theory to be grounds for taking it to be acceptable, or true, or both. Epistemic reasons I take to be reasons expressible in terms of predictive power or probability. They will become clearer shortly. The general thesis is that scientific theories are *mere* conventions, but not *arbitrary* conventions. The reasons for choosing one theory rather than another are to be epistemic in my narrow sense.

2. Generalization, Observation, and Error

We consider a set of alternative languages, **L, L′**, etc. Each language contains a set of

(a) terms,
(b) formulas,
(c) axioms, and
(d) rules of inference

in the usual way. We suppose that

(i) First-order logic and set theory are included.
(ii) "Logical" or "analytic" relations among terms are derivable.
(iii) Probability as well as provability, as relations between sentences and sets of sentences, is defined.
(iv) An inductive as well as a deductive logic is provided.

With regard to (i), the intent is merely to have on hand as much mathematics as we need to do statistics and as is involved in whatever theories we are concerned with.

With regard to (ii), the intent is merely to capture those truths, if any, that are to be regarded as characteristic of the language in question. It is the burden of the third section of this paper to elucidate this vague characterization.

With regard to (iii), we must be a little more explicit. Probability is not taken to be subjective. It is *not* to be understood as a Carnapian degree of confirmation.

It is *not* (as both of these views would have it) based on a measure defined on the sentences of the language.

Probability is:

(1) defined for a given language **L**,
(2) relativised to a corpus of knowledge **K**,

where **K** is understood to be a set of sentences in **L**,

(3) syntactically definable,
(4) based on the statistical syllogism, and
(5) interval valued.

The statistical syllogism has the form:

Between 25% and 30% of A's are B's.

Charlie is an A.

That's all that is known in **K** that counts.

Therefore the probability of "Charlie is a **B**," relative to **K**, is [.25,.30].

The trick to making this notion of probability fly, of course, is spelling out in a noncircular way the condition embodied in the third premise.[10] I will simply assume, for present purposes, that it can be done. The upshot is that probability assertions have the form

$$\text{Prob}_L(S,K) = [p,q],$$

where **L** is a specified language, S a sentence in that language, and **K** is an actual or hypothetical set of statements of that language. $[p,q]$ is just a closed subinterval of $[0,1]$.

With regard to (iv), I assume, what is somewhat controversial, that we can come to accept statements on the basis of strong but inconclusive evidence. This is what I take inductive logic to be about. Just as with probability, I shall simply assume for present purposes that it makes good sense to talk about accepting statements on the basis of strnog but inconclusive evidence. I shall furthermore assume that we can operate with a purely probabilistic rule of acceptance.

Roughly the rule has the following form: if the probability of S, relative to **K**, is at least 1 - ε, accept S. *More exactly, however, we must specify* **K**, and we must say *where S* is to be accepted. (If we look on logic as atemporal, we can't have S belong to **K**, for its probability would already be $[1,1]$, which is not our intent.)

It turns out that a useful structure for present purposes can be represented by three levels of rational corpora. We take **K*** to be the corpus of incorrigibilia: it comprises logical and mathematical truths, analytic statements, and possibly incorrigible observation statements. The next level of corpus, **K**, is the evidential corpus. It contains evidential certainties, indexed by a real number p. A statement is acceptable as an evidential certainty if its probability, relative to the set of incorrigibilia, is at least p. Finally, the corpus on the basis of which we act, and relative to which the probabilities of the various possible outcomes of our actions are defined, is the corpus of practical certainties, K'. This is indexed by a real number p' smaller than p. A sentence S is acceptable as practically certain if its minimum probability, relative to the evidential corpus, exceeds p'.

We thus have two parameters, p and p', to account for. I think a sensible story can be given, but this is not the place for it.[11]

We need just a little statistical inference. I claim, without argument here, that we can inductively infer a proportion in a population from the observation of a proportion in a sample. This makes use of a set theoretical truth:

> For perfectly reasonable n, ε, and δ, the proportion of n-membered subsets of a finite set A that exhibit a proportion of Bs that is within ε of the proportion of Bs among As in general is at least $1 - \delta$.

The argument also depends on the fact that the "third condition" of the statistical syllogism may be satisfied with regard to a given subset of A, relative to plausible bodies of knowledge K.

From these premises, we obtain: The probability that this particular sample, having its given proportion of Bs, exhibits a proportion that is close to that among As in general, is $[1 - \delta, 1.0]$.

Assume that our counting is incorrigible, so that the data is in K^*, and that $1 - \delta$ is greater than p, and we have shown that "The proportion in our sample is r and the proportion in A differs from that in the sample by less than ε" is entitled to be in our corpus K of evidential certainties. Since this statement entails "The proportion of As that are Bs is in the interval $[r - \varepsilon, r + \varepsilon]$" the latter statement, too, may be in K.

We need one more piece of machinery to examine the possibility that scientific theories may be construed as mere, but not arbitrary, conventions. We must consider also, corresponding to each language L, its metalanguage ML. In each case ML contains L as a sublanguage. In addition, ML is assumed to contain a single nonlogical primitive relation, $O(X, S, t)$, which is to be interpreted thus: the individual X, our agent, an individual, society, group, or whatever, *observes* the state of affairs or event denoted by the sentence S of L, at some time earlier than t.

With the help of this metalinguistic predicate, we can define two important classes of sentences of L:

D1 $VO\ (X) = \{S \mid (\exists t)\ (O(X, S, t)\ \&\ S)\}$

These are X's veridical observations.

D2 $EO\ (X) = \{S \mid (\exists t)(O(X, S, t)\ \&\ S)\}$

These are X's erroneous observations.

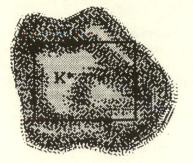

$$O(X, \text{'}Aa\text{'}, t)$$
$$\downarrow p$$

K

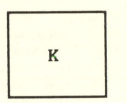

MK

Aa, Ab, Ba,...
$\downarrow p'$

$$\%(O(X), EO\,(X)) \in [p, q]$$

K'

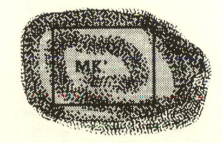

$$\%(A, B) \in [p, q], Bd \qquad\qquad \text{"}Cb\text{"} \in VO(X)$$

Figure 5.

Finally, we add one principle to our inductive logic: that if a statement in **L** has a probability greater than *p* relative to the evidential metacorpus, then it may appear in the object language corpus of evidential certainties.

Figure 5 may help to give the general idea of the structure I have in mind. The corpus of incorrigibilia in the language **L** seems to serve no useful purpose; logical truths and the like will enter the corpus of evidential certainty by being highly probable (having probability 1) relative to the metacorpus of evidential certainties. Observation statements themselves *may* appear in **K** when they are the result of observation: what is required is that they not be known to belong to any class of observations in which error is rampant. (If the observation is made under bad light, it is more likely to be erroneous—and we can come to know this in our metacorpus.)

For an example of how this works, let us consider the familiar ravens. This will also constitute an initial example of my main point. Consider two languages, L_1 and L_2, that both contain the predicates "is a raven" and "is black." L_2 contains *as a meaning postulate* the generalization that all ravens are black. This is construed as a priori, as a *linguistic* constraint, embodied in the language itself.

Let us now consider the question, "Given a set of observation reports, how do we choose between the two languages?" (This reflects Mary Hesse's distinction[12] between observation *statements* and observation *reports*. The observation report is the linguistic entity entered into the laboratory notebook. The observation statement represents the corresponding fact. It is bad form to correct an observation report; but observation statements may sometimes turn out to be wrong and need to be corrected.)

Suppose our agent X has accumulated a long list of observation reports in his incorrigible metacorpus; statements such as $O(X, "a_1$ is a raven", $t_4)$. These statements may be the same in either language. In particular, there is no inconsistency, even in **ML**$_2$, between

$O(X, "a_1$ is a raven", $t_4)$, and

$O(X, "a_1$ is not black", $t_4)$,

though of course the two mentioned statements are inconsistent in both L_2 and **ML**$_2$. In either case, our agent is simply reporting his observations.

The difference is that in L_1 we have no reason to doubt any of the mentioned statements. It seems perfectly natural for the mentioned statements to pass directly into the body of evidential certainties K_1. Given the vagaries of human observation, we may suppose that even under the best of circumstances, not all observed ravens will have been reported to be black. Most will be, and relative to the evidence embodied in K_1, it may become practically certain, in K_1' that almost all ravens are black; that is, " %(is a raven, is black) ε [.9,1.0]" may be a member of the practical corpus. (Even if, unnaturally, *all* the observed ravens have been observed to be black, this could be the strongest statement acceptable on the basis of probability in the first language.)

This would allow our agent to be pretty sure that the next raven he saw would be black, or that a random raven would be black. It does not, in itself, provide any grounds for supposing that a nonblack thing is a nonraven.

Consider the other language, in which it is a priori true that all ravens are black. Given the same set of observation reports, including some reports of non-black ravens, the situation of our agent X is quite different. Now he *knows*, on syntactical grounds alone, that some of his observation reports are erroneous. Some observation statements, corresponding to observation reports, must be rejected. Which ones? How many?

We are free, of course, whichever language we adopt, to suppose that all or almost all of our observations are erroneous. This is silly skepticism; we have no grounds for making any such assumption. So I propose that we adopt the following principle:

P_1 *The minimization principle:* Attribute no more error to your observations than your language requires you to.

This principle tells us how many observations we are required to regard as erroneous, but it doesn't specify which ones. It is not unreasonable not to be able to tell whether it is the observation report "a is a raven" or the observation report "a is not black" that is wrong. We could reject them both, but that would fly in the face of the minimization principle. The reason we want to know how many observations of each kind are in error, is that we want a statistical basis for determining long-run error rates for the various kinds of observations. A natural principle to follow (though it is not as compelling as the previous one) is this:

P_2 *The distribution principle:* Distribute the error frequencies as evenly as possible among the (syntactic) kinds of observation statements.

Given these two principles, we have a way of arriving at observed error frequencies in the incorrigible metacorpus.

Given these observed error frequencies in a sample of observation reports, we can, other things being equal, infer long-run error frequencies among the various kinds of reports. Statements asserting that in general the frequency of erroneous raven judgments lies between (say) 0.0 and 0.1 may now appear in the metacorpus of evidential certainty.

What should be the basis of our choice between the two languages in question? One natural criterion would be the predictive observational content of our corpus of practical certainties K'. That is, it seems reasonable to let past observations fend for themselves; theoretical understanding is notoriously difficult to distinguish from theological speculation; and prediction other than observational terms is hard to evaluate.

So what is "predictive observational content"? It is easy to find instances. Add to our metacorpus the sentence $O(X, "a_{17}$ is a raven", $t_{44})$, and see how many new sentences are added to our corpus of practical certainties. Generalize that idea: add to the incorrigible metacorpus a set of observation reports of each possible sort in equal numbers, and see what happens in the corpus of practical certainties. What happens depends on the values of p and p'; on what the original contents of the metacorpus is, on the language, and on the number of statements added. We will assume that this sensitivity to the levels of the evidential and practical corpora, and to the numbers of hypothetical observations added, is relatively unimportant or can be minimized. If this is so, we can adopt a third principle:

P_3 *The principle of maximizing predictive observational content:* Of two languages, adopt that that provides for the greatest predictive observational content in the corpus of practical certainties.

Applied to our example, this principle leads to the preferability of one language or another according to the relative frequencies with which various observations have been made. For example, if among our observation reports there have been a relatively large number of reports of nonblack ravens, then both "nonblack" and "raven" must be regarded as highly suspect or undependable if we were to speak L_2. In particular, the situation might be so bad that, if one were to add $O(X, "a_{17}$ is a raven", $t_{44})$ to the incorrigible metacorpus, one could not even add "a_{17} is a raven" to the evidential corpus, and therefore neither that statement nor "a_{17} is black" to the corpus of practical certainties.

On the other hand, if almost all observed ravens have been reported to be black, the addition of the observation report to the metacorpus would lead to the addition of "a_{17} is black" to the practical corpus. Furthermore, the addition (in language L_2) of $O(X, "a_{34}$ is not black", $t_{22})$ to the incorrigible metacorpus will lead (in L_2) to the inclusion of "a_{34} is not a raven" in the corpus of practical certainties.

In the case in which very few nonblack ravens have been reported, L_1 may lead to actions appropriate to the anticipation that an arbitrary new raven will be found to be black, on purely statistical grounds. (We will not in these circumstances be led to bet *against* that possibility at any odds.) But we will not be able to infer from the report that something is not black to the practical certainty that it is not a raven, and we will not be able to infer from the conjunction of reports that k objects are ravens to the conclusion that all k of them are black. Both of these inferences will go through if we adopt L_2.

The upshot, for this simple example, is this: If practically all reports conform to the generalization that all ravens are black, L_2 is the language of choice. If a significant number of reports conflict with that generalization, we must assume that our observations are so in error that L_2 offers no advantages over L_1, and indeed may suffer a loss in predictive observational content. (In fact, if the error rates are high enough, nothing may even get into the evidential corpus K_2 on the basis of past observations.)

In summary, this approach generalizes in the following way:

fewer a priori constraints	more a priori constraints
observations without error	observations prone to error
little observational prediction	more observational prediction
no loss due to error	some loss due to error

3. Scientific Change

There are four cases of scientific change to consider. There is the addition of theory, there is the deletion of theory, there is replacement of one theory by another, and there is the addition to observational vocabulary. (The last mentioned is one that has not received much attention in the literature of scientific theory change, though it is obviously an important one.)

Let us first consider the addition of a theoretical generalization to a body of scientific knowledge. The theoretical generalization embodies new constraints. If these constraints lead to changes in our statistical knowledge concerning the errors of observation (or measurement), it can only be that we have been forced to allow for an increase in observational error.

Thus we assume the same observational vocabulary; we assume greater errors of observation; but in compensation, the predictive observational content of the corpus of practical certainties under the new a priori constraints will be larger than previously. We get new predictive content at a given level of practical certainty.

Note that there is nothing inductive about this. Our new generalization is not accepted because it has, relative to the evidence, a high degree of confirmation. In fact, its confirmation, relative to the body of knowledge expressed in the old language, is 0. (Relative to L_1 the probability of "All ravens are black" is 0, as Carnap already recognized.[13]) Relative to the new language, its probability is 1.0: that is the probability of "All bachelors are unmarried," or "All ravens are black" in the language L_2. Generalizations are not (directly) supported by their instances, but by their contribution to our foreknowledge. This may, but need not, be somewhat counterbalanced by a loss of precision in our predictive statements.

Note that the situation we face is the choice between two *given* languages. There is no question of using evidence to derive a new constraint. In this regard, also, then, the acquisition of new scientific knowledge is anti-inductive. We do not learn a new theory from experience; we test a new theory against an old one, by means of experience.

The second case to consider is that in which scientific progress consists in the deletion of an old generalization or law. I suspect that this is far more rare than some philosophers, Popper, for example[14], have maintained. Rather than refuting an old generalization, refined testing most often seems to lead to the replacement of one theory by another. But refutation no doubt does occur. When it does, what happens on the model I am depicting is that a group of new observation reports leads to new statistical error distributions. These new statistical error distributions may lead to a decrease in the observational content of the corpus of evidential certainties, and do lead to a decrease in the predictive observational content of the corpus of practical certainties. At this point, it may be that the language that *lacks* the generalization or law in question may lead to a practical corpus of greater predictive content.

Note that refutation does not occur as Popper and some other philosophers suggest. We do not put a theory to the test, note that the test results do not conform to the theory, and then reject the theory. Rather, an isolated test of the theory that does not yield the predicted result contributes to the data on the basis of which we derive our distribution of errors of observation. That this effect is not often pronounced (high school laboratory tests of the law of inertia do not contribute significantly to the refutation of Newtonian mechanics, nor to the statistics of error distribution in the measurement of mass and distance) does not show that it does not exist. In some cases there is an extremely large body of data on which the error distribution is based. In such cases it can *look* as though a theory is being refuted. Note also that according to the conventional treatment of error in physics, *no* set of measurement results is ruled out as impossible by any theory. If we take error into account, no experiment can refute a theory.

The third and most common form of theoretical change is the replacement of one theory by another. Again, we assume that the observational vocabulary is the same in each case. We compare directly the error statistics and the corresponding predictive observational contents of the corpus of practical certainties of the two theories.

Only one philosopher has made a serious attempt to account in rational terms for the replacement of one theory by another. This is Isaac Levi.[15] For Levi the problem is particularly difficult, since on his view (as on mine), within the framework provided by one theoretical language, direct refutation is impossible. If an experiment is performed, and the results contradict the predictions of the theory, it is the results that must be rejected as false, not the theory. On my view, the erroneous results contribute statistically to our knowledge of the theory of error of the kinds of observations involved.

Levi's solution is that when things get bad enough, we may be led to contract to a body of knowledge in which neither the original theory nor its rival appears. When we weigh Newtonian mechanics against relativity mechanics, we first contract to a body of knowledge with no mechanics. Then we use our evidence as a basis for new expansion. Replacement is thus construed as contraction, followed by expansion.

On the view that I am advocating, this two-stage process is not necessary. We may compare directly the predictive observational contents of the corpora of practical certainties of the two cases, corresponding to the two languages.

The fourth and final case to consider is when the two languages differ in the observational vocabulary. Since I have rejected the incorrigibility of observation statements, it is necessary to say a bit about what a change in observational vocabulary comes to. What constitutes an observational term is dependent on the corresponding error theory and on what constitutes the level of evidential certainty. The basic idea is that a term is observational if an observation report making use of that term constitutes adequate evidence for including the corresponding statement among the evidential certainties of a given level of acceptance.

The kind of change in vocabulary I have in mind here includes the South Seas islander learning to distinguish dependably among the kinds of snow, or the Eskimo learning to distinguish dependably among the varieties of wave patterns. An even simpler and more direct example, however, is provided by scientists learning to see through a microscope or telescope. Galileo's inquisitors, after all, didn't need to refuse to look through his telescope. They could have looked through it, and what they would have seen were flashes of light, their own eyelashes, and difficult-to-identify shapes. It takes the student of histology a significant period of time to learn to see through a microscope. Observation, like any other skill, must be learned.

The reward of learning to observe, whether we consider the individual or the group of individuals, is the same as always: an increase in the predictive observational content of the corpus of practical certainties.

If we look at the change of observational vocabulary as learning to see something new, we may suppose that the old observational predicates are included in the new ones; what we achieve is a *refinement* of our observational vocabulary.

This case is somewhat more complicated than the preceding three cases, since we may consider two kinds of comparisons between the old and the new corpus of practical certainties. We may consider the predictive observational content of the practical corpora, constrained to sentences in the old vocabulary. Or we may look simply at the absolute numbers of predictive observational consequences in the two cases. The latter approach is not as question begging as it might appear to be, due to the requirement that *observational* predicates must be predicates that we can dependably apply. Thus one may imagine (what is no doubt the case) that phrenologists could learn to distinguish head bumps that modern people can no longer dependably detect. Well and good. But the predictive observational content of the phrenologist, given the addition of bump reports in the metacorpus of incorrigibilia, would presumably not exceed that of his modern nonphrenological counterpart, even if we include bump reports among the predictive observation statements to be taken account of.

4. Conclusions

This view of scientific knowledge and scientific change bears on a number of contemporary issues.

It bears on the question of incommensurability of alternative theories, for example. Kuhn,[16] Feyerabend,[17] and others have sometimes written as if there were no way of directly comparing theories and their successors. If the view of theories as mere conventions that I have been outlining is meritorious, it is easy to see the truth in such a view. It is quite true that two competing theories are not comparable in the sense that we can look at the evidence and decide that one of them has (or should have) a larger degree of confirmation than the other, or that one has been tested and refuted while the other has not, or that one has been tested more severely, or that one is more testable than the other. Two theories are not compared directly in the light of the evidence. But given a body of observations – that is, given a set of O-statements in the incorrigible metacorpus – we can find a direct basis of comparison of two *languages* in the predictive observational content of the corpus of practical certainties.

A serious question that might be raised concerns the apparent ubiquity of language change in the history of science. As I have sketched scientific change, all change, other than mere statistical inference, reflects a change in the language of science. Is it really the case that every time a universal generalization is confirmed (as we naively say), what is going on is that we are choosing between languages?

We must distinguish two cases. The natural history case, in which we look at ravens or swans and arrive at a universal generalization, it seems to me, does involve change of language. It is not that we decide that "almost all swans are white," and then discover that Australian swans are black. It is that we take as a matter of certainty (convention) that all swans are white; and then, given the evidence from Australia, must decide – nontrivially – whether there are adequate anatomical grounds for regarding the southern black birds as also swans. It is quite clear that there are issues in such a case as this that only an ornithologist is in a position to deal with.

The other case is the one in which we already have a body of structured scientific knowledge to use. This corresponds to what used to be called "demonstrative induction." We can decide that all samples of the compound X melt at about 45 degrees Celsius, on the basis of a small number of experiments, because it is a part of the general theory – i.e., language – that each pure compound has exactly one melting point. All we have to do is to determine that one sample of X melts at 45 degrees to determine that all samples do.

Can we decide between *any* two languages for science? For example, between

French physics and English physics? The answer is clearly negative. French physics and English physics (assuming they embody the same laws) will have the same number of predictive observational consequences in the corpus of practical certainties. Indeed, they will have the same predictive observational consequences, though expressed in a different vocabulary. So the difference between French physics and English physics is exactly a matter of *arbitrary* convention. The difference between Newtonian physics and Einsteinian physics is also a matter of convention, but while it is *mere* convention, it is not arbitrary convention. Einstein's physics allows us to predict a bit more.

This way of looking at scientific languages might be thought to have consequences for the issues of scientific realism and instrumentalism. This is true, but it might also be thought that since I am arguing that scientific theories are mere conventions, instrumentalism is the only plausible ontological view. This is mistaken. It reflects a confusion between mere and arbitrary convention. If the difference between theories quantifying over different sorts of entities were a matter of arbitrary convention, then perhaps instrumentalism would be warranted. But I have just argued that the choice between alternative scientific languages is an epistemological issue. We have epistemological grounds for choosing one theory over another. And this suggests that we can have epistemological grounds for quantifying over one set of entities rather than another.

In addition, the framework suggested weakens the distinction between "observational" and "theoretical" terms profoundly. Given modern instrumentation and modern training, it seems clear that many of the properties of cells must be regarded as observational properties. Before the invention of the microscope, some of these same properties must have been taken to be unobservable and theoretical. There are indeed complexities to be dealt with here, but the point that emerges quite clearly is that nothing definitive — in particular no form of instrumentalism — emerges as entailed by the view described.

To view scientific theories as mere but not arbitrary conventions offers a framework in which the rational succession of one scientific theory by another can be understood; it offers an alternative to the view of scientific theories as subjective, arbitrary, and subject to personal whim and social pressure, rather than the demands of rationality. What views of the world the evidence we happen to have supports is an objective, rational matter. This is not to say that experience and reason dictate what theory we should hold of the world, but that experience and reason dictate which of two *alternative* theories (languages) is preferable.

Notes

1. Ernst Mach, *The Science of Mechanics* (LaSalle, Ill.: Open Court, 1960). 1st German ed., 1883; 1st American ed., 1893.

2. Henri Poincaré, *Science and Hypothesis* (New York: Dover Publications, 1952). 1st French ed., *La Science et L'hypothese* (Paris, 1903).

3. Pierre Duhem, *The Aim and Structure of Physical Theory* (Princeton: Princeton University Press, 1954). 1st French ed., 1906.

4. W. V. O. Quine, "Two Dogmas of Empiricism," *The Philosophical Review* 60 (1951). This essay has been extensively reprinted. It appears in Sandra Harding ed., *Can Theories Be Refuted?* (Dordrecht: Reidel, 1976), 41–64. This volume contains a number of interesting essays that bear on our present concerns.

5. W. V. O. Quine and J. S. Ullian, *The Web of Belief* (New York: Random House, 1970). This is not the first appearance of Quine's web metaphor, but this volume contains the most developed statement of the metaphor.

6. William Craig, "Replacement of Auxiliary Expressions," *Philosophical Review* 65, 1956, 38–55.

7. W. V. O. Quine, *Word and Object* (Cambridge: MIT Press, 1960). The issue is related to the question of radical translation; can we sort our experiences uniquely according to the expressions of our language?

8. Quine's *Word and Object* is the locus classicus for undertermination arguments. It should be noted that Quine's own views have evolved significantly since that work.

9. David Lewis, *Convention: A Philosophical Study* (Cambridge: Harvard University Press, 1969).

10. Henry E. Kyburg, Jr., "The Reference Class," *Philosophy of Science* 50 (1983): 374–97. The issue has been addressed in a number of other publications, including my *The Logical Foundations of Statistical Inference.* (Dordrecht: Reidel, 1974).

11. Henry E. Kyburg, Jr., "Full Belief," *Theory and Decision* 25, 1988, 137–162, contains one proposal for the determination of p and p'. Others are no doubt possible.

12. Mary Hesse, *The Structure of Scientific Inference* (Berkeley and Los Angeles: University of California Press, 1974).

13. Rudolf Carnap, "On Inductive Logic," *Philosophy of Science* 12, 1945, 72–97.

14. K. R. Popper, *The Logic of Scientific Discovery* (London: Hutchinson and Co., 1959). The first German edition appeared in 1934, and Popper has had a numerous and influential following ever since.

15. Isaac Levi, *The Enterprise of Knowledge* (Cambridge: MIT Press, 1980). Recent work on this problem has also been done by Peter Gardenfors and others; see *Knowledge in Flux* (Cambridge, Mass.: Bradford Books, 1988) and references therein.

16. Thomas Kuhn, *The Structure of Scientific Revolutions* (Chicago: University of Chicago Press, 1962).

17. Paul K. Feyerabend, "Against Method: Outline of an Anarchistic Theory of Knowledge," in *Minnesota Studies on the Philosophy of Science,* Vol. 4, *Analyses of Theories and Methods of Physics and Psychology,* eds. Michael Radner and Stephen Winokur (Minneapolis: University of Minnesota Press, 1970), 3–130.

Wesley C. Salmon

Rationality and Objectivity in Science

or

Tom Kuhn Meets Tom Bayes

Twenty-five years ago, as of this writing, Thomas S. Kuhn published *The Structure of Scientific Revolutions.*[1] It has been an extraordinarily influential book. Coming at the height of the hegemony of *logical empiricism*—as espoused by such figures as R. B. Braithwaite, Rudolf Carnap, Herbert Feigl, Carl G. Hempel, and Hans Reichenbach—it posed a severe challenge to the logistic approach that they practiced.[2] It also served as an unparalleled source of inspiration to philosophers with a historical bent. For a quarter of a century there has been a deep division between the logical empiricists and those who adopt the historical approach, and Kuhn's book was undoubtedly a key document in the production and preservation of this gulf.

At a 1983 meeting of the American Philosophical Association (Eastern Division), Kuhn and Hempel—the most distinguished living advocates for their respective viewpoints—shared the platform in a symposium devoted to Hempel's philosophy.[3] I had the honor to participate in this symposium. On that occasion Kuhn chose to address certain issues pertaining to the rationality of science that he and Hempel had been discussing for several years. It struck me that a bridge could be built between the differing views of Kuhn and Hempel if Bayes's theorem were invoked to explicate the concept of scientific confirmation.[4] At the time it seemed to me that this maneuver could remove a large part of the dispute between standard logical empiricism and the historical approach to philosophy of science on this fundamental issue.

I still believe that we have the basis for a new consensus regarding the choice among scientific theories. Although such a consensus, if achieved, would not amount to total agreement on every problem, it would represent a major rapprochement on an extremely fundamental issue. The purpose of the present essay is to develop this approach more fully. As it turns out, the project is much more complex than I thought in 1983.

I should like to express my deepest gratitude to Adolf Grünbaum and Philip Kitcher for important criticism and valuable suggestions with respect to an earlier version of this paper.

175

§1. Kuhn on Scientific Rationality

A central part of Kuhn's challenge to the logical empiricist philosophy of science concerns the nature of theory choice in science. The choice between two fundamental theories (or paradigms), he maintains, raises issues that "cannot be resolved by proof." To see how they are resolved we must talk about "techniques of persuasion," or about "argument and counterargument in a situation in which there can be no proof." Such choices involve the exercise of the kind of judgment that cannot be rendered logically explicit and precise. Such statements, along with many others that are similar in spirit, led a number of critics to attribute to Kuhn the view that science is fundamentally irrational and lacking in objectivity.

Kuhn was astonished by this response, which he regarded as a serious misinterpretation. In his "Postscript—1969," in the second edition of *The Structure of Scientific Revolutions,* and in "Objectivity, Value Judgment, and Theory Choice"[5] he replies to these charges. What he had intended to convey was the claim that the decision by the community of trained scientists *constitutes* the best criterion of objectivity and rationality we can have. In order better to understand the nature of such objective and rational methods we need to look in more detail at the considerations that are actually brought to bear by scientists when they endeavor to make comparative evaluations of competing theories.

For purposes of illustration, Kuhn offers a (nonexhaustive) list of characteristics of good scientific theories that are, he claims:

> individually important and collectively sufficiently varied to indicate what is at stake. . . . These five characteristics—accuracy, consistency, scope, simplicity, and fruitfulness—are all standard criteria for evaluating the adequacy of a theory. . . . Together with others of much the same sort, they provide the shared basis for theory choice.[6]

Two sorts of problems arise when one attempts to use them.

> Individually the criteria are imprecise: individuals may legitimately differ about their applicability to concrete cases. In addition, when deployed together, they repeatedly prove to conflict with one another; accuracy may, for example, dictate the choice of one theory, scope the choice of its competitor.[7]

For reasons of these sorts—and others as well—individual scientists may, at a given moment, differ regarding a particular choice of theories. In the course of time, however, the interactions among individual members of the community of scientists produce a consensus for the group. Individual choices inevitably depend upon idiosyncratic and subjective factors; only the outcome of the group activity can be considered objective and fully rational.

One of Kuhn's major claims seems to be that observation and experiment, in conjunction with hypothetico-deductive reasoning, do not adequately account for the choice of scientific theories. This has led some philosophers to believe that theory choice is not rational. Kuhn, in contrast, has tried to locate the additional factors that are involved. These additional factors constitute a crucial aspect of scientific rationality.

§2. Bayes's Theorem

The first step in coming to grips with the problem of evaluating and choosing scientific hypotheses or theories[8] is the recognition of the inadequacy of the traditional hypothetico-deductive (H-D) schema as a characterization of the logic of science. According to this schema, we confirm a scientific hypothesis by deducing from it, in conjunction with suitable initial conditions and auxiliary hypotheses, an observational prediction that turns out to be true. The H-D method has a number of well-known shortcomings. (1) It does not take account of alternative hypotheses that might be invoked to explain the same prediction. (2) It makes no reference to the initial plausibility of the hypothesis being evaluated. (3) It cannot accommodate cases, such as the testing of statistical hypotheses, in which the observed outcome is not deducible from the hypothesis (in conjunction with the pertinent initial conditions and auxiliary hypotheses), but only rendered more or less probable.

In view of these and other considerations, many logical empiricists agreed with Kuhn regarding the inadequacy of hypothetico-deductive confirmation. A number—including Carnap and Reichenbach—appealed to Bayes's theorem, which may be written in the following form:

$$P(T|E.B) = \frac{P(T|B)P(E|B.T)}{P(T|B)P(E|B.T) + P(\sim T|B)P(E|B.\sim T)}. \tag{1}$$

Let "T" stand for the theory or hypothesis being tested, "B" for our background information, and "E" for some new evidence we have just acquired. Then the expression on the left-hand side of the equation represents the probability of our hypothesis on the basis of the background information and the new evidence. This is known as the *posterior probability*. The right-hand side of the equation contains four probability expressions. Two of these, $P(T|B)$ and $P(\sim T|B)$, are called *prior probabilities*; they represent the probability, on the basis of background information alone, without taking account of the new evidence E, that our hypothesis is true or false respectively. Obviously the two prior probabilities must add up to one; if the value of one of them is known, the value of the other can be inferred immediately. The remaining two probabilities, $P(E|T.B)$ and $P(E|\sim T.B)$, are known as *likelihoods*; they are, respectively, the probability that the new evidence would occur if our hypothesis were true and the probability that it would occur if our hypothesis were false. The two likelihoods, in contrast to the two prior probabilities, must be established independently; the value of one does not automatically determine the value of the other. To calculate the posterior probability of our hypothesis, then, we need three separate probability values to plug into the right-hand side of Bayes's theorem – a prior probability and two likelihoods.

Before attempting to resolve any important issues concerning the nature of scientific reasoning, let us look at a simple and noncontroversial application of Bayes's theorem. Consider a factory that produces can openers at the rate of 6,000 per day. This factory has two machines, a new one that produces 5,000 can openers per day and an old one that produces 1,000 per day. Among the can openers produced by the new machine 1 percent are defective; among those produced by the old machine 3 percent are defective. We pick one can opener at random from today's production and find it defective. What is the probability that it was produced by the new machine?

We can get the answer to this question via Bayes's theorem. If we let "B" stand for the class of can openers produced in this factory today, "T" for the class of can openers produced by the new machine, and "E" for a can opener that is defective, then the probability we seek is the posterior probability $P(T \mid B.E)$ – the probability that a defective can opener from today's production was produced by the new machine. The values of the prior probabilities and likelihoods have been given, namely:

$$P(T \mid B) = 5/6 \qquad\qquad P(\sim T \mid B) = 1/6$$
$$P(E \mid T.B) = 1/100 \qquad\qquad P(E \mid \sim T.B) = 3/100.$$

Plugging these values into equation (1) immediately yields $P(T \mid B.E) = 5/8$. Notice that the old machine has a greater probability of producing a defective can opener than does the new, but the probability that a defective can opener was produced by the new machine is greater than that it was produced by the old one. This results, obviously, from the fact that the new machine produces so many more can openers overall than does the old one.

One way to look at this example is to consider the hypothesis T that a given can opener was produced by the new machine. This is a causal hypothesis. Our background information B is simply that the can opener is part of today's production at this factory. On the basis of this prior information, we can evaluate the prior probability of T; it is 5/6. Now we add to our knowledge about this can opener the information E that it is defective. This knowledge is relevant to the hypothesis that it was produced by the new machine; the posterior probability is 5/8. Although one does not *need* to appeal to Bayes's theorem to establish this result,[9] the highly artificial example shows clearly just how Bayes's theorem can be used to ascertain the posterior probability of a simple causal hypothesis.

When we come to more realistic scientific cases, it is not so easy to see how to apply Bayes's theorem; the prior probabilities may seem particularly difficult. I believe that, in fact, they reflect the plausibility arguments scientists often bring to bear in their deliberations about scientific hypotheses. I shall discuss this issue in §4; indeed, in subsequent sections we shall have to take a close look at all of the probabilities that enter into Bayes's theorem.

In this section I have been concerned to present Bayes's theorem and to make a few preliminary remarks about its application to the problem of evaluating scientific hypotheses. In the next section I shall try to spell out the connections between Bayes's theorem and Kuhn's views on the nature of theory choice. Before moving on to that discussion, however, I want to present two other useful forms in which Bayes's theorem can be given. In the first place, because of the theorem on total probability:

$$P(E|B) = P(T|B)\ P(E|T.B) + P(\sim T|B)\ P(E|\sim T.B), \tag{2}$$

equation (1) can obviously be rewritten as:

$$P(T|E.B) = \frac{P(T|B) \times P(E|B.T)}{P(E|B)}. \tag{3}$$

In the second place, equation (1) can be generalized to handle several alternative hypotheses, instead of just one hypothesis and its negation, as follows:

$$P(T_i|B.E) = \frac{P(T_i|B) \times P(E|T_i.B)}{\sum\limits_{j=1}^{k} [P(T_j|B) \times P(E|T_j.B)]}, \tag{4}$$

where T_1- T_k are mutually exclusive and exhaustive alternative hypotheses and $1 \leq i \leq k$.

Strictly speaking, (4) is the form that is needed for realistic historical examples—such as the corpuscular (T_1) and wave (T_2) theories of light in the nineteenth century. In that case, although we could construe T_1 and T_2 as mutually exclusive, we could not legitimately consider them exhaustive, for we cannot be sure that one or the other is true. Therefore, we would have to introduce T_3— what Abner Shimony has called the *catchall hypothesis*—which says that T_1 and T_2 are both false. T_1 — T_3 thus constitute a mutually exclusive and exhaustive set of hypotheses. This is the sort of situation that obtains when scientists are attempting to choose a correct hypothesis from among two or more serious candidates.

§3. Kuhn and Bayes

For purposes of discussion, Kuhn is willing to admit that "each scientist chooses between competing theories by deploying some Bayesian algorithm which permits him to compute a value for $P(T|E)$, i.e., for the probability of the theory T on the evidence E available both to him and the other members of his professional group at a particular period of time."[10] He then formulates the crucial issue in terms of the question of whether there is one unique algorithm used by all rational scientists, yielding a unique value for P, or whether different scientists, though fully rational, may use different algorithms yielding different values of P. I want to suggest a third possibility to account for the phenomena of theory choice—namely, that many different scientists might use the same algorithm, but nevertheless arrive at different values of P.

When one speaks of a Bayesian algorithm, the first thought that comes to mind is Bayes's theorem itself, as embodied in any of the equations (1), (3) or (4). We have, for instance:

$$P(T \mid E.B) = \frac{P(T \mid B) \times P(E \mid B.T)}{P(E \mid B)} \ , \tag{3}$$

which constitutes an algorithm in the most straightforward sense of the term. Let us call $P(E \mid B)$ the *expectedness* of the evidence. Given values for the prior probability, likelihood, and expectedness, the value of the posterior probability can be computed by trivial arithmetical operations.[11]

If we propose to use equation (3) as an algorithm, the obvious question is how to get values for the expressions on the right-hand side. Several answers are possible in principle, depending on what interpretation of the probability concept is espoused. If one adopts a Carnapian approach to inductive logic and confirmation theory, all of the probabilities that appear in Bayes's theorem can be derived a priori from the structure of the descriptive language and the definition of degree of confirmation. Since it is extremely difficult to see how any genuine scientific case could be handled by means of the highly restricted apparatus available within that approach, not many philosophers are tempted to follow this line. Moreover, even if a rich descriptive language were available, it is not philosophically tempting to suppose that the probabilities associated with serious scientific theories are a priori semantic truths.

Two major alternatives remain. First, one might maintain that the probabilities on the right-hand side of (3)—especially the prior probability $P(T \mid B)$—are objective and empirical. I have attempted to defend the view that they refer, at bottom, to the frequencies with which various kinds of hypotheses or theories have been found successful.[12] Clearly, enormous difficulties are involved in working out that alternative; I shall return to the issue below. In the meantime, let us consider the other—far more popular—alternative.

The remaining alternative approach involves the use of personal probabilities. Personal probabilities are subjective in character; they represent subjective degrees of conviction on the part of the individual who has them, provided that they fulfill the condition of coherence.[13] Consider a somewhat idealized situation. Suppose that, in the presence of background knowledge B (which may include initial conditions, boundary conditions, and auxiliary hypotheses) theory T deductively entails evidence E. This is the situation to which the hypothetico-deductive method appears to be applicable. In this case, $P(E|T.B)$ must equal 1, and equation (3) reduces to:

$$P(T|E.B) = P(T|B)/P(E|B). \tag{5}$$

One might then ask a particular scientist for his or her plausibility rating of theory T on background knowledge B alone, quite irrespective of whether evidence E obtains or not. Likewise, the same individual may be queried regarding the degree to which evidence E is to be expected irrespective of the truth or falsity of T. According to the personalist, it should be possible—by direct questioning or by some less direct method—to elicit such *psychological* facts regarding a scientist involved in investigations concerning the theory in question. This information is sufficient to determine the degree of belief this individual should have in the theory T given the background knowledge B and the evidence E, namely, the posterior probability $P(T|E.B)$.

In the more general case, when T and B do not deductively entail E, the procedure is the same, except that the value of $P(E|T.B)$ must also be ascertained. In many contexts, where statistical significance tests can be applied, a value of the likelihood $P(E|T.B)$ can be calculated, and the personal probability will coincide with the value thus derived. In any case, whether statistical tests apply or not, there is no *new* problem in principle involved in procuring the needed degree of confidence. This reflects the standard Bayesian approach in which all of the probabilities are taken to be personal probabilities.

In any case, whether one adopts an objective or a personalistic interpretation of probability, equation (3)—or some other version of Bayes's theorem—can be taken as an algorithm for evaluating scientific hypotheses or theories. Individual scientists, using the same algorithm, may arrive at different evaluations of the same hypothesis because they plug in different values for the probabilities. If the probabilities are construed as objective, different individuals may well have different estimates of these objective values. If the probabilities are construed as personal, different individuals may well have different subjective assessments of them. Bayes's theorem provides a mechanical algorithm, but the judgments of individual scientists are involved in procuring the values that are to be fed into it. This is a general feature of algorithms; they are not responsible for the data they are given.

§4. Prior Probabilities

In §2 I remarked that the prior probabilities in Bayes's theorem can best be seen as embodying the kinds of plausibility judgments that scientists regularly make regarding the hypotheses with which they are concerned. Einstein, who was clearly aware of this consideration, contrasted two points of view from which a theory can be criticized or evaluated:

The first point of view is obvious: the theory must not contradict empirical facts. . . . [it] is concerned with the confirmation of the theoretical foundation by the available empirical facts.   The second point of view is not concerned with the relation of the material of observation but with the premises of the theory itself, with what may briefly but vaguely be characterized as the "naturalness" or "logical simplicity" of the premises. . . . The second point of view may briefly be characterized as concerning itself with the "inner perfection" of a theory, whereas the first point of view refers to the "external confirmation."[14]

Einstein's second point of view is the sort of thing I have in mind in referring to plausibility arguments or judgments concerning prior probabilities.

Plausibility considerations are pervasive in the sciences; they play a significant — indeed, *indispensable* — role. This fact provides the initial reason for appealing to Bayes's theorem as an aid to understanding the logic of evaluating scientific hypotheses. Plausibility arguments serve to enhance or diminish the probability of a given hypothesis prior to — i.e., without reference to — the outcome of a particular observation or experiment. They are designed to answer the question, "Is this the kind of hypothesis that is likely to succeed in the scientific situation in which the scientist finds himself or herself?" On the basis of their training and experience, scientists are qualified to make such judgments.

This point can best be explained, I believe, in terms of concrete examples. Since before the time of Newton, for instance, a well-known plausibility argument for the inverse square character of gravitational forces has been around. It is natural to think of the gravitational force emanating from a particle of matter as one that spreads spherically from it in a uniform manner. In the seventeenth and eighteenth centuries all competent physical scientists believed that physical space has a three-dimensional Euclidean structure. Since the surface of a Euclidean sphere increases as the square of the radius, it is reasonable to suppose that the force of gravity is diluted in just the same way, for the farther one goes from the particle, the greater the spherical surface over which the force must be spread.

A famous Canadian study of the effects of the consumption of large doses of saccharin provides another example.[15] A statistically significant association between heavy saccharin consumption and bladder cancer in a controlled experiment with rats lends considerable plausibility to the hypothesis that use of saccharin as an artificial sweetener in diet soft drinks increases the risk of bladder cancer in humans. This example, unlike the preceding one, is inherently statistical and does not have even the prima facie appearance of a hypothetico-deductive inference.

In order to come to a clearer understanding of the nature of prior probabilities, it will be necessary to look at them from the point of view of the personalist and that of the objectivist (frequency or propensity theorist).[16] The frightening thing about pure unadulterated personalism is that nothing prevents prior probabilities (and other probabilities as well) from being determined by all sorts of idiosyncratic and objectively irrelevant considerations. A given hypothesis might get an extremely low prior probability because the scientist considering it has a hangover, has had a recent fight with his or her lover, is in passionate disagreement with the politics of the scientist who first advanced the hypothesis, harbors deep prejudices against the ethnic group to which the originator of the hypothesis belongs, etc. What we want to demand is that the investigator make every effort to bring all of his or her *relevant* experience in evaluating hypotheses to bear on the question of whether the hypothesis under consideration is of a type likely to succeed, and to leave aside emotional irrelevancies.

It is rather easy to construct really perverse systems of belief that do not violate the coherence requirement. But we need to keep in mind the objectives of science. When we have a long series of events, such as tosses of fair or biased coins, or radioactive decays of unstable nuclei, we want our subjective degrees of conviction to match what either a frequency theorist or a propensity theorist would regard as the objective probability. Carnap was profoundly correct in his notion that inductive or logical or epistemic probabilities should be reasonable estimates of relative frequencies.

A sensible personalist, I would suggest, is someone who wants his or her personal probabilities to reflect objective fact. Betting on a sequence of tosses of a coin, a personalist wants not only to avoid Dutch books,[17] but also to stand a reasonable chance of winning (or of not losing too much too fast). As I read it, the whole point of F. P. Ramsey's famous article on degrees of belief is to consider what you get if your subjective degrees of belief match the relevant frequencies.[18] One of the facts recognized by the sensible personalist is that whether the coin lands heads or tails is not affected by on which side of the bed he or she got out that morning. If we grant that the personalist's aim is to do as well as possible in betting on heads and tails, it would be obviously counterproductive to allow the betting odds to be affected by such irrelevancies.

The same general sort of consideration should be brought to bear on the assignment of probabilities to hypotheses. Whether a particular scientist is dyspeptic on a given morning is irrelevant to the question of whether a physical hypothesis that is under consideration is correct or not. Much more troubling, of course, is the fact that any given scientist may be inadvertently influenced by ideological or metaphysical prejudices. It is obvious that an unconscious commitment to capitalism or racism might seriously affect theorizing in the behavioral sciences.

Similar situations may arise in the physical sciences as well; another historical example will illustrate the point. In 1800 Alessandro Volta invented the battery, thereby providing scientists with a way of producing steady electrical currents.

It was not until 1820 that Hans Christian Oersted discovered the effect of an electrical current on a magnetic needle. Why was there such a delay? One reason was the previously established fact that a static electric charge has no effect on a magnetic needle. Another reason that has been mentioned is the fact that, contrary to the expectation if there were such an effect, it aligns the needle perpendicular to the current carrying wire. As Holton and Brush remark, "But even if one has currents and compass needles available, one does not observe the effect unless the compass is placed in the right position so that the needle can respond to a force that seems to act in a direction *around* the current rather than *toward* it."[19] I found it amusing when, on one occasion, a colleague set up the demonstration with the magnetic needle oriented at right angles to the wire to show why the experiment fails if one begins with the needle in that position. When the current was turned on, the needle rotated through 180 degrees; he had neglected to take account of polarity. How many times, between 1800 and 1820, had the experiment been performed without reversing the polarity? Not many. The experiment had apparently not been tried by others because of Cartesian metaphysical commitments. It was undertaken by Oersted as a result of his proclivities toward *naturphilosophie*.

How should scientists go about evaluating the prior probabilities of hypotheses? In elaborating a view he calls *tempered personalism,* —a view that goes beyond standard Bayesian personalism by placing further constraints on personal probabilities—Shimony[20] points out that experience shows that the hypotheses seriously advanced by serious scientists stand some chance of being successful. Science has, in fact, made considerable progress over the past four or five centuries, which constitutes strong empirical evidence that the probability of success among members of this class is nonvanishing. Likewise, hard experience has also taught us to reject claims of scientific infallibility. Thus, we have good reasons for avoiding the assignment of extreme values to the priors of the hypotheses with which we are seriously concerned. Moreover, Shimony reminds us, experience has taught that science is difficult and frustrating; consequently, we ought to assign fairly low prior probabilities to the hypotheses that have been explicitly advanced, allowing a fairly high prior for the catchall hypothesis—the hypothesis that we have not yet thought of the correct hypothesis. The history of science abounds with situations of choice among theories in which the successful candidate has not even been conceived at the time.

In *The Foundations of Scientific Inference,* I proposed that the problem of prior probabilities be approached in terms of an objective interpretation of probability, in particular, the frequency interpretation. I suggested three sorts of criteria that can be brought to bear in assessing the prior probabilities of hypotheses: formal, material, and pragmatic.

Pragmatic criteria have to do with the circumstances in which a new hypothesis originates. We have already seen an example of a pragmatic criterion in Shimony's observation that hypotheses advocated by serious scientists have nonvanishing chances of success. The opposite side of the same coin is provided by Martin Gardner, who offers an enlightening characterization of scientific cranks.[21] Since it is doubtful that a single useful scientific suggestion has ever been originated by anyone in that category, hypotheses advanced by people of that ilk have negligible chances of being correct. I recall when L. Ron Hubbard's *Dianetics* was first published. A psychologist friend, asked what he thought of it, said, "I can't condemn this book before reading it, but after I have read it, I will." When competent scientists offer hypotheses outside of their areas of specialization, we have a right to wonder whether appreciable plausibility accrues to such suggestions. Hubbard was, incidentally, an engineer with no training in psychology.

The *formal criteria* have to do not only with matters of internal consistency of a new hypothesis, but also with relations of entailment or incompatibility of the new hypothesis with accepted laws and theories. The fact that Immanuel Velikovski's *Worlds in Collision* [22] contradicts many of the accepted basic laws of physics — e.g., the law of conservation of angular momentum — renders his 'explanations' of such biblically reported incidents as the parting of the waters of the Red Sea and the brief interruption of the rotation of the earth (the sun standing still) utterly implausible.

It should be recalled that among his five considerations for the evaluation of scientific theories — mentioned above — Kuhn includes consistency of the sort we are discussing. I take this as a powerful hint that one of the main issues Kuhn has raised about scientific theory choice involves the use of prior probabilities and plausibility judgments.

The *material criteria* have to do with the actual structure and content of the hypothesis or theory under consideration. The most obvious example is simplicity — another of Kuhn's five items. Simplicity strikes me as singularly important, for it has often been treated by scientists and philosophers as an a priori criterion. It has been suggested, for example, that the hypothesis that quarks are fundamental constitutents of matter loses plausibility as the number of different types of quarks increases, since it becomes less simple as a result.[23] It has also been advocated as a universal methodological maxim: *Search for the simplest possible hypothesis.* Only if the simpler hypotheses do not stand up under testing should one resort to more complex hypotheses.

Although simplicity has obviously been an important consideration in the physical sciences, its applicability in the social/behavioral sciences is problematic. In a recent article, "Slips of the Tongue," Michael T. Motley criticizes Freud's theory for being too simple — an oversimplification.

> Further still, the categorical nature of Freud's claim that all slips have hidden meanings makes it rather unattractive. It is difficult to imagine, for example, that my six-year-old daughter's mealtime request to "help cut up my meef" was

the result of repressed anxieties or anything of that kind. It seems more likely that she simply merged "meat" and "beef" into "meef." Similarly, about the only meaning one can easily read into someone's saying "roon mock" instead of "moon rock" is that the m and r got switched. Even so, how does it happen that words can merge or sounds can be switched in the course of speech production? And in the case of my "pleased to beat you" error [to a competitor for a job], might Freud have been right?[24]

The most reasonable way to look at simplicity, I think, is to regard it as a highly relevant characteristic, but one whose applicability varies from one scientific context to another. Specialists in any given branch of science make judgments about the degree of simplicity or complexity that is appropriate to the context at hand, and they do so on the basis of extensive experience in that particular area of scientific investigation. Since there is no precise measure of simplicity as applied to scientific hypotheses and theories, scientists must use their judgment concerning the degree of simplicity a given hypothesis or theory possesses and concerning the degree of simplicity that is desirable in the given context. The kind of judgment to which I refer is not spooky; it is the kind of judgment that arises on the basis of training and experience. This experience is far too rich to be the sort of thing that can be spelled out explicitly. As Patrick Suppes[25] has pointed out, the assignment of prior probability by the Bayesian can be regarded as the best estimate of the chances of success of the hypothesis or theory on the basis of all relevant experience in that particular scientific domain. The personal probability represents, not an effort to contaminate science with subjective irrelevancies, but rather an attempt to facilitate the inclusion of all relevant evidence.

Simplicity is only one among many material criteria. Another closely related criterion—frequently employed in contemporary physics—is symmetry. Perhaps the most striking historical example is de Broglie's hypothesis regarding matter waves. Since light exhibits both particle and wave behavior, which are linked in terms of linear momentum, he suggested, why should not material particles, which obviously possess linear momentum, also have such wave characteristics as wave length and frequency? Unbeknownst to de Broglie, experimental work by Davisson was, at that very time, providing positive evidence of wavelike behavior of electrons.

A third widely used material criterion is analogy, as illustrated by the saccharin study. The physiological analogy between rats and humans is sufficiently strong to lend considerable plausibility to the hypothesis that saccharin can cause bladder cancer in humans. I suspect that the use of arguments by analogy in science is almost always aimed at establishing prior probabilities. The formal criteria enable us to take account of the ways in which a given hypothesis fits *deductively* with what else we know. Analogy helps us to assess the degree to which a given hypothesis fits *inductively* with what else we know.

The moral I would draw concerning prior probabilities is that they can be understood as our best estimates of the frequencies with which certain kinds of hypotheses succeed. These estimates are rough and inexact; some philosophers might prefer to think of them in terms of intervals. If, however, one wants to construe them as personal probabilities, there is no harm in it, as long as we attribute to the subject who has them the aim of bringing to bear all his or her experience that is relevant to the success or failure of hypotheses similar to that being considered. The personalist and the frequentist need not be in any serious disagreement over the construal of prior probabilities.[26]

One point is apt to be immediately troublesome. If we are to use Bayes's theorem to compute values of posterior probabilities, it would appear that we must be prepared to furnish numerical values for the prior probabilities. Unfortunately, it seems preposterous to suppose that plausibility arguments of the kind we have considered could yield exact numerical values. The usual answer is that, because of a phenomenon known as "washing out of the priors" or "swamping of the priors," even very crude estimates of the prior probabilities will suffice for the kinds of scientific judgments we are concerned to make. Obviously, however, this sort of convergence depends upon agreement regarding the likelihoods.

§5. The Expectedness

The term "$P(E|B)$" occurring in the denominator of equation (3) is called the *expectedness* because it is the opposite of surprisingness. The smaller the value of $P(E|B)$, the more surprising E is; the larger the value of $P(E|B)$, the less surprising, and hence, the more expected E is. Since the expectedness occurs in the denominator, a smaller value tends to increase the value of the fraction. This conforms to a widely held intuition that the more surprising the predictions a theory can make, the greater is their evidential value when they come true.

A classic example of a surprising prediction that came true is the Poisson bright spot. If we ask someone who is completely naive about theories of light how probable it is that a bright spot appears in the center of the shadow of a brightly illuminated circular object (ball or disk), we would certainly anticipate the response that it is very improbable indeed. There is a good inductive basis for this answer. In our everyday lives we have all observed many shadows of opaque objects, and they do not contain bright spots at their centers. Once, when I demonstrated the Poisson bright spot to an introductory class, one student carefully scrutinized the ball bearing that cast the shadow because he strongly suspected that it had a hole through it.

Another striking example, to my mind, is the Cavendish torsion-balance experiment. If we ask someone who is totally ignorant of Newton's theory of universal gravitation how strongly they expect to find a force of attraction between a lead ball and a pith ball in a laboratory, I should think the answer, again, would be that it is very unlikely. There is, in this example as well, a sound inductive basis for the response. We are all familiar with the gravitational attraction of ordinary-size objects to the earth, but we do not have everyday experience of an attraction between two such relatively small (electrically neutral and unmagnetized) objects as those Cavendish used to perform his experiment. Newton's theory predicts, of course, that there will be a gravitational attraction between any two material objects. The trick was to figure out how to measure it.

As the foregoing two examples show, there is a possible basis for assigning a low value to the expectedness; it was made plausible by assuming that the subject was completely naive concerning the relevant physical theory. The trouble with this approach is that a person who wants to use Bayes's theorem — in the form of equation (3), say — cannot be totally innocent of the theory T that is to be evaluated, since the other terms in the equation refer explicitly to T. Consequently, we have to recognize the relationship between P(E | B) and the prior probabilities and likelihoods that appear on the right-hand side in the theorem on total probability:

$$P(E|B) = P(T|B) \ P(E|T.B) + P(\sim T|B) \ P(E| \sim T.B). \tag{2}$$

Suppose that the prior probability of T is not negligible and that T, in conjunction with suitable initial conditions, entails E. Under these circumstances E cannot be totally surprising; the expectedness cannot be vanishingly small. Moreover, to evaluate the expectedness of E we must also consider its probability if T is false. By focusing on the expectedness, we cannot really avoid dealing with likelihoods.

There is a further difficulty. Suppose, for example, that the wave theory of light is true. It is surely *true enough* in the context of the Poisson bright spot experiment. If we want to evaluate P(E | B) we must include in B the initial conditions of the experiment — the circular object illuminated by a bright light in such a way that the shadow falls upon a screen. Given the truth of the wave theory, the *objective probability* of the bright spot is one, for whenever those initial conditions are realized, the bright spot appears. It makes no difference whether we know that the wave theory is true, or believe it, or reject it, or have ever thought of it. Under the conditions specified in B the bright spot invariably occurs. Interpreted either as a frequency or a propensity, P(E | B) = 1. If we are to avoid trivialization in many important cases, the expectedness must be treated as a personal probability. To anyone who, like me, wants to base scientific theory preference or choice on objective considerations, this result poses a serious problem.

The net result is a twofold problem. First, by focusing on the expectedness, we do *not* escape the need to deal explicitly with the likelihoods. In §6 I shall discuss the difficulties that arise when we focus on the likelihoods, especially the problem of the likelihood on the catchall hypothesis. Second, the expectedness defies interpretation as an objective probability. In §7 I shall propose a strategy for avoiding involvement with either the expectedness or the likelihood on the catchall. That maneuver will, I hope, keep open the possibility of an objective basis for the evaluation of scientific hypotheses.

§6. Likelihoods

Equations (1), (3), and (4) are different forms of Bayes's theorem, and each of them contains a likelihood, $P(E|T.B)$, in the numerator. Two trivial cases can be noted at the outset. First, if the conjunction of theory T and background knowledge B are logically incompatible with evidence E, the likelihood equals zero, and the posterior probability, $P(T|E.B)$, automatically becomes zero.[27] Second, as we have already noticed, if T.B entails E, that likelihood equals one, and consequently drops out, as in equation (5).

Another easy case occurs when the hypothesis T involves various kinds of randomness assumptions, for example, the independence of a series of trials on a chance setup.[28] Consider, for example, the case of a coin that has been tossed 100 times, with the result that heads showed in 63 cases and tails in 37. We assume that the tosses are independent, but we are concerned whether the system consisting of the coin and tossing mechanism is biased. Calculation shows that the probability, given an unbiased coin and tossing mechanism, of the actual frequency of heads differing from 1/2 by 20 percent or more on 100 tosses (i.e., falling outside of the range 40 to 60) is about .05. Thus, the likelihood of the outcome on the hypothesis that the coin and mechanism are fair is less than .05. On the hypothesis that the coin has a 60 to 40 bias for heads, by contrast, the probability that the number of heads in 100 trials differs from 6/10 by less than 20 percent (i.e., lies within the 48 to 72 range) is well above .95. These are the kinds of likelihoods that would be used to compare the *null hypothesis* that the coin is fair with the hypothesis that it has a certain bias.[29] This example typifies a wide variety of cases, including the above-mentioned controlled experiment on rats and saccharin, in which statistical significance tests are applied. These yield a comparison between the probability of the observed result if the hypothesis is correct and the probability of the same result on a null hypothesis.

In still another kind of situation the likelihood $P(E|T.B)$ is straightforward. Consider, for example, the case in which a physician takes an X ray for diagnostic purposes. Let T be the hypothesis that the patient has a particular disease and let E be a certain appearance on the film. From long medical experience it may be known that E occurs in 90 percent of all cases in which that disease is present. In many cases, as this example suggests, there may be accumulated frequency data from which the value of $P(E|T.B)$ can be derived.

Unfortunately, life with likelihoods is not always as simple as the foregoing cases suggest. Consider an important case, which I will present in a highly unhistorical way. In comparing the Copernican and Ptolemaic cosmologies, it is easy to see that the phases of Venus are critical. According to the Copernican system, Venus should exhibit a broad set of phases from a narrow crescent to an almost full disk. According to the Ptolemaic system, Venus should always present nearly the same crescent-shaped appearance. One of Galileo's celebrated telescopic observations was of the phases of Venus. The likelihood of such evidence on the Copernican system is unity; on the Ptolemaic it is zero. This is the decisive sort of case that we cherish.

The Copernican system did, however, face one serious obstacle. On the Ptolemaic system, because the earth does not move, the fixed stars should not appear to change their positions. On the Copernican system, because the earth makes an annual trip around the sun, the fixed stars should appear to change their positions in the course of the year. The very best astronomical observations, including those of Tycho Brahe, failed to reveal any observable stellar parallax.[30] However, it was realized that, if the fixed stars are at a very great distance from the earth, stellar parallax, though real, would be too small to be observed. Consequently, the likelihood $P(E|T.B)$, where T is the Copernican system and E the absence of observable stellar parallax, is not zero. At the time of the scientific revolution, prior to the advent of Newtonian mechanics, there seemed no reasonable way to evaluate this likelihood. The assumption that the fixed stars are almost unimaginably distant from the earth was a highly ad hoc, and consequently implausible, auxiliary hypothesis to adopt just to save the Copernican system. Among other things, Christians did not like the idea that heaven was so very far away.

The most reasonable resolution of this anomaly was offered by Tycho Brahe, whose cosmology placed the earth at rest, with the sun and moon moving in orbits around the earth, but with all of the other planets moving in orbits around the sun. In this way both the observed phases of Venus and the absence of observable stellar parallax could be accomodated. Until Newton's dynamics came upon the scene, it seems to me, Tycho's system was clearly the best available theory.

In §2 I suggested that the following form of Bayes's theorem is the most appropriate for use in actual scientific cases in which more than one hypothesis is available for serious consideration:

$$P(T_i|B.E) = \frac{P(T_i|B) \times P(E|T_i.B)}{\sum_{j=1}^{k} [P(T_j|B) \times P(E|T_j.B)]} \ . \tag{4}$$

It certainly fits the foregoing example in which we compared the Ptolemaic, Copernican, and Tychonic systems. This equation involves a mutually exclusive and exhaustive set of hypotheses $T_1, \ldots, T_{k-1}, T_k$, where $T_1 - T_{k-1}$ are seriously entertained and T_k is the catchall. Thus, the scientist who wants to calculate the posterior probability of one particular hypothesis T_i on the basis of evidence E must ascertain likelihoods of three types: (1) the probability of evidence E given T_i, (2) the probability of that evidence on each of the other seriously considered alternatives T_j ($j \neq i, j \neq k$), and (3) the probability of that evidence on the catch-all T_k.

In considering the foregoing example, I suggested that, although likelihoods in the first two categories are sometimes straightforward, there are cases in which they turn out to be quite problematic. We shall look at more examples in which they present difficulties as our discussion proceeds. But the point to be emphasized right now is the utter intractability of the likelihood on the catchall. The reason for this difficulty is easy to see. Whereas the seriously considered candidates are bona fide hypotheses, the catchall is a hypothesis only in a Pickwickian sense. It refers to all of the hypotheses we are *not* taking seriously, including all those that have not been thought of as yet; indeed, the catchall is logically equivalent to their disjunction. These will often include brilliant discoveries in the future history of science that will eventually solve our most perplexing problems.

Among the hypotheses hidden in the catchall are some that, in conjunction with present available background information, entail the present evidence E. On such as-yet-undiscovered hypotheses the likelihood is one. Obviously, however, the fact that its probability on one particular hypothesis is unity does not entail anything about its probability on some disjunction containing that hypothesis as one of its disjuncts. These considerations suggest to me that the likelihood on the catchall is totally intractable. To try to evaluate the likelihood on the catchall involves, it seems to me, an attempt to guess the future history of science. That is something we cannot do with any reliability.

In any situation in which a small number of theories are competing for ascendency it is tempting, though quite illegitimate, simply to ignore the likelihood on the catchall. In the nineteenth century, for instance, scientists asked what the probability of a given phenomenon is on the wave theory of light and what it is on the corpuscular theory. They did not seriously consider its probability if neither of these theories is correct. Yet we see, from the various forms in which Bayes's theorem is written, that either the expectedness or the likelihood on the catchall is an indispensable ingredient. In the next section I shall offer a *legitimate* way of eliminating those probabilities from our consideration.

§7. Choosing Between Theories

Kuhn has often maintained that in actual science the problem is never to evaluate one particular hypothesis or theory in isolation; it is always a matter of choosing from among two or more viable alternatives. He has emphasized that an old theory is never completely abandoned unless there is currently available a rival to take its place. Given that circumstance, it is a matter of choosing between the old and the new. On this point I think that Kuhn is quite right, especially as regards reasonably mature sciences. And this insight provides a useful clue on how to use Bayes's theorem to explicate the logic of scientific confirmation.

Suppose that we are trying to choose between T_1 and T_2, where there may or may not be other serious alternatives in addition to the catchall. By letting i = 1 and i = 2, we can proceed to write equation (4) for each of these candidates. Noting that the denominators of the two are identical, we can form their ratio as follows:

$$\frac{P(T_1|E.B)}{P(T_2|E.B)} = \frac{P(T_1|B)\,P(E|T_1.B)}{P(T_2|B)\,P(E|T_2.B)}.\tag{6}$$

No reference to the catchall hypothesis appears in this equation. Since the catchall is not a bona fide hypothesis, it is not a contender, and we need not try to calculate its posterior probability. The use of equation (6) frees us from the need to deal either with the expectedness of E or with its probability on the catchall.

Equation (6) yields a relation that can be regarded as a *Bayesian algorithm for theory preference.* Suppose that, prior to the emergence of evidence E, you prefer T_1 to T_2; that is, $P(T_1|B) > P(T_2|B)$. Then E becomes available. You should change your preference in the light of E if and only if $P(T_2|E.B) > P(T_1|E.B)$. From (6) it follows that

$$P(T_2|E.B) > P(T_1|E.B) \text{ iff } P(E|T_2.B)/(E|T_1.B) > (T_1|B)/P(T_2|B).\tag{7}$$

In other words, you should change your preference to T_2 if the ratio of the likelihoods is greater than the reciprocal of the ratio of the respective prior probabilities. A corollary is that, if both $T_1.B$ and $T_2.B$ entail E, so that:

$$P(E|T_1.B) = P(E|T_2.B) = 1,$$

the occurrence of E can never change the preference rating between the two competing theories.

At the end of §4 I made reference to the well-known phenomenon of washing out of priors in connection with the use of Bayes's theorem. One might well ask what happens to this swamping when we switch from Bayes's theorem to the ratio embodied in equation (6).[31] The best answer, I believe, is this. If we are dealing with two hypotheses that are serious contenders in the sense that they do not differ too greatly in plausibility, the ratio of the priors will be of the order of unity. If, as the observational evidence accumulates, the likelihoods come to differ greatly, the ratio of the likelihoods will swamp the ratio of the priors. Recall the example of the tossed coin. Suppose we consider the prior probability of a fair device to be ten times as large as that of a biased device. If about the same proportion of heads occurs in 500 tosses as occurred in the aforementioned 100, the likelihood on the null hypothesis would be virtually zero and the likelihood on the hypothesis that the device has a bias approximating the observed frequency would be essentially indistinguishable from unity. The ratio of prior probabilities would obviously be completely dominated by the likelihood ratio.

§8. Plausible Scenarios

Although, by appealing to equation (6), we have eliminated the need to deal with the expectedness or the likelihood on the catchall, we cannot claim to have dealt adequately with the likelihoods on the hypotheses we are seriously considering, for their values are not always straightforwardly ascertainable. We have already mentioned one example, namely, the probability of absence of observable stellar parallax on the Copernican hypothesis. We noted that, by adding an auxiliary hypothesis to the effect that the fixed stars are located an enormous distance from the Earth, we could augment the Copernican hypothesis in such a way that the likelihood on this augmented hypothesis is one. But, for many reasons, this auxiliary assumption could hardly be considered plausible in that historical context. By now, of course, we have measured the parallax of relatively nearby stars, and from those values have calculated these distances. They are extremely far from us in comparison to the familiar objects in our solar system.

Consider another well-known example. During the seventeenth and eighteenth centuries the wave and corpuscular theories of light received considerable scientific attention. Each was able to explain certain important optical phenomena, and each faced fundamental difficulties. The corpuscular hypothesis easily explained how light could travel vast distances through empty space, and it readily explained sharp shadows. The theory of light as a longitudinal wave explained various kinds of diffraction phenomena, but failed to deal adequately with polarization. When, early in the nineteenth century, light was conceived as a transverse wave, the wave theory explained polarization as well as diffraction quite straightforwardly. And Huygens had long since shown how the wave theory could handle rectilinear propagation and sharp shadows. For most of the nineteenth century the wave theory dominated optics.

The proponent of the particle theory could still raise a serious objection. What is the likelihood of a wave propagating in empty space? Lacking a medium, the answer is zero. So wave theorists augmented their theory with the auxiliary assumption that all of space is filled with a peculiar substance known as the *luminiferous ether*. This substance was postulated to have precisely the properties required to transmit light waves.

The process I have been describing can appropriately be regarded as the discovery and introduction of *plausible scenarios*. A theory is confronted with an *anomaly*—a phenomenon that appears to have a small, possibly zero, likelihood given that theory. Proponents of the theory search for some auxiliary hypothesis that, if conjoined to the theory, renders the likelihood high, possibly unity. This move shifts the burden of the argument to the plausibility of the new auxiliary hypothesis. I mentioned two instances involved in the wave theory of light. The first was the auxiliary assumption that the wave is transverse. This modification of the theory was sufficiently plausible to be incorporated as an integral part of the theory. The second was the luminiferous ether. The plausibility of this auxiliary hypothesis was debated throughout the nineteenth, and into the twentieth, century. The ether had to be dense enough to transmit transverse waves (which require a denser medium than do longitudinal waves) and thin enough to allow astronomical bodies to move through it without noticeable diminution of speed. Attempts to detect the motion of the earth relative to the ether were unsuccessful. The Lorentz-Fitzgerald contraction hypothesis was an attempt to save the ether theory—that is, another attempt at a plausible scenario—but it was, of course, abandoned in favor of special relativity.

I am calling these auxiliaries *scenarios* because they are stories about how something could have happened, and *plausible* because they must have some degree of acceptability if they are to be of any help in handling problematic phenomena. The wave theory could handle the Poisson bright spot by deducing it from the theory. There seemed to be no plausible scenario available to the particle theory that could deal with this phenomenon. The same has been said with respect to Foucault's demonstration that the velocity of light is greater in air than it is in water.[32]

One nineteenth century optician of considerable importance who did not adopt the wave theory, but remained committed to the Newtonian emission theory, was David Brewster.[33] In a "Report on the Present State of Physical Optics," presented to the British Association for the Advancement of Science in 1831, he maintained that the undulatory theory is "still burthened with difficulties and cannot claim our implicit assent."[34] Brewster freely admitted the unparalleled explanatory and predictive success of the wave theory; nevertheless, he considered it false.

Among the difficulties Brewster found with the wave theory, two might be mentioned. First, he considered the wave theory implausible, for the reason that it required "an *ether* invisible, intangible, imponderable, inseparable from all bodies, and extending from our own eye to the remotest verge of the starry heavens."[35] History has certainly vindicated him on that issue. Second, he found the wave theory incapable of explaining a phenomenon that he had discovered himself, namely, *selective absorption*—dark lines in the spectrum of sunlight that has passed through certain gases. Brewster points out that a gas may be opaque to light of one particular index of refraction in flint glass, while transmitting freely light whose refractive indices in the same glass are only the tiniest bit higher or lower. Brewster maintained that there was no plausible scenario the wave theorists could devise that would explain why the ether permeating the gas transmits two waves of very nearly the same wave length, but does not transmit light of a very precise wave length lying in between:

> There is no fact analogous to this in the phenomena of sound, and I can form
> no conception of a simple elastic medium so modified by the particles of the

body which contains it, as to make such an extraordinary selection of the undulations which it stops or transmits. . . . [36]

Brewster never found a plausible scenario by means of which the Newtonian theory he favored could cope with absorption lines, nor could proponents of the wave theory find one to bolster their viewpoint. Dark absorption lines remained anomalous for both the wave and particle theories; neither could see a way to furnish them with high likelihood.

With hindsight we can say that the catchall hypothesis was looking very strong at this point. We recognize that the dark absorption lines in the spectrum of sunlight are closely related to the discrete lines in the emission spectra of gases, and that they, in turn, are intimately bound up with the problem of the stability of atoms. These phenomena played a major role in the overthrow of classical physics at the turn of the twentieth century.

I have introduced the notion of a plausible scenario to deal with problematic likelihoods. Likelihoods can cause trouble for a scientific theory for either of two reasons. First, if you have a pet theory that confers an extremely small—for all practical purposes zero—likelihood on some observed phenomenon, that is a problem for that favored theory. You try to come up with a plausible scenario according to which the likelihood will be larger—ideally, unity. Second, if there seems to be no way to evaluate the likelihood of a piece of evidence with respect to some hypothesis of interest, that is another sort of problem. In this case, we search for a plausible scenario that will make the likelihood manageable, whether this involves assigning it a high, medium, or low value.

What does this mean in terms of the Bayesian approach I am advocating? Let us return to:

$$\frac{P(T_1 \mid E.B)}{P(T_2 \mid E.B)} = \frac{P(T_1 \mid B) \ P(E \mid T_1.B)}{P(T_2 \mid B) \ P(E \mid T_2.B)} , \tag{6}$$

which contains two likelihoods. Suppose, as in nineteenth-century optics, that both likelihoods are problematic. As we have seen, we search for plausible scenarios A_1 and A_2 to augment T_1 and T_2 respectively. If the search has been successful, we can assess the likelihoods of E with respect to the augmented theories $A_1.T_1$ and $A_2.T_2$. Consequently, we can modify (6) so as to yield

$$\frac{P(A_1.T_1 \mid E.B)}{P(A_2.T_2 \mid E.B)} = \frac{P(A_1.T_1 \mid B) \ P(E \mid A_1.T_1.B)}{P(A_2.T_2 \mid B) \ P(E \mid A_2.T_2.B)} . \tag{8}$$

In order to use this equation to compare the posterior probabilities of the two augmented theories, we must assess the plausibilities of the scenarios, for the prior probabilities of both augmented theories—$A_1.T_1$ and $A_2.T_2$—appear in it. In §4 I tried to explain how prior probabilities can be handled—that is, how we can obtain at least rough estimates of their values. If, as suggested, the plausible scenarios have made the likelihoods ascertainable, then we can use them in conjunction with our determinations of the prior probabilities to assess the ratio of the posterior probabilities. We have, thereby, handled the central issue raised by Kuhn, namely, what is the basis for preference between two theories.[37] Equation (8) is a Bayesian algorithm.

If either augmented theory, in conjunction with background knowledge B, entails E, then the corresponding likelihood is one and it drops out of (8). If both likelihoods drop out we have the special case in which:

$$\frac{P(A_1.T_1\,|\,E.B)}{P(A_2.T_2\,|\,E.B)} = \frac{P(A_1.T_1\,|\,B)}{P(A_2.T_2\,|\,B)}\,, \tag{9}$$

thereby placing the *whole* burden on the prior probabilities — the plausibility considerations. Equation (9) represents a simplified Bayesian algorithm that is applicable in this type of special case.

Another type of special case was mentioned above. If, as in our coin tossing example, the values of the prior probabilities do not differ drastically from one another, but the likelihoods become widely divergent as the observational evidence accumulates, there will be a washing out of the priors. In this case, the ratio of the posterior probabilities equals, for practical purposes, the ratio of the likelihoods.

The use of either (8) or (9) as an algorithm for theory choice does not imply that all scientists will agree on the numerical values or prefer the same theory. The evaluation of prior probabilities clearly demands the kind of scientific judgment whose importance Kuhn has rightly insisted upon. It should also be clearly remembered that these formulas provide no evaluations of individual theories; they furnish only comparative evaluations. Thus, instead of yielding a prediction regarding the chances of one particular theory being a component of "completed science," they compare existing theories with regard to their present merits.

§9. Kuhn's Criteria

Early in this paper I quoted five criteria that Kuhn mentioned in connection with his views on the rationality and objectivity of science. The time has come to relate them explicitly to the Bayesian approach I have been attempting to elaborate. In order to appreciate the significance of these criteria it is important to distinguish three aspects of scientific theories that may be called *informational virtues, confirmational virtues,* and *economic virtues.* Up to this point we have concerned ourselves almost exclusively with confirmation, for our use of Bayes's theorem is germane only to the confirmational virtues. But since Kuhn's criteria patently refer to the other virtues as well, we must also say a little about them.

Consider, for example, the matter of *scope.* Newton's three laws of motion and his law of universal gravitation obviously have greater scope than the conjunction of Galileo's law of falling bodies and Kepler's three laws of planetary motion. This means, simply, that Newtonian mechanics contains more information than the laws of Kepler and Galileo taken together. Given a situation of this sort, we prefer the more informative theory because it is a basic goal of science to increase our knowledge as much as possible. We might, of course, hesitate to choose a highly informative theory if the evidence for it were extremely limited or shaky, because the desire to be right might overrule the desire to have more information content. But in the case at hand that consideration does not arise.

In spite of its intuitive attraction, however, the appeal to scope is not altogether unproblematic. There are two ways in which we might construe the Galileo-Kepler-Newton example of the preceding paragraph. First, we might ignore the small corrections mandated by Newton's theory in the laws of Galileo and Kepler. In that case we can clearly claim greater scope for Newton's laws than for the conjunction of Galileo's and Kepler's laws, since the latter is entailed by the former but not conversely. Where an entailment relation holds we can make good sense of comparative scope.

Kuhn, however, along with most of the historically oriented philosophers, has been at pains to deny that science progresses by finding more general theories that include earlier theories as special cases. Theory choice or preference involves *competing* theories that are *mutually incompatible* or *mutually incommensurable*. To the best of my knowledge Kuhn has not offered any precise characterization of scope; Karl Popper, in contrast, has made serious attempts to do so. In response to Popper's efforts, Adolf Grünbaum has effectively argued that none of the Popperian measures can be usefully applied to make comparisons of scope among mutually incompatible competing theories.[38] Consequently, the concept of scope requires fundamental clarification if we are to use it to understand preferences among competing theories. However, since scope refers to information rather than confirmation, it plays no role in the Bayesian program I have been endeavoring to explicate. We can thus put aside the problem of explicating that difficult concept.

Another of Kuhn's criteria is *accuracy*. It can, I think, be construed in two different ways. The first has to do with informational virtues; the second with economic. On the one hand, two theories might both make true predictions regarding the same phenomena, but one of them might give us precise predictions where the other gives only predictions that are less exact. If, for example, one theory enables us to predict that there will be a solar eclipse on a given day, and that its path of totality will cross North America, it may well be furnishing correct information about the eclipse. If another theory gives not only the day, but also the time, and not only the continent, but also the precise boundaries, the second provides much more information, at least with respect to this particular occurrence. It is not that either is incorrect; rather, the second yields more knowledge than the first. However, it should be clearly noted—as it was in the case of scope—that these theories are not incompatible or incommensurable competitors (at least with respect to this eclipse), and hence do not illustrate the interesting type of theory preference with which Kuhn is primarily concerned.

On the other hand, one theory may yield predictions that are nearly, but not quite, correct, while another theory yields predictions that are entirely correct—or, at least, more nearly correct. Newtonian astrophysics does well in ascertaining the orbit of the earth, but general relativity introduces a correction of 3.8 seconds of arc per century in the precession of its perihelion.[39] Although the Newtonian theory is literally false, it is used in contexts of this sort because its inaccuracy is small, and the economic gain involved in using it instead of general relativity (the saving in computational effort) is enormous.

The remaining three criteria are *simplicity, consistency,* and *fruitfulness;* all of them have direct bearing upon the confirmational virtues. In the treatment of prior probabilities in §4, I briefly mentioned simplicity as a factor having a significant bearing upon the plausibility of theories. More examples could be added, but I think the point is clear.

In the same section I also made passing reference to consistency, but more can profitably be said on that topic. Consistency has two aspects, internal consistency of a theory and its compatibility with other accepted theories. While scientists may be fully justified in *entertaining* collections of statements that contain contradictions, the goal of science is surely to accept only logically consistent theories.[40] The discovery of an internal inconsistency has a distinctly adverse effect on the prior probability of that theory, to wit, it must go straight to zero.

When we consider the relationships of a given theory to other accepted theories we again find two aspects. There are *deductive* relations of entailment and incompatibility, and there *inductive* relations of fittingness and incongruity. The deductive relations are quite straightforward. Incompatibility with an accepted theory makes for implausibility; being a logical consequence of an accepted theory makes for a high prior probability. Although deductive subsumption of narrower theories under broader theories is probably something of an oversimplification of actual cases, nevertheless, the ability of an overarching theory to deductively unify diverse domains furnishes a strong plausibility argument.

When it comes to the inductive relations among theories, analogy is, I think, the chief consideration. I have already mentioned the use of analogy in inductively transferring results of experiments from rats to humans. In archaeology, the method of ethnographic analogy, which exploits similarities between extant primitive societies and prehistoric societies, is widely used. In physics, the analogy between the inverse square law of electrostatics and the inverse square law of gravitation provides an example of an important plausibility consideration.

Kuhn's criteria of consistency (broadly construed) and simplicity seem clearly to pertain to assessments of the prior probabilities of theories. They cry out for a Bayesian interpretation.

The final criterion in Kuhn's list is *fruitfulness*; it has many aspects. Some theories prove fruitful by unifying a great many apparently different phenomena in terms of a few simple principles. The Newtonian synthesis is, perhaps, the outstanding example; Maxwellian electrodynamics is also an excellent case. As I suggested above, this ability to accommodate a wide variety of facts tends to enhance the prior probability of a given theory. To attribute diverse success to happenstance, rather than basic correctness, is implausible.

Another sort of fertility involves the predictability of theretofore unknown phenomena. We might mention as familiar illustrations the prediction of the Poisson bright spot by the wave theory of light and the prediction of time dilation by special relativity. These are the kinds of instances in which, in an important sense, the expectedness is low. As we have noted, a small expectedness tends to increase the posterior probability of a hypotheses.

A further type of fertility relates directly to plausible scenarios; a theory is fruitful in this way if it successfully copes with difficulties with the aid of suitable auxiliary assumptions. Newtonian mechanics again provides an excellent example. The perturbations of Uranus were explained by postulating Neptune. The perturbations of Neptune were explained by postulating Pluto.[41] The motions of stars within galaxies and of galaxies within clusters are explained in terms of *dark matter*, concerning which there are many current theories. A theory that readily gives rise to plausible scenarios to deal with problematic likelihoods can boast this sort of fertility.

The discussion of Kuhn's criteria in this section is intended to show how adequately they can be understood within a Bayesian framework — insofar as they are germane to confirmation. If it is sound, we have constructed a fairly substantial bridge connecting Kuhn's views on theory choice with those of the logical empiricists — at least, those who find in Bayes's theorem a suitable schema for characterizing the confirmation of hypotheses and theories.

§10. Rationality vs. Objectivity

In the title of this essay I have used both the concept of *rationality* and that of *objectivity*. It is time to say something about their relationship. Perhaps the best way to approach the distinction between them is to enumerate various grades of rationality. In a certain sense one can be rational without paying any heed at all to objectivity. It is essentially a matter of good housekeeping as far as one's beliefs and degrees of confidence are concerned. As Bayesians have often emphasized, it is important to avoid logical contradictions in one's beliefs and to avoid probabilistic incoherence in one's degrees of conviction. If contradiction or incoherence are discovered, they must somehow be eliminated; the presence of either constitutes a form of irrationality. But the removal of such elements of irrationality can be accomplished without any appeal to facts outside of the subject's corpus of beliefs and degrees of confidence. To achieve this sort of rationality is to achieve a minimal standard that I have elsewhere called *static* rationality.[42]

One way in which additional facts may enter the picture is via Bayes's theorem. We have a theory T in which we have a particular degree of confidence. A new piece of evidence turns up—some objective fact E of which we were previously unaware—and we use Bayes's theorem to calculate a posterior probability of T. To accept this value of the posterior probability as one's degree of confidence in T is known as *Bayesian conditionalization*. Use of Bayes's theorem does not, however, guarantee objectivity. If the resulting posterior probability of T is one we are not willing to accept, we can make adjustments elsewhere to avoid incoherence. After all, the prior probabilities and likelihoods are simply personal probabilities, so they can be adjusted to achieve the desired result. If, however, the requirement of *Bayesian conditionalization* is added to those of static rationality we have a stronger type of rationality that I have called *kinematic*. [43]

The highest grade of rationality—what I have called *dynamic rationality* — requires much fuller reference to objective fact than is demanded by advocates of personalism. The most obvious way to inject a substantial degree of objectivity into our deliberations regarding choices of scientific theories is to provide an objective interpretation of the probabilities in Bayes's theorem. Throughout this discussion I have adopted that approach as thoroughly as possible. For instance, I have argued that prior probabilities can be given an objective interpretation in terms of frequencies of success. I have tried to show how likelihoods could be objective—by virtue of entailment relations, tests of statistical significance, or observed frequencies. When the likelihoods created major difficulties, I appealed to plausible scenarios. The result was that an intractable likelihood could be exchanged for a tractable prior probability—namely, the prior probability of a theory in conjunction with an auxiliary assumption.

We noted that the denominators of the right-hand sides of the various versions of Bayes's theorem—equations (1), (3), and (4)—contain either an expectedness or a likelihood on the catchall. It seems to me futile to try to construe either of these probabilities objectively. Consequently, in §7 I introduced equation (6), which involves a ratio of two instances of Bayes's theorem, and from which the expectedness and the likelihood on the catchall drop out. Confining our attention, as Kuhn recommends, to comparing the merits of competing theories, rather than offering absolute evaluations of individual theories, we were able to eliminate the probabilities that most seriously defy objective interpretation.

§11. Conclusions

For many years I have been convinced that plausibility arguments in science have constituted a major stumbling block to an understanding of the logic of scientific inference. Kuhn was not alone, I believe, in recognizing that considerations of plausibility constitute an essential aspect of scientific reasoning, without seeing where they fit into the logic of science. If one sees confirmation solely in terms of the crude hypothetico-deductive method, there is no place for them. There is, consequently, an obvious incentive for relegating plausibility considerations to heuristics. If one accepts the traditional distinction between the *context of discovery* and the *context of justification,* it is tempting to place them in the former context. But Kuhn recognized, I think, that plausibility arguments enter into the justifications of choices of theories, with the result that he became skeptical of the value of that distinction. If, as I believe, plausibility considerations are simply evaluations of prior probabilities of hypotheses or theories, then it becomes apparent via Bayes's theorem that they play an indispensable role in the context of justification. We do not need to give up that important distinction.

At several places in this paper I have spoken of Bayesian algorithms, mainly because Kuhn introduced that notion into the discussion. I have claimed that such algorithms exist—and attempted to exhibit them—but I accord *very little* significance to that claim. The algorithms are trivial; what is important is the scientific judgment involved in assessing the probabilities that are fed into the equations. The algorithms give frameworks in terms of which to understand the role of the sort of judgment upon which Kuhn rightly placed great emphasis.

The history of science chronicles the successes and failures of attempts at scientific theorizing. If the Bayesian analysis I have been offering is at all sound, history of science—in addition to contemporary scientific experience, of course—provides a rich source of information relevant to the prior probabilities of the theories among which we are at present concerned to make objective and rational choices. This viewpoint captures, I believe, the point Kuhn made at the beginning of his first book:

> But an age as dominated by science as our own does need a perspective from which to examine the scientific beliefs which it takes so much for granted, and history provides one important source of such perspective. If we can discover the origins of some modern scientific concepts and the way in which they supplanted the concepts of an earlier age, we are more likely to evaluate intelligently their chances for survival.[44]

I suggested at the outset that an appeal to Bayesian principles could provide some aid in bridging the gap between Hempel's logical-empiricist approach and Kuhn's historical approach. I hope I have offered a convincing case. However that may be, there remain many unresolved issues. For instance, I have not even broached the problem of incommensurability of paradigms or theories. This is a major issue. For another example, I have assumed uncritically throughout the discussion that the various parties to disputes about theories share a common body B of background knowledge. It is by no means obvious that this is a tenable assumption. No doubt other points for controversy remain. I do not for a moment maintain that complete consensus would be in the offing even if both camps were to buy the Bayesian line I have been peddling. But I do hope that some areas of misunderstanding have been clarified.

Notes

1. Thomas S. Kuhn, *The Structure of Scientific Revolutions* (Chicago: University of Chicago Press, 1962; 2d ed., 1970). "Postscript—1969," added to the second edition, contains discussions of some of the major topics that are treated in the present essay.

2. Such philosophers are often characterized by their opponents as *logical positivists,* but this is an egregious historical inaccuracy. Although some of them had been members of or closely associated with the Vienna Circle in their earlier years, none of them retained the early positivistic commitment to phenomenalism and/or instrumentalism in their more mature writings. Reichenbach and Feigl, for example, were outspoken realists, and Carnap regarded physicalism as a tenable philosophical framework. Reichenbach never associated himself with positivism; indeed, he regarded his 1938 book, *Experience and Prediction* (Chicago: University of Chicago Press), as a refutation of logical positivism. I could go on and on . . .

3. "Symposium: The Philosophy of Carl G. Hempel," *Journal of Philosophy* LXXX, 10 (Oct., 1983): 555–72.

4. I had offered a similar suggestion in "Bayes's Theorem and the History of Science," in *Minnesota Studies in the Philosophy of Science,* vol. 5 *Historical and Philosophical Perspectives of Science,* ed. Roger H. Stuewer (Minneapolis: University of Minnesota Press, 1970), 68–86.

5. Thomas S. Kuhn, *The Essential Tension* (Chicago: University of Chicago Press, 1977), 320–39. The response is given in greater detail in this article than it is in the Postscript.

6. "Objectivity, Value Judgment, and Theory Choice," 321–22.

7. Ibid., p. 322.

8. Throughout this paper I shall use the terms "hypothesis" and "theory" more or less interchangeably. Kuhn tends to prefer "theory," while I tend to prefer "hypothesis," but nothing of importance hinges on this usage here.

9. As Adolf Grünbaum pointed out to me, if we assume that in a given day the actual frequency of defective can openers produced by the two machines matches precisely the respective probabilities, we can calculate the result as follows. The new machine produces 50 defective can openers and the old machine produces 30, so that 50 out of a total of 80 are produced by the new machine. However, it would be *incorrect* to assume that the frequencies match the probabilities each day; in fact, the probability of an exact match is quite small.

10. "Objectivity, Value Judgment, and Theory Choice," 328.

11. I remarked above that three probabilities are required to calculate the posterior probability—a prior probability and two likelihoods. Obviously, in view of (2), the theorem on total probability, if we have a prior probability, one of the likelihoods, and the expectedness, we can compute the other likelihood; likewise, if we have one prior probability and both likelihoods, we can compute the expectedness.

12. *The Foundations of Scientific Inference* (Pittsburgh: University of Pittsburgh Press, 1967), chap. 7.

13. A set of degrees of conviction is coherent provided that its members do not violate any of the conditions embodied in the mathematical calculus of probability.

14. From "Autobiographical Notes," in Paul A. Schilpp, ed., *Albert Einstein: Philosopher-Scientist* (Evanston, Ill: Library of Living Philosophers, 1949), 21–22.

15. This example is discussed in Ronald N. Giere, *Understanding Scientific Reasoning,* 2d ed. (New York: Holt, Rinehart, and Winston, 1984), 274–76.

16. I reject the so-called propensity interpretation of probability because, as Paul Humphreys

pointed out, the probability calculus accommodates inverse probabilities of the type that occur in Bayes's theorem, but the corresponding inverse propensities do not exist. In the example of the can opener factory, each machine has a certain propensity to produce defective can openers, but it does not make sense to speak of the propensity of a given defective can opener to have been produced by the new machine.

17. A so-called Dutch book is a combination of bets such that, no matter what the outcome of the event upon which the wagers are made, the subject is bound to suffer a net loss.

18. "Truth and Probability," in Frank Plumpton Ramsey, *The Foundations of Mathematics* ed. R. B. Braithwaite (New York: Humanities Press, 1950).

19. Gerald Holton and Stephen G. Brush, *Introduction to Concepts and Theories in Physical Science,* 2d ed. (Reading, Mass.: Addison-Wesley, 1973), 416, italics in original.

20. Abner Shimony, "Scientific Inference," in Robert G. Colodny, ed., *The Nature and Function of Scientific Theories* (Pittsburgh: University of Pittsburgh Press, 1970), 79–172.

21. Martin Gardner, *Fads and Fallacies in the Name of Science* (New York: Dover Publications, 1957), 7–15.

22. Doubleday & Company, 1950.

23. Haim Harari, "The Structure of Quarks and Leptons," *Scientific American* 248, (April 1983): 56–68.

24. *Scientific American,* 253 (Sept. 1985): 116. Adolf Grünbaum, *The Foundations of Psychoanalysis* (Berkeley/Los Angeles/London: University of Califronia Press, 1984), 202–4, criticizes Motley's account of Freud's theory; he considers Motley's version a distortion, and points out that Freud's motivational explanations were explicitly confined to a very circumscribed set of slips. He defends Freud against Motley's criticism on the grounds that Freud's actual account has greater complexity than Motley gives it credit for.

25."A Bayesian Approach to the Paradoxes of Confirmation," in Jaakko Hintikka and Patrick Suppes, eds., *Aspects of Inductive Logic* (Amsterdam: North Holland, 1966), 202–3.

26. I have discussed the relations between personal probabilities and objective probabilities in "Dynamic Rationality: Propensity, Probability, and Credence," in James H. Fetzer, ed., *Probability and Causality* (Dordrecht: Reidel 1988), 3–40.

27. As Duhem has made abundantly clear, in such cases we may be led to reexamine our background knowledge B, which normally involves auxiliary hypotheses, to see whether it remains acceptable in the light of the negative outcome E. Consequently, refutation of T is not usually as automatic as it appears in the simplified account just given. Nevertheless, the probability relation just stated is correct.

28. Exchangeability is the personalist's surrogate for randomness; it means that the subject would draw the same conclusion regardless of the order in which the members of an observed sample occurred.

29. Note that, in order to get the posterior probability—the probability that the observed results were produced by a biased device—the prior probabilities have to be taken into account.

30. Indeed, stellar parallax was not detected until the nineteenth century.

31. This question was, in fact, raised by Adolf Grünbaum in a private communication.

32. See, for example, Gerald Holton and Stephen G. Brush, *Introduction to Concepts and Theories in Physical Science,* 2d ed. (Reading, Mass.: Addison-Wesley, 1973), 392–93.

33. An excellent account of Brewster's position can be found in John Worrall, "Scientific Revolutions and Scientific Rationality: The Case of the Elderly Holdout," this volume, pp. 319–36.

34. Quoted by Worrall, p. 321.

35. Quoted by Worral, p. 322.

36. Quoted by Worral, p. 323.

37. If more than two theories are serious candidates, the pairwise comparison can be repeated as many times as necessary.

38. Adolf Grünbaum, "Can a Theory Answer More Questions Than One of Its Rivals," *British Journal for the Philosophy of Science* 27 (1976): 1-23.

39. Steven Weinberg, *Gravitation and Cosmology* (New York: John Wiley & Sons, 1972), 198. Note that this correction is smaller by an order of magnitude than the correction of 43 seconds of arc per century for Mercury.

40. See Joel Smith, *The Status of Inconsistent Statements in Scientific Inquiry* (doctoral dissertation, University of Pittsburgh, 1987).

41. Unfortunately, recent evidence regarding the mass of Pluto strongly suggests that Pluto is not sufficiently massive to explain the perturbations of Neptune. A different plausible scenario is needed, but I do not know of any serious candidates that have been offered.

42. See "Dynamic Rationality" (note 26 above), 5-12, for a more detailed discussion of various grades of rationality. The term "static" was chosen to indicate the lack of any principled method for changing personal probabilities in the face of inconsistency or incoherence.

43. Ibid., esp. pp. 11-12

44. Thomas S. Kuhn, *The Copernican Revolution* (Cambridge: Harvard University Press, 1957), 3-4.

Bayesian Problems of Old Evidence

1. Introduction

According to "standard" or "classical" Bayesian confirmation theory, a piece of evidence E confirms a theory or hypothesis T, for a given person, if and only if $Pr(T/E) > Pr(T)$, where Pr is the relevant person's subjective (personal) probability function, representing this individual's degrees of belief. $Pr(T/E)$ is the individual's probability for T conditional on E, $Pr(T\&E) / Pr(E)$, which Bayesians argue is equal to the degree of confidence the individual should have in T if, and after, E is learned. In defenses and further "articulations" of this theory, Bayesians have assumed that these "rational degrees of belief" satisfy the standard probability axioms, where the appropriateness of this assumption has been defended in various ways, including "Dutch book" and decision theoretical approaches.[1]

Clark Glymour (1980, 85–93) has recently raised an interesting problem for this theory. It is not uncommon for scientists to find support for a theory in evidence known long before the theory was even introduced, so that, intuitively, there are cases of already known, or "old," evidence confirming "new" theories or hypotheses. Glymour cites the examples of the support for Copernicus' theory derived from previous astronomical observations, the support for Newton's theory of gravitation derived from the already established second and third laws of Kepler, and the support for Einstein's gravitational field equations derived from the already known anomalous advance of the perihelion of Mercury. But if evidence E is already known before theory or hypothesis T is invented, then $Pr(E)$ already equals 1 at that later time, so that, at that later time, $Pr(T/E)$ *must equal*

I thank the American Council of Learned Societies for financial support during the time most of this paper was written. And I thank the students in my seminar on confirmation theory at the University of Wisconsin-Madison during the fall semester of 1984 – especially Martin Barrett and Mark Bauder, as well as my colleague Mike Byrd, who also attended – for some very good discussions in the meetings on the old evidence problem. And I thank Elliott Sober for helpful comments on an earlier draft.

$Pr(T)$; this follows from the usual axioms of probability and the definition of $Pr(T / E)$. Thus, Bayesian confirmation theory seems to imply that already known evidence *cannot* support newly invented theories, contrary to what seems true in the cases Glymour cites.

After some further clarification of this problem for Bayesian confirmation theory in section 2 below — which will help isolate the version of the problem of old evidence that poses the most potent threat to the theory — I shall in section 3 consider the principal line of defense that has been offered in support of Bayesian confirmation theory in light of the problem. I believe that the general idea embodied in such responses is sound. Roughly, the strategy is: (1) to suppose, contrary to standard or classical Bayesianism, that rational scientists are *not* "logically omniscient" at all times over the set of propositions entertained (as explained below), and then, (2) to argue that, in cases of the kind Glymour cites, there really *is* new evidence, namely, the discovery of some *logical relation* between E and T (which relation might suggest, for example, that the truth of T would *explain E*). In sections 4 and 5, however, I will argue that the ways in which the two components of this strategy have been developed are not adequate, and I will suggest more promising ways of developing them.

2. Clarification of the Problem

Based on a suggestion by Brian Skyrms, Daniel Garber (1983) has suggested that there is an ambiguity in Glymour's problem, and Garber divides the problem into two separate problems, which he calls the "historical" and "ahistorical" problems of old evidence. Garber argues (quickly) that his ahistorical problem can be solved by means of some variant of what he calls a "counterfactual strategy." This strategy has also been discussed, and criticized, by Glymour (1980); the strategy and some of its difficulties will be briefly described later in this section. Since I wish to avoid this strategy, I shall here segment the problem somewhat differently, in a way that I think will allow for more plausible "quick" resolutions of several of its variants. Below I'll also describe the relation between my way of dividing up the problem and Garber's.

In all versions of the problem to be described, with two exceptions to be noted later, I shall assume that the correct assessment of the relation between the evidence E and theory or hypothesis T in question is that E is in fact positively evidentially relevant to T. (I hope that the rationale for the perhaps somewhat awkward terminology I shall employ for labeling the three main problems I shall distinguish will become evident as the discussion unfolds; also, see the outline below.) What I shall call "The Problem Of *Old New* Evidence" arises in cases in which one *first* formulates a theory or hypothesis T and *subsequently* discovers evidence E, which is thus "new" in relation to the time of the formulation of T. In cases of the problem of this first kind, at the point in time at which E is learned,

E in fact *does* increase one's confidence in *T*. However, later on, when *E* becomes "old" in relation to the time of its discovery, *E* can no longer thus increase one's confidence in *T*. Nevertheless, there seems to be a valid sense in which *E* is *still* good evidence for *T*. So, at such later times, *E* would, in this sense, seem to confirm *T*—it is in this sense still good evidence for *T*—even though $Pr(T \mid E)$ = $Pr(T)$ so that the Bayesian theory says that *E* does *not* confirm *T*.

What I shall call (henceforth) simply "The Problem Of *Old* Evidence" arises in cases in which *E* is learned first, and *T* is formulated subsequent to the learning of *E*. This problem can be subdivided in turn. "The Problem Of *Old Old* Evidence" arises at times *after* the formulation of *T*: if *E* somehow confirmed *T* at or around the time of the formulation of *T* (even if *E* had probability 1 at or around the time of the formulation of *T*), then it would seem that, even later on, *E* is still, in a valid sense, good evidence for *T*. And "The Problem Of *New Old* Evidence" concerns how, at the time of the formulation of *T*, *E* can confirm *T* even though *E* already had subjective probability 1. Finally, it will be helpful to divide each of these latter two problems into two cases. In Case 1, *T* was specifically designed to explain *E*; and in Case 2, it was not.

For easier reference, here in outline form are the versions of Glymour's problem to be discussed:[2]

I.   The Problem of *Old New* Evidence: *T* formulated before the discovery of *E*; but it is now later and $Pr(E)$ = 1, so that $Pr(T \mid E)$ = $Pr(T)$

II.   The Problem of *Old* Evidence: *E* known before the formulation of *T*
    A.   The Problem of *Old Old* Evidence: It is now some time subsequent to the formulation of *T*.
        1.   *T* originally designed to explain *E*
        2.   *T* not originally designed to explain *E*
    B.   The Problem of *New Old* Evidence: It is now the time (or barely after it) of the formulation of *T*
        1.   *T* originally designed to explain *E*
        2.   *T* not originally designed to explain *E*

Garber's ahistorical problem of old evidence, as I understand it, arises in cases I and IIA above, and the historical problem arises in case IIB. Roughly, the counterfactual strategy endorsed by Garber is to argue that, in cases I and IIA, if, now (sometime after both the discovery of *E* and the formulation of *T*), we *had not* known *E* (e.g., if our education in the history of science had been incomplete), then the probability $Pr(T \mid E)$ *would have been* greater than the probability $Pr(T)$. Thus, on the modification of Bayesian confirmation theory suggested by Garber, appeal is made to *counterfactual degrees of belief*.

Garber admits that there are "some details to be worked out here" (1983, 103), and, as noted above, Glymour (1980) has criticized the strategy. For example, Glymour and Garber both point out that there will not necessarily be any *particu-*

lar degrees of belief that we can say a person would have had in *E*, or in *T* given *E*, if this person's degree of belief in *E* had been less than 1. Indeed, it seems plausible that in some cases *T would not even have been formulated* had *E* not been learned. And surely there also will be cases in which the person's knowledge of *E saved his life* at some time in the past, so that had the individual's degree of belief in *E* been less than 1, the person would be *dead* now. Also, there are of course the well-known difficulties attending the proper interpretation of counterfactual conditionals that would befall any such modification of Bayesian confirmation theory.

Brian Skyrms (1983) has discussed ways of "giving a probability assignment a memory," so that one may, in one way or another, retain information about one's old *actual* degrees of belief in *E*, and in *T* given *E*. This kind of approach is more promising, I think, and it is closer to the approach I shall describe below. To apply this kind of approach, however, it is necessary first to divide Garber's ahistorical problem into problems I and IIA above, for such an approach works only for the first of these.

The problem of old new evidence (problem I) can be handled quite plausibly, I think, as follows. One of the central tenets of Bayesian confirmation theory is that confirmation is a relation between three things: a piece of evidence, a hypothesis or theory, *and a set of background beliefs*. As background beliefs change over time (as well as from person to person), *so does what confirms what* – where, of course, our background beliefs, of various degrees, are given by our subjective probability assignment. At the time (or just before) *E* was learned, *E* was not one of our background beliefs of degree 1. At that time – that is, relative to the set of background beliefs that includes *E* only to some intermediate degree – our degree of belief in *T* is less than our degree of belief in *T* conditional on *E*, so that at that time, *E* does confirm *T* according to the Bayesian theory. However, after *E* is learned, and we face problem I above, *our background beliefs have changed*. At *this* time – that is, relative to the new set of background beliefs – *E* (or being again told that *E*, or pondering the discovery of *E*) can no longer increase our degree of confidence in *T*; although it did so once, it cannot do so again. I think it is quite natural to say simply that, because of the change in our background beliefs, *E* simply does not confirm *T* at the later time, after its evidential impact on *T* has already been "absorbed." The Bayesian theory of confirmation was *designed* to reflect this possibility: what confirms what, depends on one's background beliefs. What now remains to be explicated in Bayesian terms, however, is that sense in which *E* may *remain good evidence* for *T*.

I think it is good idea to distinguish *E*'s *actually confirming T* from *E*'s *being (actual) evidence in favor of T*, as follows. Recall that Bayesian *confirmation* is a relation that obtains between evidence *E*, theory or hypothesis *T*, and degrees of belief *Pr* if and only if $Pr(T/E) > Pr(T)$. Whether or not this relation obtains is independent of whether or not *T* ever actually gets confirmed: the relation may obtain, or not, independently of whether or not E will ever get discovered. But we may say that *actual confirmation* is an *event*, involving E, T, and a person with a subjective probability assignment Pr, that takes place when the three-part relation $Pr(T / E) > Pr(T)$ obtains and E is learned. In cases of the problem of old new evidence, such an event happened in the past, once, and it cannot recur (with the same piece of evidence and the same ideal Bayesian agent). As to the second idea, it seems appropriate to say that E *is (actual) evidence for* T, for a given individual, if, at some time in the past, the event of its confirming T, for that individual, took place. Indeed, this seems to be exactly what it means for E to constitute part of our current body of evidence for T: at some time in the past E raised our rational degree of confidence in T. In cases of the problem of old new evidence, therefore, it is clearly consistent and appropriate to say *both* that E confirmed T but no longer does *and* that E is now, but before its discovery was not, part of our body of evidence in favor of T.

To be more precise about the idea of E's being evidence for T, we have to say that it is a relation that may obtain among *four* things: (1) a piece of evidence, (2) a hypothesis or theory, (3) a *time*, and (4) a *history of a set of background beliefs (i.e., a sequence of subjective probability functions indexed by times)*. Roughly: *E is, at time t, part of the body of evidence in favor of theory T relative to history H of background beliefs* if and only if, at some time prior to t in the history H (of a set of background beliefs), the confirmation event took place between E, T, and the state of H (the relevant probability assignment) at that earlier time. Relativity to a history of a set of background beliefs is essential. For it is not difficult to invent cases (or find actual cases) in which, relative to one history, E is now evidence for T, but, relative to an alternative history, E is now evidence *against* T (i.e., *for* $\sim T$). For example, if initially $Pr(T) = 0.5$, $Pr(T / E) = 0.7$, $Pr(T / F) = 0.9$, and $Pr(T / E\&F) = 0.8$, then whether E is evidence for or against T will depend on whether E is learned before or after F. Thus the relativity of *confirmation* to a set of background beliefs shows up in this Bayesian conception of *evidence*. There is no univocal matter of fact about whether or not an E confirms a T or about whether a known E is evidence for or against a T: the first depends on a set of background beliefs, and the second depends also on the history of a set of background beliefs. On this conception of confirmation and evidence, whether E confirms, disconfirms, or is neutral for T, and whether a known E is evidence for, against, or neutral towards T, both involve an element of historical accident pertaining to what our background beliefs, and their history, happen to be.

It might be objected that it is inappropriate to let whether or not E is part of our body of evidence in favor of T depend on *when* it was learned in relation to other evidence (as in the numerical example above). I am not entirely sympathetic (or entirely unsympathetic) with such an objection, but in any case there is a natural framework in which it can be accommodated (suggested to me by Brian

Skyrms in correspondence). Suppose we begin with an initial probability assignment Pr_0, and as evidence $E_1, E_2, \ldots, E_n$ (n = "now") comes in, we update: $Pr_{i+1}(-) = Pr_i(-/E_{i+1})$. Then it is natural to say that E ($= E_i$, say) is now part of our body of evidence in favor of T if $Pr_n(T) = Pr_0(T / E_1 \& \ldots \& E_n) > Pr_0(T / E_1 \& \ldots \& E_{i-1} \& E_{i+1} \& \ldots \& E_n)$. On this explication, whether or not E is part of our body of evidence in favor of T does not depend on the order in which the evidence comes in: if we permute the subscripts on the E_is, the verdict remains the same. Note also that the "prior" used need not be actual.

As to the problem of *old* evidence (problem II above), I think we can quickly dispense with versions A1 and B1. If our knowledge of E inspired and guided our formulation of theory T, where the intention was to give an explanation of E, then it would seem that E does not confirm T in the first place, and E does not constitute evidence in favor of T. In this case, the Bayesian theory gives the right answer: for both IIA1 and IIB1, we have $Pr(T / E) = Pr(T)$. (Compare Garber 1983, 104.) This is not to say that T must be completely *without support*, however. T may derive support from the fact (if it is a fact) that it has some of the virtues usually thought to attach to good theories, such as simplicity, analogy with other well-confirmed theories, independent evidence, and the very fact (if it is a fact) that its truth *would* successfully explain E, and so on. Indeed, it would seem that if T had none of these virtues, then it *would* be entirely without support, despite the truth of E; therefore, if T does have support, it is not from E itself.

This leaves problems IIA2 and IIB2. But if problem IIB2 can be solved, so can IIA2, by employing the very same considerations employed in resolving problem I earlier. If it can be shown, in resolving IIB2, that *at the time of the formulation of T*, T gets confirmation, in the Bayesian sense, in virtue of some aspect of its relation to E, then, for problem IIA2, we should say that whatever exactly it was that confirmed T at the earlier time *no longer does so* at the later time, even though it will still be, in the sense clarified earlier, a part of our body of *evidence* in favor of T. Let us thus turn to problem IIB2.

3. The Basic Bayesian Defense

If, in cases of problem IIB2, T *does* receive confirmation in the Bayesian sense explained above, then, *when* it does, there must be some proposition F such that (1) one's subjective probability of F increases from some value short of 1 to 1, and (2) the prior probability of T is less than the posterior probability of T (i.e., the prior probability of T conditional on F). For reasons already explained, F cannot be the same as E. Daniel Garber (1983) and Richard Jeffrey (1983) have recently sought to show how some proposition other than E might plausibly play the role of F.

The *total* strategy in these defenses involves basically three steps; and the two authors concentrate on different steps of the total strategy. First, it is argued that classical Bayesianism's assumption of "logical omniscience" is clearly unrealistic, and this first step includes some "nonclassical" formulation of Bayesianism that is intended to be more realistic yet at the same time appropriately logically restrictive on rational agents' degrees of belief. Classical Bayesianism's assumption of logical omniscience may be formulated as follows: All logical truths have subjective probability 1; and if propositions A and B are logically incompatible, then $Pr(A \vee B) = Pr(A) + Pr(B)$. These two conditions, together with the condition that all subjective probabilities are greater than or equal to 0, are just the usual axioms of subjective probability theory. Of course, given the "richness" of the languages we use, it is very unrealistic to suppose that any rational scientist's degrees of belief will be sensitive to *all* the logical facts encompassed by these axioms. So it is desirable, even independently of the problem Glymour has raised for Bayesianism, to formulate weaker versions of the usual axioms of subjective probability.

The second step of the total strategy is to describe a *logical relation* that holds between E and T — or between E, T, and others of one's beliefs — whose obtaining perhaps suggests that the truth of T (or the truth of T in the presence of one's background beliefs) would explain E. It is then argued, in the third step, that it is not E itself that confirms T in cases of problem IIB2, but rather the discovery of this logical relation between E and T. That the obtaining of this relation *can be discovered*, and need not have been known all along, should now be a possibility in light of the successful completion of step one of the strategy. Thus, where "$T \vdash E$" expresses the proposition that E and T are so logically related to each other (and perhaps also to one's background beliefs), step one of the strategy makes it theoretically possible that $Pr(T / T \vdash E) > Pr(T)$. It is the task of step three to argue that this inequality *should* indeed obtain.

Roughly speaking, Garber focuses on step one, without providing much in the way of argument for step three, while Jeffrey concentrates on step three; for various reasons (good ones I think, as explained below), neither provides much detail in the way of carrying out step two.

As to step one, Garber advances a theory of what might be called "limited logical omniscience." He begins with a truth-functional language L, with atomic sentences a_i, and builds language L^* from L by adding new atomic sentences $A \vdash B$, where A and B are any sentences of L (i.e., truth-functional compounds of the a_is). Treating sentences $A \vdash B$ as atomic has the effect of making each of them *formally*, in L^*, logically independent of all the other atomic sentences: L^*-atomic sentences neither L^*-imply nor are L^*-implied by other L^*-atomic sentences. Thus, there are no axiomatic constraints on what subjective probabilities (between 0 and 1) may be assigned to atomic sentences: any sprinkling of numbers between 0 and 1 (inclusive) on the atomic sentences is allowed.

"Extrasystematically," however, we will want to understand "$\vdash$" as meaning, say, *implies*, or *explains*, and this makes it desirable to put some sort of formal constraint on the relation. Recall that step two of the basic Bayesian response to

the problem is to identify the appropriate logical relation that is discovered between T and E; "$\vdash$" is intended to be interpreted as denoting that appropriate relation. Let us thus briefly digress from step one and consider Garber's treatment
of step two. Garber doesn't actually insist on any particular interpretation of "$\vdash$".
He states:

> Depending on the context of investigation, "$\vdash$" may be understood as truth-
> functional implication, or implication in . . . the global language of science.
> We can even read "$h_i \vdash e_i$" as "e_i is a positive instance of h_i," or as "e_i bootstrap
> confirms h_i with respect to some appropriate theory," as Glymour demands.
> (1983, 112)

I agree with Garber that it need not be part of an adequate solution to the problem
Glymour has raised that a particular interpretation of "$\vdash$" be specified, i.e., that
a particular logical relation between T and E be described. For, as Garber suggests, different interpretations of "$\vdash$" may be appropriate in different particular,
actual cases of problem IIB2. What relation will be appropriate depends, as
Garber suggests, on the context of investigation. Indeed, in different actual investigations, different relations (of the appropriate kind) between an E and a T *will
be* discovered, assuming that Garber's general approach to problem IIB2 parallels
what transpires in actual cases of IIB2. Of course, the Bayesian defense would
be strengthened if analyses of actual, historical cases of problem IIB2 could be
given, supplying particular appropriate interpretations of "$\vdash$". For the most part,
in any case, both Garber and Jeffrey use the interpretation of "$\vdash$" as "logically
implies" to guide some of their intuitions and support various moves in their analyses; hence, one would suppose, the symbolism.

In any case, some formal constraint or other on $\vdash$ would seem to be desirable,
if plausible, given the more or less vague intended interpretation. Garber
assumes:

$$(K^*) \quad Pr(A \ \& \ B \ \& \ A \vdash B) = Pr(A \ \& \ A \vdash B).$$

This principle guarantees that if a person assigns subjective probability 1 to A and
to $A \vdash B$, then this individual will also assign probability 1 to B. Thus, $\vdash$ will
behave somewhat like implication in classical Bayesianism, although, as Garber
points out, K^* by itself doesn't rule out its being interpreted as conjunction, for
example, or biconditionalization.

Garber distinguishes what he calls "global Bayesianism" from what he calls
"local Bayesianism," and the distinction shows up when we look, *extrasystematically*, into the structures of the L^*-atomic sentences a_i and $A \vdash B$. As Garber explains, while these are atomic sentences of L^*, they may, from a broader perspective, have complex logical structure: truth-functional structure, quantificational
structure, modal logical structure, and so on—including of course a special interpretation of "$\vdash$" in the case of the $A \vdash B$s. In virtue of the internal structures

of the L^*-atomic sentences, some of them may be logically true from such a broader perspective. Global Bayesianism includes the thesis that those that are logically true, whether from the broad perspective or just from the point of view of L^*, must be assigned subjective probability 1. Local Bayesianism requires only that those that are logically true from the perspective of L^* − *i.e., those that are tautological truth-functional compounds of the a_is and the $A \vdash Bs$* − be assigned subjective probability 1. (A similar argument applies to extrasystematically logically incompatible L^*-atomic sentences.)

Garber proposes that local Bayesianism can handle problem IIB2, since although it may be *true* − even logically true − that $T \vdash E$ (when "$\vdash$" is interpreted), it is nevertheless allowable that $Pr(T \vdash E) < 1$, so that it may also be the case that $Pr(T/T \vdash E) > Pr(T)$. In cases of problem IIB2, according to Garber, it is $T \vdash E$, and not E, that actually confirms T; and although E is old, $T \vdash E$ may not be old, for "local Bayesians."

I agree that Garber's defense does all that needs to be done in the way of carrying out step two of the strategy, as explained above. However, there are serious problems with his approach to step one, as I shall argue in the next section. In addition to taking a more critical look at Garber's theory of local Bayesianism, I shall also in the next section examine the approach to a more realistic Bayesianism offered by Hacking (1967). I then will try to characterize a more adequate kind of approach to step one (without, however, attempting actually to carry out such an approach). Finally, in section 5, I shall turn to step three of the strategy.

4. Bayesianism and Logical Fallibility

Although I sympathize with the idea that axioms of subjective probability theory should be weakened to allow for the failure of logical omniscience in rational individuals, I think Garber's approach does not go nearly far enough in the way of allowing logical fallibility. To extend Garber's analogy, his method is to draw a "line" demarcating the "local" from the "nonlocal," and then to insist on logical omniscience only on the local side of the line. The line Garber draws (as an example of the approach at least) is, so to speak, between truth-functional logic and logic that attends to more features of the logical form of statements than truth-functional logic does. That is, Garber's approach requires rational locally Bayesian agents to assign probability 1 to all *tautologies* (of L^*), and to recognize all cases of pairs of sentences that are logically incompatible in virtue of their (L^*) truth-functional logical structure (in the sense that the subjective probability of their disjunction will in each such case equal the sum of their subjective probabilities). The same need not hold for cases of sentences (namely, some "atomic" sentences of L^*) that are logically true in virtue of non-truth-functional (or even perhaps, as far as L^* is concerned, "extrasystematic" *truth-functional*) features of their form, or for pairs of sentences that are logically incompatible in virtue of (*L**) non-truth-functional features of their forms.

But this seems to be an inappropriate place to draw the relevant "line." For there are extremely complex tautologies (of *L**), so complex that it would be more difficult to recognize them as logically true than to recognize as logically true certain simple sentences that are logically true in virtue of their (say) quantificational logical form. Simple sentences of the form "For all *x*, if *Fx* then *Fx*," would be atomic in *L**. And it seems completely inappropriate not to require an agent to assign probability 1 to such sentences, while at the same time insisting that the agent assign probability 1 to arbitrarily complex tautologies of *L**.

Of course the choice of making *L* and *L** *truth-functional* languages is just an example. They could instead be first-order languages, where sentences containing modal logical structure, second-order quantifiers, and so on, would be considered atomic. However, the same objection would apply to any such proposal: There will always be extremely complex logically true sentences of the local language, and extremely simple logically true sentences "outside" the local language, where it will be inappropriate to insist on probability 1 for the former while not so insisting in the case of the latter. Thus, in one way, Garber's theory of local Bayesianism requires too much, and in another, too little.[3]

Of course there may be some atomic sentences a_i of *L** that, from an extrasystematic point of view, are extremely *truth-functionally* complex. Some of these may be tautologies from the extrasystematic point of view. Local Bayesianism does *not* require subjective probability 1 for them. But this seems arbitrary in view of the fact that local Bayesianism *does* require subjective probability 1 for tautologies of *L** that have *exactly the same truth-functional form* from the point of view of *L** as extrasystematically extremely complex tautological *L**-atomic sentences have from the extrasystematic point of view. That is, an *L**-atomic sentence a_i may be an extremely complex tautology from an extrasystematic point of view, and have the same complex form from that point of view as a sentence *A* has from the point of view of *L**. Yet local Bayesianism requires subjective probability 1 for *A* and allows any subjective probability for a_i.

Suppose some evidence statement *E* and some theory *T* are, extrasystematically, quite complex, but complex *only with respect to the kind of logical form to which the local language is sensitive.* As an example, let's say *E* and *T* are quite complex *truth-functionally*, from the extrasystematic point of view. Suppose also that *T* truth-functionally implies *E* from the extrasystematic point of view. From the point of view of the local language, however, *E*, *T*, and *T⊢E* are all atomic. Suppose that we are in a case of problem IIB2, so that $Pr(E) = 1$. Now it must be true that $Pr(T⊢E) < 1$ if Garber's approach is to work. But it seems quite arbitrary to think that $Pr(T⊢E)$ may be less than 1, while there are sentences *A* and *B* of the local language that have the same forms from the point of view of the local language as *T* and *E* have from the extrasystematic point of view, respectively, so that $Pr(A \supset B)$ has to equal 1. In this case, $T \supset E$ *has the same truth table* from the extrasystematic point of view as $A \supset B$ has from the point of view of the local language! And if one assigns subjective probability 1 to the conditional $A \supset B$ out of local logical omniscience — in virtue of having perceived the truth-functional logical connection between A and B — then it is hard to see why this individual would miss the (identical kind of) logical connection in the case of T and E.[4]

The point here, of course, is that Garber's local Bayesianism makes the prospects for successfully carrying out step three of the general strategy look very bleak. In the case described in the previous paragraph, for example, it seems very implausible that $Pr(T \mid T \vdash E) > Pr(T)$, if it is required that the agent be logically omniscient with respect to the parallel logical relations between A and B.

It is worthwhile considering an alternative way of "drawing the line," advanced by Ian Hacking in his well-known article "Slightly More Realistic Personal Probability" (1967). Hacking first notes that the axioms of subjective probability theory can be stated in terms of the idea of logical possibility, rather than the ideas of logical truth and incompatibility. Thus, the usual axioms can be restated as follows (this is not the particular axiomatization considered by Hacking in his article):

For all propositions A and B:

(1) $Pr(A) \geq 0$;
(2) $Pr(A) = 1$, if not $-A$ is not *logically possible*; and
(3) $Pr(A \lor B) = Pr(A) + Pr(B)$, if $A \& B$ is not *logically possible*.

Hacking then suggests that it is unrealistic to assume these axioms for subjective probability, since it is unrealistic to assume that a rational agent will always be able to recognize cases of logical possibility and logical impossibility as such. Instead, we should assume axioms stated in terms of *personal possibility* and *sentences*, where a sentence (see below) is personally possible if the relevant individual *does not know* (in a special sense of "know," see below) that the sentence is false. Thus, a sentence is only required to have subjective probability 1 if it is *known* to be true (rather than: if it is logically true), and the probability of a disjunction is only required to have subjective probability equal to the sum of the probabilities of its disjuncts if it is *known* that the conjunction of the two disjuncts is false.

The reason for stating the axioms in terms of sentences rather than propositions is that "proposition" is usually understood in such a way that "two logically equivalent propositions" are really the same proposition (Hacking 1967, 318, cites Carnap 1947, 27). Thus, it would be absurd to entertain the possibility that an agent knows one proposition but fails to know a *different but logically equivalent* proposition. However, we want to allow the possibility that an agent can know one thing, but — in part because the agent does not know the relevant logical equivalence – does not know a second thing that is logically equivalent to the first. So it is natural to say that what the agent knows to be true is one *formulation* of a proposition and what the agent fails to know to be true is a different but logically equivalent *formulation* of the same proposition. Thus, (unambiguous) sentences are a natural choice for the objects of personal probability, sentences that *express* propositions, in a personal language closed under the linguistic connectives of negation, conjunction, disjunction, and conditionalization, for example. (For presumably the same reason, Garber's theory of local Bayesianism is also formalized in such a way that the objects of subjective probability are sentences – of formal languages, in fact.)

Hacking's special sense of knowledge – called "the examiner's view of knowledge" – is one in which certain traditional closure conditions for knowledge are explicitly rejected, e.g., "a man can know how to use *modus ponens*, can know that the rule is valid, can know *p*, and can know *p* ⊃ *q*, and yet not know *q*, simply because he has not thought of putting them together" (1967, 319). It is clear that such closure conditions must fail, if we relax the classical assumption of logical omniscience. Otherwise (for example), the agent must assign probability 1 to all first-order logical truths if (*roughly*) probability 1 is assigned to all the axioms of a (complete) deductive calculus in which the only rule is *modus ponens*.

I think Hacking's way of making subjective probability theory more realistic – of abandoning the assumption of logical omniscience – goes too far. The axioms are much too weak. However, one of the ways in which it has been argued that classical personalism – the classical axioms assuming logical omniscience – are reasonable is by way of "Dutch book arguments." And Hacking suggests a revised Dutch book argument to show the reasonableness of his weaker axioms.

One of the ways in which it has been attempted to justify classical personalism is to prove that if a person's subjective probabilities do *not* satisfy the classical axioms, and if the person is willing to accept any bet whatsoever whose odds are determined in the natural way from the person's subjective probabilities, then it is possible for a clever betting opponent to offer bets, all acceptable to the agent, such that *no matter which propositions bet on turn out to be true and which false*, the agent is assured of a loss.[5] The clever betting opponent need know no more than the agent *about matters of fact* in order to identify a series of bets (called a "Dutch book"), each acceptable to such an agent, but that will assure a net loss to the agent. All that is necessary is that the opponent be able to detect the "incoherence" (violation of the classical axioms); then, using simple mathematical techniques, a Dutch book can be found. The ability to detect an incoherence requires only logical and mathematical sophistication – and not knowledge of matters of fact.

Hacking, however, wants to put knowledge of matters of fact and knowledge of logical facts on a par, for the purpose of assessing a person's rationality.[6] Thus,

Hacking suggests that an appropriate betting opponent for carrying out a Dutch book argument must be *one who knows no more than the agent in the examiner's sense of knowledge.* But this has the consequence that if the agent is *unaware of an incoherence* in his subjective probabilities, then so must be an appropriate betting opponent. But this means that a person will turn out to be rational in the Bayesian sense as long as the person is *not aware of an incoherence.*

Put in other terms, if it is personally possible to *you* that no Dutch book can be made against you, then this must also be personally possible to any appropriate betting opponent, so that the opponent couldn't know of a Dutch book against you either. So it seems that Hacking's slightly more realistic personal probability requires only that you not know that you're not coherent. But this seems to be too severe a weakening of classical personalism.

Thus, while Garber's approach seems to require too much of a rational agent in one sense, and too little in another, Hacking's approach simply requires too little. Recall that the central idea behind all the difficulties raised for Garber's local Bayesianism is that of *complexity* of sentences, logical relations, and inferences. The counterexamples all suggested (roughly) either that "local Bayesianism requires that the agent perceive such-and-such enormously complex logical fact," or that "local Bayesianism does not require the agent to perceive such-and-such extremely simple logical fact." From this point of view, the problem with Hacking's slightly more realistic personalism is that it *allows the agent to set the standard* governing how complex a logical fact has to be in order for him not to be required to perceive it, and its implications, and yet still be considered to be rational on the theory. All this suggests, to borrow Garber's analogy again, that the appropriate place to "draw the line" between the logical facts an agent has to perceive and those that one need not perceive, in order to be considered rational, should correspond to the *complexity* of the facts on the two sides of the line. Or perhaps we should conceive of rationality as coming in (objective) degrees, corresponding to where the line in fact falls for particular agents. Although the development of such a measure is, of course, beyond the scope of this paper, I think we have seen plenty of considerations indicating that, if such a measure *could* be defined, then it would be the appropriate tool for use in developing a version of Bayesianism that is truly more realistic than classical Bayesianism and yet at the same time still reasonably restrictive in the right way.

5. The Evidential Significance of $T \vdash E$

I turn now to part three of the basic Bayesian defense — an argument to the effect that $Pr(T \mid T \vdash E)$ should be greater than $Pr(T)$ before the discovery of $T \vdash E$. The main thing Garber does in this connection is show that it is *possible* that $Pr(T \mid T \vdash E) > Pr(T)$. That is, he proves that, under some fairly general conditions, there *are* probability functions such that $Pr(T \mid T \vdash E) > Pr(T)$ (and

($K*$) is satisfied). No argument is given that one's subjective probability function *should* satisfy this inequality in the relevant kind of situation, although Garber points out in a note (1983, 131) that the discovery of $T \vdash E$ will increase the probability of T if and only if the probability of $T \vdash E$ is higher given T than given not-T. Richard Jeffrey (1983), in the part of his article about the problem of old evidence, basically assumes the adequacy of Garber's steps one and two, and, taking a hint from Garber's note, provides an argument for step three.

Here is a simplified version of Jeffrey's main result. Jeffrey proves that $Pr(T \mid T \vdash E) > Pr(T)$ if the following four conditions obtain:

(1) $Pr(E) = 1$ and $Pr(T) > 0$;

(2) $Pr(T \vdash E)$ and $Pr(T \vdash \sim E)$ are both strictly between 0 and 1, and $Pr(T \vdash E \ \& \ T \vdash \sim E) = 0$;

(3) $Pr(T \mid T \vdash E \lor T \vdash \sim E) \geq Pr(T)$; and

(4)   Garber's condition ($K*$) (see above), in particular, $Pr(T \ \& \ T \vdash \sim E)$ $= Pr(T \ \& \ \sim E \ \& \ T \vdash \sim E)$.[7]

The proof of this version is simple. In view of (3), it suffices to establish

$$Pr(T \mid T \vdash E) > Pr(T \mid T \vdash E \lor T \vdash \sim E),$$

which, given (2), is true if and only if

$$Pr(T \mid T \vdash E) > Pr(T \vdash E \mid T \vdash E \lor T \vdash \sim E)Pr(T \mid T \vdash E)$$
$$+ \ Pr(T \vdash \sim E \mid T \vdash E \lor T \vdash \sim E)Pr(T \mid T \vdash \sim E).$$

(1), and an application of (4), implies that the second term on the right-hand side equals 0. (1) and (3) imply that the right-hand side is greater than 0, so that $Pr(T \mid T \vdash E) > 0$. And (2) implies that $Pr(T \vdash E \mid T \vdash E \lor T \vdash \sim E) < 1$, giving us the desired inequality.

Let us consider the four conditions. Condition (1) is part of the specification of the problem (we assume, of course, that T initially enjoys *some* credence). Condition (2) is plausible in light of the agent's *logical nonomniscience* and the intended interpretation of "$\vdash$", as long as we assume that the agent fully believes that T is not inconsistent. Condition (3) will be discussed below; and condition (4) just specifies part of the intended interpretation of "$\vdash$".

Condition (3) expresses the idea that "your confidence in [T] would not be weakened by discovering that it implies something about [the relevant phenomenon]" (Jeffrey, 1983, 150). Conversely, in order for (3) to be true, it must also be the case that your confidence in T *would* be weakened (or left unchanged) by the discovery that it does *not* imply either E or $\sim E$. There is, however, the intuition that the more a hypothesis or theory implies (the "stronger" it is logically), the less chance it has of being true — an intuition that says more, I think, than just that the *probability* of a hypothesis can be no greater than propositions it implies.

And it seems odd that a theory should be *disconfirmed* just by the fact that it is *silent* on a certain issue. Here, given that T implies *something* about the relevant phenomenon, there is, of course, always the chance that it implies *something false* about it. Thus, in order for (3) to be true, we must be antecedently relatively more confident that, *if* T implies *something* (E or $\sim E$) about the relevant phenomenon, then it implies a *truth* about it (i.e., E), than we are that T would imply *a falsehood* about it (i.e., $\sim E$). But this seems to run against the spirit of the idea of allowing the agent to be *logically nonomniscient*, as I shall presently explain.

Condition (3) (in the presence of the other conditions) implies that:

$$\frac{Pr(T\vdash E)}{Pr(T\vdash E) + Pr(T\vdash \sim E)} \times Pr(T \mid T\vdash E) \geq Pr(T).$$

Suppose now that the agent is "so logically nonomniscient" that $Pr(T\vdash E) = Pr(T\vdash \sim E)$. This is not implausible if T and E are sufficiently complex in the right way. Now, given that T was not designed just with the intention of explaining E, T may already have been confirmed somewhat; say $Pr(T)$ is equal to 0.6. In that case, it is clear from the last displayed inequality that it is impossible for condition (3) to be satisfied. Note also that the agent's having the same degree of confidence in $T\vdash E$ as in $T\vdash \sim E$, while at the same time assigning probability 1 to E, is not incompatible with (4), together with a high degree of confidence in T, for the agent's degree of confidence in $T\vdash E$ and in $T\vdash \sim E$ may be quite low.

Thus, in order for (3) to be satisfied, the agent must *either* assign a high probability to $T\vdash E$, compared with the probability assigned to $T\vdash \sim E$, *or* have a relatively low degree of confidence in T - or both. When would we expect this to be true? To me, this disjunctive condition strongly suggests (though strictly speaking it doesn't imply) that T *was designed to explain* E. This hypothesis would certainly *explain* why $Pr(T\vdash E)$ is much greater than $Pr(T\vdash \sim E)$, if it is: the agent thinks he is pretty good at coming up with theories that would, if true, explain things. And it would also explain a low initial degree of confidence in T: the theory hasn't been around very long and thus has not received *independent* confirmation. But if T was designed to explain E, then we have only a solution to problem IIB1, above, and not to IIB2. (Earlier it was noted that, for IIA1 and IIB1 situations, T may derive support from its explaining E, but not from E itself.)

Regardless of whatever connection there may be between the disjunctive condition just discussed and cases in which T was invented with the intention of explaining E, we have seen that there is what would seem to be an important class of cases outside the scope of Jeffrey's approach, as further suggested below.

Failure of condition (3) is, of course, compatible with $Pr(T \mid T\vdash E) > Pr(T)$: it could be true that finding out merely that T implies *something* about the relevant phenomenon (i.e., that it either implies E or implies $\sim E$) would *decrease* one's confidence in T, while finding out that T implies *a truth* (i.e., E) about the relevant phenomenon would *increase* one's confidence in T. For example, one can imagine an investigator's having somehow hit upon the idea that Einstein's equations *might* have precise implications pertaining to the apparent orbit of Mercury, *without* having actually *gone through any calculations* to determine what the precise implications might be; and the investigator might conceivably, pessimistically, think it unlikely that any such precise consequences of the equations would match the previous (in this hypothetical example) precise observations of the orbit. In this case, nevertheless, a match would, for this investigator, provide striking confirmation.

More formally, note that conditions (1) and (4) above imply that $Pr(T \mid T \vdash \sim E) = 0$. Thus, $Pr(T \mid T \vdash E) > Pr(T)$ if and only if

$$(*) \quad Pr(T \mid T \vdash E) > Pr[\sim(T \vdash E) \ \& \ \sim[T \vdash \sim E) \mid \sim(T \vdash E)]$$
$$\times Pr[T \mid \sim(T \vdash E) \ \& \ \sim(T \vdash \sim E)].$$

And it is clear that this relation may be satisfied even if Jeffrey's condition (3), which, given his others, is equivalent to

$$(3') \quad Pr(T \mid T \vdash E \text{ v } T \vdash \sim E) > Pr[T \mid \sim(T \vdash E) \ \& \ \sim(T \vdash \sim E)],$$

is not. Clearly (when Jeffrey's other conditions are met), the left-hand side of (*) is greater than the left-hand side of (3') (see the derivation of this above), and the right-hand side of (*) is less than the right-hand side of (3'), thus making (*) "easier to satisfy" than (3'). And, as suggested in the last paragraph with an example, it seems that there are genuine cases of confirmation involving old evidence in which (*) holds but (3'), i.e., (3), does not.

But how to "justify" (*)? Of course, we should not hope for a universal justification of (*), applicable in all cases of theories T and evidence statements E. For example, as Jeffrey points out,

> a purported theory of acupuncture that implies the true value of the gravitational red shift would be undermined thereby: [for example,] its implying *that* is likely testimony to its implying everything, i.e., to its inconsistency. (147)

Also, note that we are virtually "back to square one." Condition (*) is *equivalent* to $Pr(T \mid T \vdash E) > Pr(T)$, given Jeffrey's conditions (1) and (2) and Garber's (K^*), all of which are quite plausible given the specification of our problem, given logical nonomniscience, and given the intended interpretation of "$\vdash$"!

There are good reasons to think that there can be no *single* justification of (*); there *are* no "more primitive" assumptions that will justify (*) and be satisfied in all and only those situations in which $T \vdash E$ should be taken as confirming T. As Jeffrey's acupuncture/red shift example shows, whether or not there is a rational increase in confidence in T as a result of discovering that $T \vdash E$ will depend on what T and E *are about*, on the relationships between what T is about and what

E is about, and on our background beliefs. And of course these relevant items differ quite a lot from case to case; we should not expect them to be amenable to one single, systematic, formal treatment in the form of "more primitive" assumptions.

The case is parallel, I think, to the Bayesian explication of "E confirms T." The explication is "$Pr(T \mid E) > Pr(T)$", thus taking into account one's background beliefs, codified in Pr. Any justification of a statement of the form "$Pr(T \mid E) > Pr(T)$" will have to work from more or less particular information about the relation between T, E, and one's background beliefs, such information as "$T \vdash E$," or "E is a positive instance of T where T is subjectively probabilistically independent of the relevant object's satisfying the antecedent of T," or "E bootstrap confirms T relative to an appropriate theory," and so on. Such information as this pertains to particular cases, and no particular such piece of information will apply generally.

Similarly, it seems to me, the Bayesian should simply *explicate* "$T \vdash E$ confirms T (relative to Pr)" as "$Pr(T \mid T \vdash E) > Pr(T)$," without expecting there to be any single formal kind of *justification* of the latter for exactly the cases in which $T \vdash E$ *should* be taken as confirming T. Jeffrey's analysis will shed light in many cases; but in other cases, there may be other formal conditions, incompatible with his condition (3), that will imply $Pr(T \mid T \vdash E) > Pr(T)$. And it is conceivable that in other cases, it will be a formally "intractable" feature of Pr that the inequality holds; i.e., there is no simple, more primitive, relation holding between various items in virtue of which the inequality holds.

This latter kind of possibility, is, incidentally, closely connected with part of the Bayesian rationale for appeal to subjective probability distributions in the first place. As Charles Chihara has put it,

> To take account of heterogeneous information and evidence obtained from a variety of sources, all of different degrees of reliability and relevance, as well as of intuitive hunches and even vague memories, the Bayesian theory provides us with a subjective "prior probability distribution," which functions as a sort of systematic summary of such items. (1981, 433)

Of course, the prior probability distribution does not literally *summarize* the relevant *items* (hence the phrase "sort of" in the quotation), but rather summarizes the *effects* that the agent's absorption of such items has on the evidential situation at hand—the point being, in part, that there may be no simple formal relation between the relevant items themselves, the expression of which would accomplish the same task. And it seems that in some cases, there will not even be any simpler *subjective probabilistic* relations between items that justify or explain why $Pr(T \mid E) > Pr(T)$, or why $Pr(T \mid T \vdash E) > Pr(T)$.

As in the case of step two of the basic Bayesian strategy, I think the best hope for step three is (at least for starters) in the analysis of concrete, particular, historical cases of confirmation, which come complete with particular sets of background beliefs, particular theories, and evidence statements, and what they are about.

Notes

1. For more on Bayesian confirmation theory, and the Dutch book and decision theoretical approaches alluded to, and for further references, see, for example, Savage (1972), Jeffrey (1983), Hesse (1974), Chihara (1981), and Eells (1982). For criticisms, see, for example, Kennedy and Chihara (1979), Kyburg (1983), and Glymour (1980).

2. Note that to complete a classification of *situations* in which an E can confirm a T, we would have to add the case of "*New New* Evidence": the case in which E is new relative to the theory T (i.e., learned after T is formulated), and in which we are presently at, or just subsequent to, the moment of the discovery of E. Of course, this situation does not present a problem for Bayesian confirmation theory of the kind under discussion. For, as required by the theory, if E actually confirms T, then E increases our confidence in T at the time of E's discovery, the time specified in the description of this kind of situation. It is, of course, central to Bayesian confirmation theory that confirmation of a hypothesis or theory implies "rational increase in confidence" in the hypothesis or theory, given one's background beliefs (full and partial), which are supposed to be systematically codified in a subjective probability (degree of confidence) function. Current controversy concerning the idea that confirmation coincides with confidence increase involves, I think, mainly (1) Glymour's problem of old evidence and (2) the plausibility of Glymour's (1980) "bootstrap" conception of evidence, which seems to be inconsistent with the idea of confirmation's implying increase in confidence. As to (1), that problem, and the possibility of a Bayesian resolution to it, is the focus of this paper. As to (2), see Horwich (1983) for discussion.

3. Martin Barrett made essentially these points in my seminar on confirmation theory.

4. Garber doesn't require what he calls condition (*), that $Pr(A \vdash B) = 1$ if $A \supset B$ is a tautology of L^* and must thus itself be assigned subjective probability 1. This leaves open the possibility that *even though locally Bayesian agents assign probability 1 to all tautologies*, they may do so, even in the infinity of complex cases, on grounds other than a perception (intuitive or otherwise) of truth-functional logical structure. But how might this possibility "realistically" be realized (even setting aside the implausibility of assigning probability 1 to all tautologies of L^*? Surely we cannot suppose that, for example, for very many cases of tautological conditionals, their probability 1 status is secured by the agent's being told by a source believed to be totally reliable that they are true. Garber acknowledges "a kind of *informal* contradiction in requiring that S be certain of $A \supset B$ when A truth-functionally entails B in $[L^*]$, while at the same time allowing him to be uncertain of $A \vdash B$" (p. 118), while at the same time insisting (correctly) that this is no formal contradiction. Despite the intuitive implausibility of failure of (*) *given that* all tautologies must have subjective probability 1, Garber seems, in the end, neutral (or noncommittal), with respect to the condition (118).

5. For details on the Dutch book argument, originally due to de Finetti (1964), see, for example, Shimony (1955), Kemeny (1955), or Skyrms (1984, chapter 2).

6. Indeed, Hacking has emphasized (1967, 312-13), plausibly I think, the appropriateness of investigating certain logical and mathematical issues by empirical methods. Though not cited by Hacking, one plausible example of this is the question of which of a number of blackjack ("21") strategies yields a player the highest expectation against the house, which plays a fixed strategy. Though mathematically intractable, Monte Carlo techniques using computer simulation have been successfully employed. The point is that questions pertaining to how one can "in principle" come to know such-and-such are irrelevant to rationality and confirmation, where what is relevant is what one in fact knows, and the reliability and efficiency of one's actual methods of coming to know such-and-such, given one's background belief and knowledge.

7. In Jeffrey's version, E is contrasted not just with $\sim E$ but is, more generally, a member of a

set of mutually exclusive propositions about some phenomenon, such as the tides. The version just given, however, would seem to be an instance of this, since if E is about some phenomenon, then $\sim E$ would seem to be about the same phenomenon. Also, Jeffrey uses "H" rather than "T."

References

Carnap, Rudolf. 1947. *Meaning and Necessity*. Chicago: University of Chicago Press.

Chihara, Charles. 1981. "Quine and the Confirmational Paradoxes." In *Midwest Studies in Philosophy*. Vol. 6, *The Foundations of Analytic Philosophy*, eds. Peter A. French, Theodore E. Uehling, Jr., and Howard K. Wettstein. Minneapolis: University of Minnesota Press.

de Finetti, Bruno. 1964. "Foresight: Its Logical Laws, Its Subjective Sources." Trans. Henry E. Kyburg, Jr., from the 1937 French. In *Studies in Subjective Probability*, eds. Henry E. Kyburg, Jr., and Howard E. Smokler. New York: John Wiley & Sons, Inc.

Eells, Ellery. 1982. *Rational Decision and Causality*. New York: Cambridge University Press.

Garber, Daniel. 1983. "Old Evidence and Logical Omniscience in Bayesian Confirmation Theory." In *Minnesota Studies in the Philosophy of Science*, Vol. 10, *Testing Scientific Theories*, ed. John Earman. Minneapolis: University of Minnesota Press.

Glymour, Clark. 1980. *Theory and Evidence*, Princeton: Princeton University Press.

Hacking, Ian. 1967. Slightly More Realistic Personal Probability, *Philosophy of Science* 34:311–25.

Hesse, Mary. 1974. *The Structure of Scientific Inference*. Berkeley and Los Angeles: University of California Press.

Horwich, Paul. 1982. *Probability and Evidence*. New York: Cambridge University Press.

——. 1983. "Explanations of Irrelevance." In *Minnesota Studies in the Philosophy of Science*, Vol. 10, *Testing Scientific Theories*, ed. John Earman. Minneapolis: University of Minnesota Press.

Jeffrey, Richard C. 1983a. *The Logic of Decision*, 2d ed. Chicago: University of Chicago Press.

——. 1983b. "Bayesianism with a Human Face." In *Minnesota Studies in the Philosophy of Science*, Vol. 10, *Testing Scientific Theories*, ed. John Earman. Minneapolis: University of Minnesota Press.

Kemeny, John. 1955. Fair Bets and Inductive Probabilities. *Journal of Symbolic Logic* 20:263–73.

Kennedy, Ralph, and Chihara, Charles. 1979. The Dutch Book Argument: Its Logical Flaws, Its Subjective Sources. *Philosophical Studies* 36:19–33.

Kyburg, Henry E., Jr.. 1978. Subjective Probability: Criticisms, Reflections, and Problems. *Journal of Philosophical Logic* 7:157–180. Reprinted, with modifications, in his *Epistemology and Inference*, Minneapolis: University of Minnesota Press, 1983.

Savage, Leonard J. 1972. *The Foundations of Statistics*, 2d ed. New York: Dover Publications.

Shimony, Abner. 1955. Coherence and the Axioms of Probability. *Journal of Symbolic Logic* 20:1–28.

Skyrms, Brian. 1983. "Three Ways to Give a Probability Assignment a Memory." In *Minnesota Studies in the Philosophy of Science* Vol. 10, *Testing Scientific Theories*, ed. John Earman. Minneapolis: University of Minnesota Press.

——. 1984. *Pragmatics and Empiricism*. New Haven: Yale University Press.

Fitting Your Theory to the Facts: Probably Not Such a Bad Thing After All

1. Introduction

In the following pages I shall try to show that the variety of Bayesian confirmation theory based on so-called personal probabilities provides an intuitively correct solution to an outstanding problem in a controversial area of methodology, and that it does so in an entirely natural and unforced way. In the course of the discussion it will, I hope, become apparent that the objections usually thought to be decisive against Bayesian theories in general, and especially against this one, are in reality nowhere near as damaging as they seem; and I shall end by making the possibly surprising claim that Personalist Bayesianism offers just the sort of theory of confirmation, fallibilistic in temper, founded on deductive principles only, yet fruitful in methodological information, with which even Popper himself should not be able to find fault (though he would, and does, of course, because it is 'subjective'; but more on that later).

2. The Null-Support Thesis

The outstanding problem is the ancient one of what epistemological distinction if any we are entitled to draw between a theory that has independently predicted an observed effect and one that has been deliberately constructed to yield the effect as a consequence. One answer to that question, and an extremely popular one, is to deny that the second theory derives any support at all from its 'prediction' of the effect in question. An apparently powerful argument in support of this position is the following. If we were to concede that the second theory is supported by the known effect, we should, or so it appears, be faced with the awkward, if not intolerable, consequence that it becomes a simple matter to generate arbitrarily many theories that are supported by any given piece of data, but that are, intuitively speaking, not really supported at all. For example, suppose that A's heuristic deliberations result in a hypothesis, call it H, that x and y stand in a specific functional relation $y = f(x)$, and that A then tests H with a large number

n of joint observations $(x_1,y_1), \ldots ,(x_n,y_n)$ of x and y. Suppose also that, within the given error bounds, the observations all lie on that curve. While we will probably agree that the observations confirm H, we should equally probably be reluctant to concede that they also support each of the infinitely many hypotheses determined by a particular choice of g:

$$y = f(x) + (x-x_1)(x-x_2) \ldots (x-x_n)g(x) \tag{1}$$

(These ad hoc variants, essentially precursors of Goodman's 'grue' hypothesis, were introduced into the literature by Jeffreys [1948] 3).

Non-Bayesian theories have difficulty with hypotheses like (1); that the observations do not by themselves appear to discriminate between them and H is after all precisely the rub of Goodman's paradox. Bayesian theories at least have the formal capacity to discriminate by means of an appropriate distribution of prior probabilities: to what extent this is successful is a question I shall answer later. Without leave to appeal to considerations of plausibility prior to the data (possibly, as in Jeffreys's Bayesian theory, which I shall discuss briefly later, based on considerations of simplicity, in some one of the numerous explicata of that troublesome notion), there seems to be no intrinsic difference between H and its variants (1) that justifies denying them all equal status in explaining and being supported by the observations. But there is an extrinsic, but apparently quite objective, difference: the hypotheses (1) are constructed specifically to fit the data, while H was not. According to what I shall henceforward call the null-support thesis, the variants (1) are not supported by the n observations, precisely because they were deliberately contrived to explain those observations.

The null-support thesis has a long and respectable pedigree: it is to be found in Bacon, Descartes, and Leibniz; it is a principle incorporated into Popper's theory of corroboration, and we find it recently endorsed by Zahar (1989, 16), Redhead (1986), Giere (1984, 159–61), and Worrall (1978, 48). In what follows I hope to show that, despite this powerful advocacy, the null-support thesis is false. Indeed, I shall produce some hypotheses that well-established canons of scientific procedure pronounce very strongly supported by the data they were constructed to explain. It follows that the reason we tend to depreciate the ad hoc alternatives (1) cannot be that they are constructed from the data. What is it then? It is, I shall argue, just because they are irremediably ad hoc; because, in other words, the structural model implicit in the parametric hypothesis: there are $a_1, \ldots ,a_n$ such that for all x,

$$y = f(x) + (x-a_1) \ldots (x-a_n)g(x),$$

where g is not identically zero, is thought not to correspond to the actual state of affairs; we just don't think it true. And in general, hypotheses simply contrived to fit the data will be looked askance at, not because they have been made to fit the data, but because they have been contrived *simply* to fit the data; which is just another way of saying that we have, as far as we can tell, no independent reason to believe them true.

These remarks are not as empty of explanatory value as they might appear. The falsity of the null-support thesis means that a more flexible account of support is required, which, in the course of explaining when and for what reasons we ascribe support to hypotheses generally, explains also why we discriminate *within* the class of hypotheses constrained to fit specific pieces of data, as to which merit support therefrom and which do not. I shall provide such an account. It is, I shall argue, implicit in the Personalist Bayesian theory, which does indeed say that the variants (1) will be assigned vanishingly low support if the structural models implicit in them are sufficiently strongly disbelieved—as indeed, by assumption, they are. But the most important consequence of the Bayesian theory is the fundamental principle of all inductive inference: evidence supports a hypothesis h the more, the less it is explicable by any plausible alternative compared with its explicability by h. This is the criterion by which in practice we decide which data support a hypothesis and which do not, and it is a criterion that can just as easily be satisfied by a hypothesis constructed from that evidence as by one that was not.

Much of the motivation for the null-support thesis comes, I think, from a desire to legitimate a preference, which it is alleged that a study of the history of science reveals, for hypotheses that independently predict facts over those that merely accommodate them. Whatever else this preference reflects, however, it is not that the former are always better supported by those facts than the latter. For there are convincing counterexamples to that doctrine too. Lest it should be thought that things are now going too far in a direction away from the null-support thesis, I shall show that the Bayesian account explains why, of two rival theories, initially equally well supported, but differing in that one independently predicts data that the other merely absorbs into the evaluation of a free parameter, the former receives the greater support from those data. This result is, I think, the germ of truth in the generally false thesis that the independent prediction of facts invariably merits greater rewards of support than their post hoc explanation.

These conclusions will be shown to follow straightforwardly. No arbitrary assignments of prior probabilities, no fiddling or gerrymandering the formulas are necessary; what will emerge from the discussion is just how essentially Bayesian our informal reasoning is. I recognize that any such conclusion is bound to be greeted with strong reservations in many quarters, and I shall conclude this paper by trying to show that the usual objections advanced against allowing the Bayesian theory a role in 'objective' methodology are unsound.

Some preliminary words are in order about the Bayesian theory that is being credited with these virtues. That theory is, as I have already said, the one that, following L. J. Savage, has come to be called Personalist Bayesianism. It exploits the fact, explicitly stated and proved only in this century, but taken for granted from the outset, that the probability calculus furnishes the fundamental laws of fair odds, or to be more precise, fair betting quotients (betting quotients p are related to odds x by the equation $p = x/(1+x)$, with inverse $x = p/(1-p)$). Fair odds are odds that would give no advantage to either side of a bet were one to be called; whether any such odds exist, however, or what they are if they do, is, except in a very restricted class of cases, a matter on which the theory eschews an opinion.

The significance of setting out the general laws that any system of fair odds must obey is that from the very first, when people started talking about the probabilities of hypotheses, they habitually glossed those probabilities as what would be, relative to the contemporary state of knowledge, the fair betting quotients in the ideal situation that the bets could be unambiguously settled. Now a famous result, proved in this century independently by Frank Ramsey and Bruno de Finetti, implies that if any set of betting quotients fails to satisfy the calculus of probabilities, then those betting quotients cannot all be fair. What Ramsey and de Finetti in fact showed was that if an opponent is free to dictate which side of a bet you will take, and the stakes on each, then were you to engage in simultaneous bets with betting quotients that do not satisfy the probability calculus, you could be forced to make a loss (or gain) come what may. But if a betting quotient is fair, the advantage to taking a given side of a bet should be zero; and the net advantage relative to a system of bets at fair betting quotients should also be zero, since it is a sum of zeros. So if someone could tell in advance that a particular betting strategy would entail a positive loss or gain, and hence presumably that the net advantage at those odds cannot be zero, then it follows that the odds are not all fair. Ramsey's and de Finetti's theorem tells us, therefore, that the rules of the probability calculus are nothing more than consistency constraints on the construction of a set of subjectively fair betting quotients, i.e., betting quotients that you believe, in the light of your available information, determine fair odds.

The fundamental notion in the application of this theory to the problem of inductive inference is that of the conditional probability, $P(h/e)$, of h relative to e. Your conditional probability of h on e is what you would assess the probability of h as, were you to come to know e (but nothing more). $P(h/e)$ is, by Bayes's theorem, equal to $P(e/h)P(h)(P(e))^{-1}$. $P(e/h)$ is called the likelihood of h on e, and is equal to one if h entails e (modulo initial conditions and the more or less extensive quantity of other background information you are equipped with). Where h describes a hypothetical physical probability distribution over a set of data points, some measurable subset of which is defined by e, then $P(e/h)$ is set equal to the probability h ascribes to e. $P(h)$ and $P(e)$ are the so-called prior probabilities of h and e respectively. e is reckoned to support h if $P(h/e) > P(h)$, and the difference between the two is a useful measure of the extent of this support, which clearly can be negative. The fact that support is defined in terms only of the relation between the conditional probability $P(h/e)$ and the prior $P(h)$ implies that the so-called 'dynamic assumption', that after receipt of e (and no stronger information), your degree of belief in h should be P(h/e), is unnecessary, at any rate from the point of view of the Bayesian account of inductive inference.

The only endogenously determined quantities in the Personalistic Bayesian theory are the likelihood terms P(e/h) in the conditions stated above, and the probabilities of necessary truths and falsehoods relative to the individual's background information. The prior probabilities in Bayes's theorem are not usually of this type and are, therefore, parameters undetermined within the theory. This fact often leads people to believe that the account Personalist Bayesians give of inductive inference is subject to no significant constraints at all, and is consequently irremediably 'subjective', a mere branch of psychology, and not very good psychology at that. I shall simply ask the reader to wait until the final section for the discussion—and rebuttal—of this well-worn accusation.

3. The Falsity of the Null-Support Thesis

Since arguments have been put forward for the null-support thesis based on what appear to be unobjectionable methodological precepts, we perhaps ought first to see where these break down. There seem to be two such arguments: (i), advanced by Giere and Zahar, is that if h was designed to explain, and at the very least to be consistent with e, then h stood no chance of being refuted by those facts described by e. Hence, it is concluded, h cannot be supported by e, since support by e allegedly implies the possibility of refutation by e. The second argument, (ii), is due to Giere and Redhead, who assert that when the data are used as an explicit constraint in the process of constructing a hypothesis, then the probability of the data given the falsity of the hypothesis cannot be small, but on the contrary must be unity. But evidence, they contend, supports a hypothesis that predicts it when and only when it would be improbable if the hypothesis were false.

This criterion embodies a quite fundamental intuition. It is the intuition behind significance testing, for example: the null hypothesis (i.e., the negation of the proposed causal hypothesis, usually identified merely with the hypothesis that the observed data are due 'to chance') is rejected, and the causal hypothesis correspondingly confirmed, if those data are very improbable on the assumption that the null hypothesis is true; and—for future reference—it follows immediately from Bayes's theorem, in the form

$$P(h/e) = \frac{P(h)}{P(h) + \dfrac{P(e/\sim h)P(\sim h)}{P(e/h)}},$$

as entailing a high posterior probability of h when, as is assumed in Giere's and Redhead's discussion, e is probable given h. In general, though, as the expression above makes clear, it is merely the likelihood ratio $P(e/\sim h):P(e/h)$ that has to be small to confer a high posterior probability on h.

I shall not quarrel with either the improbability criterion, or the principle enounced in (i), that support for a hypothesis h from an experiment (a term I shall use to describe any data source) requires the possibility that the experiment is capable in principle of generating also refuting outcomes. Indeed, I shall be claiming that both these are naturally explained by the Bayesian theory. What I quarrel with is the claim that from either or both follows the null-support thesis, and I shall show that that claim is false.

Let us start with (i). Giere states that

> if the known facts were used in constructing the model and were thus built into the resulting hypothesis . . . then the fit between these facts and the hypothesis provides no evidence that the hypothesis is true [since] these facts had no chance of refuting the hypothesis. (1984, 161)

Glymour (1980, 114) voices a substantially identical opinion, and much the same occurs, in a slightly more elaborate way, in Zahar (1983, 245) (incidentally, ignore the fact that statistical theories are not strictly refutable; we can take these authors to be using the word 'refute' in a sense that accommodates weaker criteria than the purely deductive). Plausible though it may sound, the argument is quite specious. How can *facts* ever have a chance of refuting anything? If e is a factual statement and h a hypothesis, then it is simply false to say that e has a chance of refuting h; it either refutes h or it doesn't, and it does so or doesn't whether h was designed to explain e or not. Giere has confused what is in effect a random variable (the experimental setup or data source E together with its set of distinct possible outcomes) with one of its values (the outcome e). It is only E, not e, that has the chance of refuting any particular hypothesis. Moreover, it makes perfectly good sense to say that E might well have produced an outcome other than the one, e, it did as a matter of fact produce. It follows that whether or not h was deliberately designed to explain e but nevertheless does so, E (in general) *could*, on the occasion on which it produced e, have generated another outcome inconsistent with h. Obviously, once E is performed and e results, there is no chance of that performance of E refuting h; but, as we have seen, this would be true whether h was designed to explain e or not. Either way, (i) collapses.

(ii) fares no better. A propos Mendel's use of the observed ratio of tall to dwarf pea plants in the second filial generation of his famous experiment, Giere remarks that "fitting this case was . . . a necessary requirement for any model to be seriously entertained. So there seems no way [in which the data could be improbable given the negation of Mendel's factorial hypothesis]" (1984, 118). Let us not dispute Giere's rather doubtful assumption that Mendel invented his theory to account for the data (though too exact agreement with those data caused Fisher (1936), in a famous paper, to conclude that the theory was in fact constructed first). But we should certainly dispute the validity of his argument, for the conclusion is a non sequitur. I at any rate can see no reason, and Giere provides none, why the data should not have been regarded as improbable on the supposition that Mendel's hypothesis was false. It might of course be argued that as the data are by assumption already known to have occurred, their probability must of necessity be equal to one, and hence equal to one conditional on the negation of Mendel's hypothesis. Glymour (1980), in a well-known argument, does indeed charge the Bayesian with having to accept just this conclusion.

I shall discuss Glymour's claim later, and argue that it is false. However, even were it true, it would not be a good strategy for supporters of the null-support thesis to adopt, since it would mean that there could be no way of distinguishing the support of hypotheses by data already known at the time both hypotheses were proposed, and that one hypothesis was designed to explain and the other explains independently. In either case, the probability of the data relative to the negation of both the hypotheses would be one, and neither would be supported according to the criterion — contrary to the declared opinion of virtually all the advocates of the thesis.

Writing, however, from an allegedly Bayesian position, Redhead (1986) seems to provide the linking argument Giere needs, but does so only by transforming Giere's premise, that constructing a hypothesis to explain e is tantamount to making the explanation of e a necessary condition for any hypothesis to be seriously entertained, into the explicitly Bayesian, and very strong, condition that in such cases e acts as a 'filter' allowing only those hypotheses nonzero prior probability which, modulo a set of auxiliary hypotheses and other statements asserting that suitable initial conditions have been satisfied, entail e. From this considerably strengthened form of Giere's premise we do indeed infer that

$$P(e/\sim h \& a) = 1.$$

(I have followed Redhead here in writing a explicitly in the form of a condition in the probabilities), since for any partition $[h_i]$,

$$P(e/a) = \Sigma P(e/h_i \& a)P(h_i/a) = \Sigma P(h_i/a) = 1,$$

where the filter condition ensures that the only h_i contributing to the sum is such that $P(e/h_i \& a) = 1$. It more or less immediately follows that $P(e/\sim h \& a) = 1$. Hence it does seem, when h is constructed in order to explain e, that the condition that e be improbable relative to a and the denial of h is never satisfied.

But it only seems that way. First of all, Redhead's filter condition is, as it stands, an impossibly strong condition, for it assigns a tautology zero prior probability (a tautology does not imply e). Even if tautologies are excluded by fiat, then if h implies e one can always find a nontautologous consequence of h that doesn't, and that must apparently then be assigned a zero probability where h is assigned a positive one, contradicting the probability calculus also. Presumably, the filter condition is intended only to apply to the hypotheses in some partition. Whichever partition is chosen, however, it is easy to see that the filter condition still has the consequence that P(e/a) = 1. But this implies that P(h'/e&a) = P(h'/a), where h' is any hypothesis that entails e modulo a, whether h' was constructed to do so or not, so that h' cannot be confirmed by e either. In other words, Redhead's filter condition, even relativized to an 'appropriate' partition, is still too strong, for it cannot make the discrimination between the two types of hypothesis he wishes to make.

To sum up: nobody has made out a tenable case for supposing that when h is constructed to explain e, e cannot therefore be regarded as improbable on the supposition that h is false. On the contrary, there seems no reason at all why this condition should not be satisfied. Take the Mendel case, for example. Mendel observed a fairly exact and stable ratio of tall to dwarf peas, whose occurrence in just those conditions corresponding to his careful selection of and mating the parent and first generation plants, for which only his theory of inherited, independently, and equiprobably segregated factors seemed to offer an explanation. In other words, the probability of obtaining such data, were Mendel's account not the correct one, should certainly not be unity. Of course, Mendel's theory was not widely regarded at that time as receiving great support from the data. But the explanation is not in P(e/ ~ h) being unity, but in the contemporary implausibility of the particulate model that contradicted the favored blending theory.

In fact, the null-support thesis explains nothing, for it is false. Counterexamples abound, and we do not even have to go to the history of science to find them: they can be invented ad lib. The following two, one statistical and the other deterministic, should suffice. They are are extremely simple in structure (one almost laughably so); and this is an advantage from more than the purely expository point of view, for it means that their salient characteristic can be diagnosed immediately.

(a) An urn contains an unknown number of black and white tickets, where the proportion p of black tickets is also unknown. The data consists simply in a report of the relative frequency r/k of black tickets in a large number k of draws with replacement from the urn. In the light of the data we propose the hypothesis that p = (r/k)+ε for some suitable ε depending on k. This hypothesis is, according to standard statistical lore, very well supported by the data from which it is clearly constructed. It is worth briefly going into the reasons why we take it to be well supported. They are that we are employing as a background theory the hypothesis that the 'experiment' has the structure of a sequence of so-called Bernoulli trials, in which p is the binomial parameter, which is to say that the draws are assumed to be independent with constant probability p of getting a black ticket. r/k determines a confidence interval of length 2ε for p, where the confidence level together with k determines ε. I am not particularly concerned here with the ultimate epistemic rightness or wrongness of regarding these intervals as actually justifying confidence of the relevant degree. My concern is simply with what is actually and uniformly regarded as legitimate practice, and there is no question but that confidence interval estimates of physical parameters, derived via some background theory involving assumptions about the form of the error distribution, are the empirical bedrock upon which practically all quantitative science is built.

But we can point to a feature of the hypothesis we have derived about p, which is, I submit, highly germane to an explanation of its epistemic merit. This is that the probability of the sample data relative to the same background distribution, but on the assumption that the parameter p lies somewhere outside the specified interval, is very much smaller than its small probability on the assumption that p does in fact lie in that interval. Recall that this is just the condition, endorsed by the Bayesian theory, for a high posterior probability for that interval to contain p. Let us leave the discussion in abeyance and now look at (b).

(b) The urn remains the same, but instead of sampling with replacement we now sample without replacement, and continue until the urn is empty. The proportion p of black tickets we discover to be p_0, and this now becomes our hypothesis about the value of p. Surely in this case the sample data support, since together with background information to the effect that the urn has remained the same throughout, they entail the hypothesis that $p = p_0$ (and a fortiori the probability of the sample data on the assumption that that hypothesis is false is zero, given the background information).

While it is difficult to maintain that (b) is representative of much of quotidian scientific inference, both it and the more representative (a) are nevertheless very instructive. They are both cases where background theory supplies a model of the experiment that leaves only a parameter to be calculated from the data, and that background theory is sufficiently firmly entrenched to be taken more or less for granted (though additional data may conceivably lead to its being questioned, nonetheless). Continuity considerations would therefore seem to suggest that in general, the support of a hypothesis $h(a_0)$ obtained from a parametric hypothesis h, whose adjustable parameter(s) is (are) evaluated as a_0 from the data, should depend on the prior plausibility of the parametric model h.

I shall argue that this is indeed the case, with the help of an example that might be thought a rather surprising choice. In this rather simple and idealized example, the data from which a_0 was evaluated are going to be data that are quite uninformative about the truth or falsity of the model h itself. So: let h be the hypothesis that two observable variables x and y are related linearly, so that $y = cx + d$, for some c,d that are to be determined by observation; and suppose also that we have no reason to believe that, within an interval determined by background information, any one set of values is any more likely than any other. Let e consist of two independent joint observations of x and y. Thus e determines, up to some interval depending on the error distribution over the observations, values of c and d, which we shall represent by a_0.

Computing the degree to which e supports $h(a_0)$ as the difference e makes to your assessment of the likelihood of h (not using that term in its specialized statistical sense), it is easy to see, even without assuming that these 'quantities' are represented by numbers as opposed to the members of some arbitrary additive semigroup, that this support may be considerable, though it is bounded above by the prior credibility of the model h. For the credibility of $h(a_0)$ in the light of e is, given the assumption of the mutual independence of h and e, no more and no less than the prior credibility of h, since all e then does is evaluate the parameters c and d. But the credibility of $h(a_0)$ independently of e is negligible, since (c,d) can, we may assume, take any values in the plane (usually there will be some prior restriction on their possible values, but these may well be very broad indeed). It follows that were we to possess exactly the same information as we do now, with the exception of a knowledge of e itself, then the adjunction of e would usually make some, and possibly a considerable, difference to our evaluation of the credibility of $h(a_0)$. In other words, the potential of e to alter the credibility of $h(a_0)$ in an otherwise identical knowledge situation will vary with the prior plausibility of h.

But surely this proves too much — for how could e possibly support $h(a_0)$ when e by hypothesis provides no information relevant to the truth of h? The answer simply is that e *does* support $h(a_0)$ to the extent that it raises its probability in general. I suspect that the apparent force of the objection derives from covertly assuming the truth of the so-called Consequence Condition, that if evidence supports a hypothesis then it must support every logical consequence of that hypothesis. I have in effect just presented an argument that I believe shows the Consequence Condition to be false. Indeed, as Popper and Miller have shown (1983), if we make increase in probability the criterion for support, then every hypothesis supported by some data has a logical consequence that is actually countersupported, in the sense that its probability is decreased, by that data. The Consequence Condition is in conflict also with more basic intuitions. We do believe, I think, that the approximate constancy of the acceleration induced in falling apples supported Newton's gravitational theory, but also that it did not support the bare hypothesis that the gravitational force is not everywhere constant.

It is, I believe, just because support for $h(a_0)$ tends to be conflated with support for h that the null-support thesis is so firmly entrenched: what seems to happen is that the null support for h in cases like the above gets illicitly transferred to $h(a_0)$. The conflation is apparent, for example, in John Worrall's grounding his conclusion that "of the empirically accepted logical consequences of a theory those, and only those, used in the construction of the theory fail to count in its support," on the alleged fact that "Mercury's perihelion [advance] is not regarded as supporting classical theory," although it is predicted by versions of that theory

(1978, 48). Classical theory may or may not be supported; but even were it not supported, it certainly would not follow, as Worrall's inference presupposes, that the versions of classical theory that predict the perihelion advance are not supported by it.

Now let us observe that the informal, intuitive reasoning above is perfectly mirrored in the Bayesian theory. The asumption of the mutual irrelevance of h and e translates into the condition of probabilistic independence: $P(h\&e) = P(h)P(e)$. Note also that, modulo initial conditions, $h(a_0) \langle = \rangle h\&e$; and substituting appropriately into Bayes's theorem we obtain $P(h(a_0)/e) - P(h(a_0)) = P(h)[1 - P(e)]$ (Howson 1984, 248–49). Thus $P(h)$ is an upper bound on the support, which is positive so long as $P(e) < 1$ (we assume that $P(h) > 0$).

The dependence of the support of $h(a_0)$ by e on the prior probability of h (since the support depends on the prior probability of $h(a_0)$, which implies h) is quite general in the Bayesian theory, and is reflected in the judgments of working scientists. The statistician and biometrician Karl Pearson discovered a family of density curves (his Type I, II, III, IV, and V curves), which he was prone to fit to a great variety of data, in a way that to many of his contemporaries seemed frankly ad hoc: on one such occasion the economist Edgeworth pointedly asked "what weight should be attached to this correspondence by one who does not perceive any theoretical reason for those formulas?" (Edgeworth 1895). Kepler fitted ellipses to Tycho's data on planetary orbits, but he also thought it necessary to present independent reasons for that type of orbit. Nearer to home we find Kitcher (1985) castigating a piece of sociobiologists' parameter adjustment on the ground that "the model gives absolutely no insight into the reasons behind the periodicity [the adjusted parameter] . . . the choice of a periodic function for the probability bears no relation to any psychological mechanisms" (375). And so on; anyone can find a host of examples.

The application to the initial problem of discriminating the support of the ad hoc variants (1) from that of $y = f(x)$ is now clear. The introduction of the parameters $a_1, \ldots, a_n$, which are subsequently evaluated from the data, corresponds to postulating models

$$y = f(x) + (x - a_1) \ldots (x - a_n)g(x)$$

of the experimental situation that we simply have no reason to believe true – and because the set of these 'models' is infinite, and for none of them is there the remotest reason to believe it true; the credibility of each is literally zero. Thus the situation with the hypotheses (1) falls under the general case of evaluating the support of $h(a_0)$, where the prior credibility of the model h is zero, or effectively so. This conclusion may seem a little disappointing: the hypotheses (1) are not supported by the data, because in effect we don't think the sort of structure they postulate is the true one. Isn't an assessment of support supposed to justify rather than be justified by our convictions as to what is likely to be true and what isn't?

We have seen many attempts to construct theories of confirmation that claim a priori status. But nothing comes out of nothing; and all these theories incorporate just such convictions, often, as in Carnap's systems, in a disguised form, as basic principles. Jeffreys's theory (expounded in his [1948]) delivers the judgment that the parametric hypotheses of which (1) are instances are a priori less likely to be true than $y = f(x)$. This judgment is an immediate consequence of his Simplicity Postulate, which asserts as a general principle that of the hypotheses advanced within science, those with fewer undetermined parameters (simpler in Jeffreys's sense) are a priori more likely to be true than those with more. However, Jeffreys himself did not see the Simplicity Postulate as an a priori valid principle; for him it was rather an explicit recognition of scientific practice: the simplest equations are as a matter of fact preferred when fitting curves to data (1948, 10).

In the unlikely event of absolutely no background information about the experimental source, simplicity by itself may well play a role in determining levels of support among the uncountably many possible functional relationships consistent with the data. Clearly, if a proposed curve fits some initial set e_1 of observations, and continues to fit all subsequent sets, we should want to say that in so doing it becomes increasingly well supported; but this is possible only if its probability prior to all the observations were positive. Since the curve is likely to have been relatively simple among all the possibilities, we are in such circumstances implicitly taking simplicity to be a ground for assigning a moderate or at any rate nonzero prior probability. But in general, where there is a body of background information constraining the plausible candidates, simplicity and prior probability may well not march in step. So the Simplicity Postulate must be rejected as a principle of general scope. (Its rejection has often been urged on the grounds of alleged inconsistency, most recently by Watkins [1985, 110–16]; that charge, I have argued elsewhere [1988], is incorrect.)

Popper's well-known reversal of Jeffreys's probability ordering must also be rejected. Popper's reason for adopting the converse ordering is that being more easily tested, simpler hypotheses are less probable than more complex ones. But this is to confuse pragmatics with epistemology: we simply have no ground a priori for believing that more easily testable hypotheses are less – or for that matter more – likely to be true, and we should certainly not allow strong and, in principle, ungroundable epistemological assumptions, which these in fact are, to play the role of logical axioms.

Questions of ultimate justification are, however, beside the point of this exercise, which is the much more limited one of diagnosing the differential status we accord the hypotheses (1) compared with the initial $y=f(x)$; and we have shown that the fact that the former were generated using the data and the former is not in itself the cause. The crucial feature of the variants (1), which accounts for their comparatively low status, is that the *introduction* of the parameters $a_1, \ldots, a_n$

has no justification in terms of what we think likely to be true (the same point is made in Nickles [1985, 200; and 1987]).

4. Prediction and Accommodation; the Bayesian Analysis

I have argued that, depending on circumstances, hypotheses can be and often are regarded as supported by data employed as constraints in their construction. The circumstances can be summarized in the condition that $P(e/\sim h)/P(e/h)$ be small. This condition is, of course, as I pointed out earlier, just the Bayesian condition for a high posterior probability of h.

The null-support thesis is false. The motivation for it was a desire to disqualify certain types of patently ad hoc accommodation by a theory of otherwise adverse, or at best neutral, data. A doctrine weaker than the null-support thesis, but similarly motivated, concedes that while accommodated data may give some support, it is nevertheless never to the same extent as if the data in question had been independently predicted. This is also false, and again it is not difficult to manufacture informal counterexamples to it. Consider the following (in essence due to Peter Urbach). A numerologist employs a number of algorithms, in a manner claimed to represent the vagaries of divine will, to predict dates of major earthquakes in California, where 'major' means exceeding some given Richter value. This goes on year after year, failures of the phenomenon to occur to order being explained away suitably, until eventually one such prediction comes true. Established geophysical theory predicts (let us suppose; the moral does not depend on factual accuracy) that earthquakes of such magnitude occur when and only when the strain along a fault line exceeds by some specified quantity a critical value. There is no independent way of estimating when this value will be exceeded. I think that there are few people who would credit the numerologist's theory with greater support from the observed phenomenon than the assertion that on the date at which that phenomenon was observed the strain exceeded the critical value by at least the amount specified by standard theory. And the reason why we do not regard the independently predicting hypothesis as having no support here is the same as the reason why we regard the data-constructed hypotheses (1) as having no support in the circumstances in which they arose: we simply don't believe that they can be true. Being constructed or not from the data has nothing to do with it.

Although the general thesis that an independently predicting hypothesis is always better supported by the data so predicted than is one that is deliberately constructed to explain them is false, there is an important residuum that is not. Because it lends itself to a perspicuous treatment, let us again consider the example of three hypotheses h', h, and h(a_0). h as before contains an undetermined parameter that is evaluated from e. h is inconsistent with h', which independently predicts e; h(a_0) only 'predicts' e after the event (a_0 is the parameter in h evaluated from e). Finally, we shall suppose that the prior probabilities of h and h' are equal. In more idiomatic language, h and h' are rival explanatory frameworks; h' predicts the effect e, and so does $h(a_0)$, but only as a consequence of e's having antecedently been used to calculate a free parameter in h: h' predicted, while $h(a_0)$ is merely an accommodation of, the data. In these circumstances it seems correct to say that h' picks up more support from e than does $h(a_0)$.

This is—or at any rate seems to be—the residuum of truth in the false general thesis that independent prediction invariably scores higher in terms of support than accommodation. It is certainly regarded as true from the point of view of the Bayesian theory; indeed, it is very easily generated as a consequence of that theory, where support S is measured as simply the difference of posterior and prior probabilities. For where the initial conditions are regarded as being part of background knowledge, we have $S(h'/e) = P(h')[1-P(e)]/P(e)$, and $S(h(a_0),e)$ $= P(h(a_0))[1-P(e)]/P(e)$. But by assumption $P(h') = P(h) \geq P(h(a_0))$ and so $S(h',e) \geq S(h(a_0),e)$. Admittedly, the inequality is weak, and so to say that h' picks up strictly more support from e is strictly incorrect. Never mind; the result is good enough.

The sufficient condition for the inequality must not be forgotten; it is that h has at most the prior probability of h'. If for example the parameter in h is introduced purely ad hoc to yield the desired effect, this ad hocness will be registered in the prior probability of h being small if not negligible (such would have presumably been the case with the de Sitter modification of Poincaré's Lorentz-invariant gravitational theory, which contained a parameter specifically introduced to explain the annual shift in Mercury's perihelion). It must be emphasized that this is merely a special case, though I suspect that the apparent plausibility of the thesis that independent prediction always gleans more support than accommodation rests on nothing more than invalidly generalizing from it.

5. The Objections to Personalistic Bayesianism

The Bayesian theory seems to offer a most promising formal reconstruction of our intuitive reasoning in the contexts we have discussed. It is, furthermore, to my knowledge the only methodological theory that is capable of making sense of our intuitions, to say nothing of canonical practice, in those, and other, contexts. But objections, and on the face of it powerful ones, have been brought against its credentials to perform such a reconstructive role.

One such objection is highly relevant to the problem we started with, of assessing the supportive power of known facts relative to theories whose construction is carried out with those facts employed as explicit constraints. The objection, which I alluded to earlier, in section 2, is that if e is already known then both P(e) and P(e/h) ought to be set equal to one, and not only then do all the Bayesian formulas we have written down adopt trivial forms, but in particular the support (2) simply goes to zero, so that the support of hypotheses by known data, whether they were designed to satisfy the data or not, is uniformly zero.

The objection originates with Glymour (1980, Chap. 3). It is, I believe, based on a misunderstanding of how the Bayesian formulas are intended to be interpreted, and it has a straightforward and natural answer (and one, we shall see, that Glymour himself anticipates). This answer is as follows. The Bayesian claims that the support that e gives a hypothesis h is to be evaluated by the extent to which the observation of e alters the prior credibility you attach to h. What does this mean when e is already known? It cannot mean that that the P(e) and P(e/h) terms are trivially one, since e is *always* known at the time you compute the support function: you would be guilty of a simple misapplication of the theory, rather analogous to dividing both sides of an equation by zero, if you *therefore* made those probabilities unity. What P(e) is intended to convey, whether e is known or not, is how likely you think e *would* be were (i) h true and (ii) h false; and P(e) is simply a weighted sum of these two magnitudes.

To the extent, then, that support is a function of P(e), it is a function not of an actual probability, but of a subjunctively characterized one; and when e is known, the subjunctive conditional characterizing it becomes counterfactual: how probable do you think e would be if you didn't already know it to be the case relative to the suppositions, respectively, that (i) h is true, and (ii) h is false? There is absolutely nothing ad hoc or in conflict with core Bayesian principles in defining support in this way. On the contrary, it seems a very natural way of proceeding. Certainly the presence of subjunctives in the definition of the constituent probabilities is nothing new: the definition of conditional probabilities is also cast in the subjunctive mood, as we observed earlier.

Glymour himself considers this response to his objection quite sympathetically, but he is doubtful as to whether adequate consistent procedures exist for computing these probabilities. He considers various methods for calculating such values and concludes that they do not work. I am quite willing to concede that it is difficult if not impossible to come up with a sharp value for the probability I would attribute to a stone's falling to the ground when dropped if current gravitation theory were false, but this is just one of those occasions when the deliverances of the Bayesian rule are going to be very imprecise. Moreover, the existence or otherwise of algorithms or general criteria of evaluation is beside the point for the sort of Bayesianism I am considering here, which is not a source of rules for computing all the probabilities figuring in Bayes's theorem, but simply an attempted reconstruction of a type of intuitive reasoning, in which more or less rough estimates of various probabilities are made. Some of these can be analyzed as involving the application of a particular rule, and others cannot, and have to be treated simply as exogenously determined data, among which typically are the prior probabilities P(e) and P(h).

The relegation of P(e) and P(h) to the status of exogenous parameters whose values in many cases seem not to be susceptible of any even moderately precise determination might well seem to invite the charge of triviality. Take the mere presence of the undetermined parameters P(e) and P(h) first. These have frequently been identified as a source of weakness in the Personalistic Bayesian theory, allegedly undermining any claim it might make either to explanatory status (those parameters can be adjusted to ensure consistency with practically any historical judgment of what has supported what: due to their presence the theory is allegedly no more than "a soft and rubberlike system which is easy to manipulate") or to objectivity (they can be adjusted in a way that, e.g., "allows [people] to assign zero probability to a promising rival hypothesis that threatens ones they personally favour"). (Both quotations are from Watkins [1985, 308].)

Let us address these objections in turn. First, the presence of undetermined parameters does not preclude a hypothesis's either being tested or having the capacity to explain phenomena. Every scientific theory of note has some parameters undetermined within the theory. It has to be conceded that it is often not possible to arrive at anything like precise measurements of individuals' prior probabilities; but this does not distinguish that account from other quite respectable explanatory theories, where sometimes the estimated values of parameters amount to nothing more than a well-grounded qualitative assumption. How many times, for example, do we see explanations in the physical sciences and elsewhere prefixed with remarks like "the masses may be considered to be so small that the potential energy of interaction is zero," or "suppose a $<<$ b; then . . . "; and so forth. Qualitative assessments of belief for which there is independent evidence can support just as good explanations in the Bayesian account.

But what about the charge of extreme subjectivism? It is, after all, the inescapable subjectivity of Bayesian assessments of support, depending as they do on the individual's priors, that alarms people most, and that inspired Fisher's famous and influential verdict that those assessments are measures "of merely psychological tendencies, theorem concerning which are useless for scientific purposes" (Fisher 1947, 6–7), echoed more recently in Jaynes's verdict that "personalistic probability belongs to the field of psychology and has no place in applied statistics" (Jaynes 1968, 231). These dismal conclusions are quite unwarranted, however. Partly they result from a simple non sequitur. The subject matter of the Personalist Bayesian theory is beliefs, and beliefs are, of course, psychological. But the theory is not psychology: it sets out to describe not beliefs as such, but the structure of beliefs regulated by consistency constraints that are completely objective. The Personalist Bayesian theory is, in other words, a *logic* of beliefs; it consists in setting out conditions of consistency that are no less objective than those of deductive logic. To be fair, even the defenders of the Personalistic theory do not emphasise—or even appear to recognize—its unimpeachably logical status. They almost invariably concede the charge of outright subjectivism, and try to mitigate it by appeal to the very general phenomenon of asymptotic convergence of the posterior probabilities relative to the same data. The charge should not, I stress, be conceded in the first place.

But it must be conceded that judgments of support do, according to the Personalistic theory, reflect to a greater or lesser extent the influence of one's own prior belief distribution. Before one condemns this as amounting to a betrayal of objective standards, one should ponder the status of principles that affect to determine substantive inductive judgments. These principles, however 'objective' they purport to be, are inevitably assumptions, and moreover somebody's assumptions, and honesty compels that they should be presented as such. The Personalistic theory merely calls a spade a spade. If anybody still doubts that it is necessary to give explicit recognition to the role of undefended prior belief in inductive inference, they should examine with care (as Peter Urbach and I do [1988]) those theories, like classical statistics, for example, that claim to dispense with it. I suspect that they will eventually, though possibly reluctantly, recognize that those theories do not deliver the goods. They should also recall that they do not dismiss deductive logic because it refrains from supplying criteria for justifying premises as well as inferences. Deductive logic contents itself with judgments of the form: "If you wish to remain consistent, then if you believe this set of statements to be true, you must also accept that statement as true." The same acceptance of the limitations of the power of human reason implicit in the restriction of scope here is unfortunately still not evident in 'objective' discussions of inductive inference.

Let us now return to the ability of the Bayesian theory to explain characteristic modes of inductive inference. It is raised in an apparently acute form by some empirical studies of subjects' evaluations of statistical data, and in a well-known survey Kahneman and Tversky express a strongly negative conclusion:

> The usefulness of the normative Bayesian approach to the analysis and the modeling of subjective probability depends primarily not on the accuracy of the subjective estimates, but rather on whether the model captures the essential determinants of the judgment process . . . In his evaluation of evidence [however] man is apparently . . . not Bayesian at all. (Kahneman and Tversky 1982, 46)

I cannot go into the detailed evidence that is taken to support this conclusion; it would take far too long. I shall simply concede that there are areas in which popular modes of reasoning fail to satisfy the Bayesian constraints (so much, incidentally, for the claim that the theory imposes none). So what? Statistical inference is an area in which popular reasoning is notoriously subject to simple fallacies. Wason's celebrated card paradox (Wason 1966) exhibits an area where popular deductive intuitions also fail in just such uniform ways. Kahneman and Tversky might wish to conclude from this that people are not deductive logicians either; and to an extent they would be right. But that is not a conclusion that should disturb those who claim that the canons of deductive logic exercise visible constraints on the way people reason, especially after some reflection, in a great variety of circumstances. We know that they recognize, at least in principle, that the characteristic of deductive inferences that makes them valuable is that they preserve truth, and it would be strange indeed if they were unable to exercise this knowledge in cases that make not too great demands on their powers of reasoning.

The same ought to be true of probabilistic reasoning, and the criteria of consistency that apply there. The mathematical theory of probability, as a matter of historical fact, started life as the theory of fair odds, and as such was immediately applied to the problem of inductive inference by the seventeenth- and eighteenth-century mathematicians. The fact that the criteria developed by the theory are not uniformly applied by everybody does not mean that people in general are not capable of recognizing the authority of those criteria, nor that in many simple cases they are incapable of applying them.

I am going to consider one final objection to this exercise in explanatory Bayesianism (a more complete discussion of all these, together with other objections, is to be found in Howson and Urbach [1988]). This is not to say that there are not others, but the list is a long one, and a criterion of importance has to be exercised if this paper is not to get too tail heavy. I shall not, therefore, discuss the status of Jeffreys-conditionalization, or other proposed modifications of the 'dynamic assumption', since they do not really conflict with any principle I have invoked here, nor shall I discuss David Miller's (1966) charge, now known as Miller's paradox, that the way the Bayesian evaluates likelihoods in terms of the values of a given physical probability distribution is inconsistent. (Graham Oddie and I have, I think, shown in Howson and Oddie [1979] that Miller's reasoning rests on a standard type of fallacy.)

This final objection is one with which readers of the earlier discussion will now be familiar. It is that to whatever extent we may invoke estimates of credibility in our inductive reasoning, we do so qualitatively, and it is beyond question that we could never honestly refine these to points in the real continuum. Yet the Bayesian theory is a theory of point-valued subjective probabilities. It has been proposed (e.g., by Koopman [1940], Good [1962], Smith [1961], Dempster [1968], and Williams [1976]), in response to such observations, that a theory of interval-valued probabilities would furnish a more realistic foundation for a theory of subjective uncertainty, and these authors have indeed developed such a theory, which now tends to go under the name of the theory of upper and lower probabilities (the upper and lower probabilities are the end points of the intervals). The theory is a generalization of the point-valued theory; the latter is obtained when upper and lower probabilities are identical. It is not clear, however, that more realism is imported into such a theory, since the upper and lower probabilities themselves are point valued, and it seems no easier in principle to arrive at nonarbitrary values for these than for point probabilities themselves.

Should we attempt to find more realistic weakenings of the point-probability model? Personally, I do not think so. Certainly the objection that we cannot realistically claim to make point-valued probability assessments should not force us to abandon the point model. For that objection is not at all as strong as it sounds; it is, on the contrary, very frail. In the first place, there is nothing in the Personalist theory that says that anyone does make point estimates: that theory is quite compatible with people's actual evaluations being as crude and as qualitative as you like. Moreover, the people who take this objection seriously commit themselves thereby to a position on the use of real number theory in empirical science that they ought, on reflection, to find wholly untenable. For if the fact that point values are in principle incapable of being arrived at is taken to invalidate such explanatory claims, then we should have to give up obedience to the laws of real arithmetic as the explanation of why we employ our customary modes of calculation where any *physical* magnitude is concerned. Although we are quite happy with the usual mathematical theory of length, volume, etc., that makes these quantities real-valued, it is nevertheless a fact that the length of, e.g., a room, is not, even discounting the practical impossibility of arbitrarily precise measurement, a real number, nor even an exact nondegenerate interval of real numbers. Our estimates of such magnitudes are made only to within a non-degenerate interval possessing no exact upper and lower limits, which *in principle* cannot be refined to a single point. This does not invalidate a theory of these magnitudes that postulates real-number values, or render meaningless calculations based on real arithmetic; we should not, therefore, feel obliged either to regard the theory of point-valued probabilities as invalidated by similar facts about the nature of subjective probability.

6. Conclusion

Methodological theories wax and wane, like the scientific hypotheses that are their subject matter. The mathematical treatment of uncertainty based on the probability calculus was one such; announced in the pages of the Port Royal Logic and some writings of Leibniz, it rapidly developed in scope and sophistication, and reached its apogee in the early years of the nineteenth century. It enabled you to compute the odds in games of chance, the odds on the sun rising the next day, and the odds on coincident testimonies of witnesses being the result of their telling the truth. By the early twentieth century the theory was virtually dead, the victim of untoward facts—not empirical facts, to be sure, but logical ones. It was inconsistent, and the inconsistency was, so it appeared, at the very heart of the theory. Yet half a century later it is back again, risen like Lazarus, and fast recruiting disillusioned members of more recent faiths.

Amended to give the theory now known as Personalist Bayesianism, it is demonstrably consistent. The cost of consistency was abandoning the famous method, called by Keynes the "Principle of Indifference" and by von Kries the

"Principle of Insufficient Reason," for obtaining ostensibly 'informationless' prior probabilities. The Principle of Indifference prescribes uniform probability or probability-density distributions over bounded parameter-spaces as the mathematical representation of ignorance, but it became apparent that these distributions are far from invariant under different ways of representing the hypothesis space, with the result that the same hypothesis may be assigned different values depending on the space in which it is embedded. Worse, one and the same hypothesis may be assigned different values under logically (or, to be more precise, logico-mathematically) *equivalent* representations of the hypothesis space.

People have tried to bring back suitably watered-down versions of the principle (they are known as Objective Bayesians). I have argued in the foregoing sections, however, that no loss in desirable strength results from simply declining to prescribe any criteria for determining prior probabilities. The addition of such criteria merely burdens the theory with indefensible assumptions, even where they are consistent. Without them, I have argued, we possess a logic of confirmation that admits that our theories are no more than theories, and seeks to discover what can be constructed according to the sole criterion of consistency. And that, I have tried to show, is quite a lot. Had Popper not shown such animosity toward the enterprise of bringing subjective probabilities into epistemology, and had he not laid down a priori his own stultifying criterion – a criterion that I have argued at length (Howson 1973, 1987) is just as indefensible as the Principle of Indifference that it much resembles – for prior probabilities, that in all interesting cases they should be zero, he might therefore have recognized in Personalist Bayesianism a genuine, purely deductive logic of confirmation, yielding nontrivial information, yet free of the sort of synthetic inductive principles he rightly declared should have no place in methodology.

References

Dempster, A. P. 1968. A Generalisation of Bayesian Inference. *Journal of the Royal Statistical Society* B, 30: 205–32.

Edgeworth, F. Y. 1895. On Some Recent Contributions to the Theory of Statistics. *Journal of the Royal Statistical Society* 58: 505–15.

Edwards, W. 1968. "Conservatism in Human Information Processing." In *Formal Representation of Human Judgement*, ed. Kleinmuntz. New York: John Wiley.

Fisher, R. A. 1936. Has Mendel's Work Been Rediscovered? *Annals of Science* 1: 115–37.

———. 1947. *The Design of Experiments*, 4th ed. Edinburgh: Oliver and Boyd.

Giere, R. N. 1984. *Understanding Scientific Reasoning*, 2d ed. New York: Holt, Rinehart and Winston.

Glymour, C. 1980. *Theory and Evidence*. Princeton: Princeton University Press.

Good, I. J. 1962. "Probability as the Measure of a Non-Measurable Set." In *Logic, Methodology and Philosophy of Science*, eds. Nagel, Suppes, and Tarski. Stanford: Stanford University Press.

Howson, C. 1984. Bayesianism and Support by Novel Facts. *British Journal for the Philosophy of Science* 35: 245–51.

——. 1973. Must the Logical Probability of Laws be Zero? *British Journal for the Philosophy of Science* 24: 133–163.

——. 1987. Popper, Prior Probabilities and Inductive Inference. *British Journal for the Philosophy of Science* 38: 207–24.

——. 1988. The Consistency of Jeffreys's Simplicity Postulate, and its Role in Bayesian Inference. *Philosophical Quarterly* 38: 68–83.

Howson, C., and Oddie, G. 1979. Miller's So-Called Paradox of Information. *British Journal for the Philosophy of Science* 30: 253–78.

Howson, C., and Urbach, P. M. 1988. *Scientific Reasoning: The Bayesian Approach*. LaSalle, Ill.: Open Court.

Jaynes, E. T. 1968. Prior Probabilities. *Institute of Electricical and Electronic Engineers Transactions on Systems Science and Cybernetics* SSC-4: 227–41.

Jeffreys, H. 1948. *Theory of Probability*, 2d. ed. Cambridge: Cambridge University Press.

Kahneman, D., and Tversky, A. 1972. Subjective Probability: A Judgement of Representativeness. *Cognitive Psychology* 3:430–54.

Koopman, B. O. 1940. The Bases of Probability. *Bulletin of the American Mathematical Society* 46: 763–74.

Kitcher, P. 1985. *Vaulting Ambition*. Cambridge: MIT Press.

Miller, D. W. 1966. A Pardox of Information. *British Journal for the Philosophy of Science* 17: 59–61.

Nickles, T. 1985. Beyond Divorce: Current Status of the Discovery Debate. *Philosophy of Science* 52: 117–207.

——. 1987. Lakatosian Heuristics and Epistemic Support. *British Journal for the Philosophy of Science* 38: 181–205.

Popper, K. R., and Miller, D. W. 1983. A Proof of the Impossibility of Inductive Probability *Nature* 302: 687f.

Redhead, M. L. G. 1986. Novelty and Confirmation. *British Journal for the Philosophy of Science* 37: 115–18.

Smith, C. A. B. 1961. Consistency in Statistical Inference. *Journal of the Royal Statistical Society* B, 23: 1–25.

Wason, P. 1966. Reasoning. In *New Horizons in Psychology*, ed. B. M. Foss. Harmondsworth, U.K.: Penguin.

Watkins, J. 1985. *Science and Scepticism*. Princeton: Princeton University Press.

Williams, P. M. 1976. Indeterminate Probabilities. *Formal Methods in the Methodology of Empirical Sciences*, eds. Przlecki, Szaniawski, and Wojcicki. Dordrecht: Ossolineum and Reidel.

Worrall, J. 1978. "The Ways in Which the Methodology of Scientific Research Programmes Improve on Popper's Methodology." In *Progress and Rationality in Science*, eds. Radnitzky and Anderson. Dordrecht: Reidel.

Zahar, E. G. 1983. Logic of Discovery or Psychology of Invention? *British Journal for the Philosophy of Science* 34: 243–61.

——. 1989. *Einstein's Revolution*. LaSalle, Ill.: Open Court.

The Value of Knowledge

Why is it better to know something rather than nothing? Perhaps because knowledge is an end in itself. But we also seek knowledge in order to make informed decisions. Informed decisions, we believe, are better than uninformed ones. What is the relevant sense of "better" and what basis is there for the belief? There is a Bayesian answer to these questions. It is that coherence *requires* you to believe in the value of knowledge.

It is a theorem in the context of the decision theory of L. J. Savage (1954) and a standard treatment of learning by conditionalization that the prior expected utility of making an informed decision is always at least as great as that of making an uninformed decision, and is strictly greater if there is any probability that the information may affect the decision. The proof is stated in a characteristically trenchant note by I. J. Good (1967).

This paper will investigate the extent to which this account can be generalized to various models of the acquisition of information and to variant theories of expected utility, its import for the theory of deliberation, and its connection with payoff in the long run.

I. Good Thinking

Suppose that one can either choose between n acts, $A_1, \ldots, A_n$ now or perform a cost-free experiment, E, with possible results $[e_k]$ and then decide. The value of choosing now is the expected value of the act with the greatest expected utility (the prior Bayes act):

$$\mathrm{MAX}_j \, \Sigma_i \, \mathrm{Pr}(K_i) \, \mathrm{U}(A_j \,\&\, K_i)$$
$$= \quad \mathrm{MAX}_j \, \Sigma_k \, \Sigma_i \, \mathrm{Pr}(K_i) \, \mathrm{Pr}(e_k | K_i) \, \mathrm{U}(A_j \,\&\, K_i).$$

The value of making an informed decision conditional on experimental result e is the expected utility conditional on e of the act that has the highest expected utility after assimilating the information e (the posterior Bayes act associated with e):

$$\mathrm{MAX}_j \, \Sigma_i \, \mathrm{Pr}(K_i | e) \, \mathrm{U}(A_j \,\&\, K_i).$$

The present value of making an informed decision is the expectation:

$$\Sigma_k \; Pr(e_k) \; MAX_j \; \Sigma_i \; Pr(K_i | e_k) \; U(A_j \; \& \; K_i)$$
$$= \quad \Sigma_k \; Pr(e_k) \; MAX_j \; \Sigma_i \; Pr(e_k | K_i) \; Pr(K_i)/Pr(e_k) \; U(A_j \; \& \; K_i)$$
(by Bayes's Theorem)
$$= \quad \Sigma_k \; MAX_j \; \Sigma_i \; Pr(e_k | K_i) \; Pr(K_i) \; U(A_j \; \& \; K_i).$$

The formulas for the present value of making an informed decision and an uninformed decision differ only in the order of the first two operations. But it is true on general mathematical grounds that $\Sigma_k \; MAX_j \; g(k,j)$ is greater than or equal to $MAX_j \; \Sigma_k \; g(k,j)$, with strict inequality if $MAX_j \; g(k,j)$ is not the same for all k, q.e.d.

The situation is easy to visualize when there are only two possible experimental results: e_1, e_2. Suppose that there are three possible acts: A_1, A_2, A_3, whose expected utility is graphed as a function of $Pr(e_2)$ in figure 1.

If the experiment is performed and e_2 is the result, then $Pr(e_2)$ will equal 1, and A_1 will have the greatest expected utility. If the experiment is performed and e_1 is the result, then $Pr(e_2)$ will equal zero and A_2 will be the optimal act. If prior to the experiment $Pr(e_2) = 1/3$, then A_3 will be the prior Bayes act. The expected utility of the prior Bayes act is thus the convex function indicated by the bold line in figure 2:

The expected utility of an informed decision is plotted as the dotted line connecting the expected utility of A_1 at $Pr(e_2) = 1$ with the expected utility of A_2 at $Pr(e_2) = 0$. The vertical difference between the bold and dotted lines is the net gain in prior expected utility resulting from the determination to make an informed decision. This is zero only if both experimental outcomes lead to the same posterior Bayes act or if one is already certain of the experimental outcome; otherwise it is positive.

I want to take explicit notice here of some of the features of the system within which the proof proceeds. In the first place the expected utility theory assumed is statistical decision theory as found in Savage (1954), where a distinction is made between states of the world, acts, and consequences; where states of the world together with acts determine the consequences; and where the relevant expectation is unconditional expected utility:

$$\Sigma_i \; Pr(K_i) \; U(A \; \& \; K_i).$$

We will see in section V that in the decision theory of Jeffrey (1965), where the value of an act goes by its conditional expected utility, Good's theorem fails.

Secondly, by using the same notation for acts and states pre- and postexperiment, we are assuming that performance of the experiment itself cannot affect the state in any relevant way and that, so far as they affect consequences, the generic acts available postexperiment are equivalent to those available preexperiment. This assumption is violated, for example, in an Abbott and Costello movie where

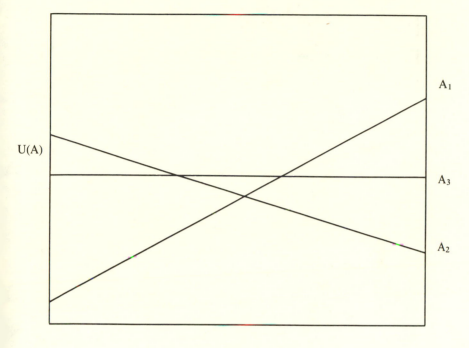

$Pr(e_2)$

Figure 1. Expected Utility as a Function of Pr (e_2).

Costello is charged with keeping a match dry until a crucial time. Abbott keeps asking Costello if the match is dry and Costello keeps replying "Yes." Finally it is time to light the match. "Are you sure it's dry?" Abbott asks for the final time. "Yes, I lit it a little while ago to make sure," Costello replies.

In the third place, the proof implicitly assumes not only that the decision maker is a Bayesian but also that he knows that he will act as one. The decision maker believes with probability one that if he performs the experiment he will (i) update by conditionalization and (ii) choose the posterior Bayes act. For an example where (i) fails, consider the case of the wrath of Khan. You are a prisoner of Khan, who (you firmly believe) is about to insert a mindworm into your brain, which will cause you to update by conditionalization on the denial of the experimental result. For an example where (ii) fails, consider the compulsive loser who conditionalizes correctly on the experimental result, but then proceeds to choose the act with minimum posterior expected utility.

I am not taking issue here with any of the assumptions, but merely making them explicit. As it stands, the theorem is a beautifully lucid answer to a fun-

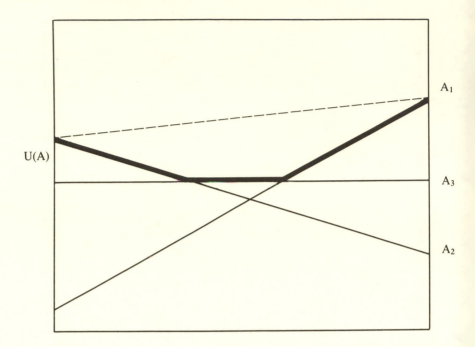

Pr(e₂)

Figure 2. The Value of Information.

damental epistemological question. It will be of some interest, however, to see
to what extent the assumptions can be relaxed and the theorem retained.

II. Probable Knowledge

Does knowledge always come nicely packaged as a proposition in one's subjective probability space? To attempt an affirmative answer would be either to defend
a form of "the myth of the given" of unprecedented strength, or to relapse into
skepticism. But the standard Bayesian theory of learning from experience by conditionalization and, in particular, the analysis of the last section appear to tacitly
make just this assumption. This is not because Bayesians have been ignorant of
the problem, but rather because it is much easier to raise the difficulty than to suggest any constructive treatment.

One positive suggestion put forward by Richard Jeffrey (1965, 1968) is to
generalize Bayes's rule of conditionalization. Suppose that an observational interaction falls short of making proposition p certain, but makes it highly probable.
Suppose also that the only effect on the observer's subjective probability space is

through the effect on p; i.e., the probabilities conditional on p and on its negation remain constant. Then we have what Jeffrey calls belief change by probability kinematics on the partition [p,-p]. Conditionalization on p and on its negation are extreme cases. Jeffrey extends the notion to any partition, all of whose members have positive probability in the obvious way. We have belief change by probability kinematics on that partition, just in case posterior probabilities conditional on members of the partition (where defined) remain unchanged.

Does the analysis of the last section extend to the case of probable knowledge? Paul Graves has recently shown that it does (Graves, forthcoming). Here is a sketch of Graves's analysis.

Suppose that you can either choose now among n acts, or perform a cost-free experiment whose result will be a belief change by probability kinematics on partition Γ. There may be no proposition in your language capturing just the phenomonological "feel" of the possible observational inputs. But are there entertainable propositions that capture the possible effects on your probability space of the observational interaction? Well, you could just *describe* the possible final probability measures that could come about. There is no reason why you could not think about these possible outcomes now, expanding your probability space to allow final probability, pr_f, to enter as a random variable.

You believe now that your observational interaction is a legitimate way of acquiring information, and so you have now:

(M) $PR(q \mid pr_f = pr^*) = pr^*(q)$.

You believe that your belief change will be by probability kinematics on partition Γ, so for any final probability pr^*, any proposition q that is "first order" (i.e., does not involve pr_f), and any member γ of Γ, you have:

(PK) $pr^*(q \mid \gamma) = PR(q \mid \gamma)$,

from the definition of probability kinematics on Γ. By (M) this is equivalent to:

(S) $PR(q \mid pr_f = pr^* \,\&\, \gamma) = PR(q \mid \gamma)$.

Since we are sure that the belief change is by probability kinematics on Γ, it is sufficient, to specify the possible final probabilities, that we specify just the final probabilities of members of the partition thus:

$$\wedge_i \, pr_f(\gamma_i) = \alpha_i,$$

since only one final probability can meet this specification and come from the initial probability by probability kinematics on Γ. So (S) becomes:

(M') $PR(q \mid \wedge_i \, pr_f(\gamma_i) = \alpha_i \,\&\, \gamma) = PR(q \mid \gamma)$.

Now the foregoing is all done in terms of what your present probabilities are about the way that your final probabilities will be after the ineffable observational interaction, without speculation as to the nature of that interaction. Nevertheless, it implies that your probabilities are structured *as if* your experimental result consisted in learning $\wedge_i \, pr_f(\gamma_i) = \alpha_i$, and conditionalizing on the result. Then Good's theorem goes through just as in the last section, with these sentences in place of the e_ks.

Graves's treatment assumes for simplicity that the prior probabilities are concentrated on a finite number of possible combinations of posterior probabilities of members of the partition, but the analysis generalizes in a straightforward way to the continuous case. Consider belief change by probability kinematics on the partition $[e_1, e_2]$ where the prior for $prf(e_1)$ is given by a continuous probability density. Here we need to strengthen (M) to:

(M+) $PR[e_1 \,|\, pr_f(e_1)\epsilon I] \; \epsilon I$,

where I is any closed interval such that $PR[pr_f(e_1)\epsilon I] > 0$. An immediate consequence of (M+) is the expectation principle:

(E) $PR\,(e_1) = E\,[pr_f\,(e_1)]$.

This together with the fact that the the expected utility of the Bayes act is a convex function of $pr(e_1)$ leads immediately to the Good theorem. Let $\emptyset$ be the expected utility of the Bayes act. From the convexity of $\emptyset$ it follows on general mathematical grounds (Jensen's inequality: Royden 1968, 110) that:

$$E\,[\emptyset\,\{pr_f(e_1)\}] \; \geq \; \emptyset\,[E\,\{pr_f(e_1)\}],$$

so by (E):

$$E\,[\emptyset\,\{pr_f(e_1)\} \; \geq \; \emptyset[\,PR(e_1)];$$

but by (M+) and the theorem on total probability, the prior expectation of the expected utility of the posterior Bayes act is equal to the prior expected utility of the decision to make an informed decision, q.e.d.

III. Ramsey's Anticipation

Good makes no great claims of originality. He cites treatments of the value of evidence by Lindley (1965) and Raiffa and Schlaifer (1961) as partial anticipations. To this list one must surely add Savage himself, who discusses the value of observation and indeed proves a form of the Good theorem in Chapter 7 (and appendix 2) of *The Foundations of Statistics* (1954). It would not be suprising if one could trace the basic idea back a little further. But it is worth reporting that Frank Ramsey had it back in the 1920s.

In two manuscript pages on "Weight or the Value of Knowledge," Ramsey (unpublished) sketches a version of the theorem. He proceeds in a rich setting where your act consists in choosing the values of a list, $x_1, x_2, \ldots$; of "control vari-

EU(x)

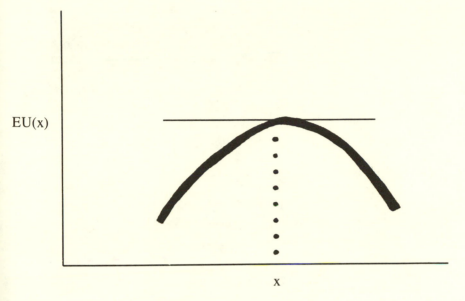

X

Figure 3. EU(x) Assuming a Maximum.

ables." He supposes that there is an unknown proposition, a, such that the expected utility of x_i considered as a function of x_i [for a fixed value of pr(a)] is continuous and twice differentiable and assumes its maximum at a nonextreme value of x_i; so that at the maximum $\delta EU(x_i|a)/\delta x_i = 0$ with the second derivative negative. This situation is illustrated in figure 3. It is also assumed that different acts are optimal for pr(a)=1 and pr(a)=0.

In this context Ramsey considers a function, $\varnothing(p)$, which he calls the "expectation of advantage in regard to a if I expect it with probability p." This is what we called in section 1 "the expected utility of the prior Bayes act." He argues that the second derivative of this function must be everywhere positive; i.e., that the function must be strictly convex. I reproduce as figure 4 Ramsey's own illustration of the situation.[1] It can be compared with the case of a finite number of acts shown in figure 2 of section 1.

It follows immediately that the expected value of an experiment whose possible results are a, not-a is positive. If Ramsey had simply noted this, it would have been enough for the theorem. Instead, however, he did something much more interesting. He considered the case in which the experimental result was not the truth value of a, but the truth value of another proposition, k, whose only effect on expected utility of the acts is in its alteration of the probability of the

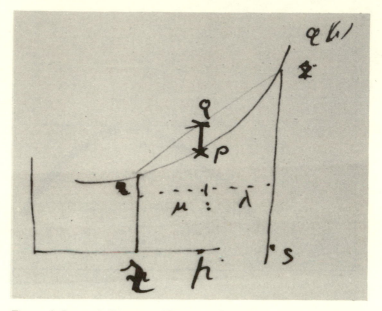

Figure 4. Ramsey's Diagram. (Courtesy of the University of Pittsburgh).

experiment. (In other words, probability of a conditional on the experimental result considered as a random variable is a sufficient statistic for the experiment.) In Ramsey's illustration, r is the probability of a conditional on k (together with background knowledge) and s is the probability of a conditional on the denial of k. As Ramsey notes, it is evident (from the strict convexity of $\varnothing$) that the gain in expected utility associated with undertaking the experiment (the length of the line segment PQ) must be positive unless k is irrelevant to the probability of a, and p=r=s. This can be seen as a special case of the problem discussed in the last section, where we have belief change by probability kinematics on a, not-a with only two possible results.

This is of special interest because we have other evidence that probability kinematical ideas were not foreign to Ramsey. There is a partial anticipation of probability kinematics in an 1851 paper by W.F. Donkin. Ramsey took a page of notes from this article. The manuscript is in the Pittsburgh archives. I quote Ramsey quoting Donkin:

> If there be any number of mutually exclusive hypotheses h_1, h_2, h_3 -; of which the probabilities relative to a particular state of information are p_1, p_2, p_3 -, and if new information is gained which changes the probabilities of some of them suppose of $h_{m=1}$ and all that follow, *without having otherwise any reference to the rest*, then the probabilities of these latter have the *same ratios* to

one another after the new information that they had *before* — . . . (Emphasis is Ramsey's.)

IV. Dynamic Probability and Other Forms of Generalized Learning

Suppose that you are in an even more amorphous learning situation than the kind that motivates Jeffrey's ideas. There is no nontrivial partition that you expect with probability one to be sufficient for your belief change. Perhaps you are in a novel situation where you expect the unexpected observational input. Perhaps there is to be no external observational input, and you are in the realm of what Good calls "dynamic probability." You are just going to think about some subject matter and the input, if you are lucky enough to have one, will be the "aha erlebniss." How unstructured can the setting for belief change be while the Good theorem is retained?

Reflection on section 2 suggests that the theorem does not really depend on any restrictions imposed by probability kinematics. After all, in the case of a discrete space where all the atoms have positive probability, *any* belief change is by probability kinematics on the partition of unit sets of the atoms. And the theorem did not depend on the belief change being by probability kinematics on any special partition. We do, however, need some form of principle M to assure you that you regard the upcoming belief-changing process as a learning experience.

Consider the case in which there are a finite number of relevant states of the world and a finite number of acts. Let the states of the world be $K_1, \ldots K_n$. The set of probability measures over these states can be represented as a k-1 dimensional simplex in k dimensional Euclidian space. The expected utility of the Bayes act is a function of these probabilities. Now suppose that the you are about to undergo what you expect will be a learning experience in that your prior probabilities satisfy an appropriate form of principle M:

$$(M++): \inf PROB\ (q) \leq PR\ [q\,|\,pr_f\ \varepsilon\ PROB] \leq \sup PROB\ (q),$$

for any q, where PROB is any rectangle (hyperinterval) of probability measures, such that $PR[pr_f \varepsilon PROB]$ is positive. This is all we assume about the learning experience. M++ guarantees that:

(E): $PR = E\ [pr_f]$.

$\varnothing$ is still convex. By the appropriate form of Jensen's inequality (Loeve 1963, 159):

$$E\ [\varnothing(pr_f)] \geq \varnothing\ [E(pr_f)],$$

and by (E):

$$E\ [\varnothing(pr_f) \geq \varnothing\ [PR].$$

The argument can be generalized under appropriate regularity conditions to cover infinite state spaces and infinite numbers of acts, but this involves some mathematical complications.[2]

Here is Good thinking reduced to its bare essentials. Nothing at all about the nature of the event that is to occasion your belief change has been specified, excepting your belief in the epistemological legitimacy of the impending belief change as embodied in (M). Bayesians have found it difficult to say anything informative about belief revision situations with so little structure, but in the presence of higher-order probabilities and condition (M) Good's theorem emerges from the mists unscathed.

V. Conflicting Expectations

I noted at the onset that Good's theorem is proved in the context of a Savage-type expected utility theory. The remark was not idle. In the variant form of expected utility theory best known to philosophers—that developed in Richard Jeffrey's *Logic of Decision* (1st ed.)—Good's theorem fails (Adams and Rosenkrantz 1980; Skyrms 1982). And for a number of forms of "causal decision theory" introduced by philosophers as alternatives to Jeffrey's theory, the question has not been adequately discussed (Skyrms 1982).

In Jeffrey's (1965) theory, there is no distinction between states, acts, and consequences. Probabilities and utilities are defined on a Boolean σ-algebra of propositions, and what we might intuitively take to be states, acts, and consequences are represented as propositions. The expected value of a proposition, A, is a conditional expected utility:

(Jeffrey) $V(A) = \Sigma_i \Pr(A \mid B_i) \, V(A \, \& \, B_i)$,

for *any* partition $[B_i]$. The probability in question is subjective degree-of-belief. Note, in particular, that A may have positive probabilistic relevance to B because A is *evidence* for B, rather than a causal factor favorable to B. For this reason we call the theory, which holds that the choice-worthiness of an act is measured by this conditional epistemic expected value, "evidential decision theory." Although Jeffrey (1965) is, perhaps, the most thorough development of evidential decision theory, it is by no means the only endorsement of this type of theory. Evidential decision theory, in various forms, has a number of advocates.

Jeffrey's theory has as a stated aim the elimination of the causal concepts implicit in Ramsey's notion of a gamble and in Savage's distinction between acts, states, and consequences from the logic of decision. This gives rise to anomalous results in certain situations in which the act is symptomatic rather than causative.

Newcomb's problem (see Nosick [1969]) is such a case. You have just taken a psychological examination that predicts fairly well how people behave in the impending decision situation. There are two boxes, one transparent and one opaque. Your choice is either to take the opaque box and get whatever is under it, or to take both boxes (yes, both!) and get everything under the opaque box, together with $1,000, which is visible under the transparent box. The experimenter has put $1,000,000 under the opaque box if her test predicted that you will take only it; nothing otherwise. The money is either there already or not, and you are convinced that there will be no cheating on the part of the experimenter. Your subjective probabilities that the experimenter will make the right prediction given that you take one box and given that you take both are both greater than .6. Taking one box is evidence that the million is there, but can in no way be a cause of its being there. A straightforward application of Jeffrey's theory leads to the recommendation that you take only the opaque box.

In order to avoid such results, Robert Stalnaker (1972) suggested that the Jeffrey expectation be modified by replacing the conditional probability with the probability of a subjunctive conditional. This idea was systematically developed by Gibbard and Harper (1980), who define expected utility of an act, A, as:

(SGH) Σ_i Pr [If I were to do A, then O_i] $D(O_i)$,

where $[O_i]$ is a partition of ultimate consequences and D is desirability.

The Stalnaker-Gibbard-Harper theory is, like that of Savage, made for conditions of determinism, for it is assumed that the subjunctive conditionals involved themselves form a partition. David Lewis (1980) suggested an extension of Stalnaker's approach to indeterministic situations by replacing the probability of the conditional with the expectation of the value of chance in a conditional with chance consequent:

(Lewis) Σ_{ij} x_j Pr [If I were to do A, CHANCE $(O_i) = x_j$] $V(O_i)$.

I sought the same generality in an extension of the Savage approach to conditions of indeterminism. My proposal was to take expected utility as:

(Skyrms) $U(A) = \Sigma_{ij}$ $Pr(K_i)$ $Pr(C_j | A \& K_i)$ $U(C_j \& K_i \& A)$,

where $[K_i]$ is a partition of states interpreted as causal preconditions of the decision, $[C_j]$ is a partition of consequences, and A is the act. The foregoing three proposals go by the (possibly misleading) name "causal decision theory."

Now Good's theorem fails for evidential decision theory. We can use a variation on Newcomb's problem to make the point. Suppose the experimenter offers you a peek under the opaque box before you make your decision, but cautions you that the accuracy of prediction holds up for subjects who are offered this option. An evidential decision theorist will presumably refuse the offer, reasoning that whatever he sees, he will subsequently take both boxes (on evidential theoretic grounds), which will be bad news (i.e., evidence that there is nothing under the opaque box).

What is the status of Good's theorem in causal decision theory? The answer is simplest with respect to the Stalnaker-Gibbard-Harper theory, for that theory can be viewed as a reformulation of the theory of Savage. In Savage's theory, the act together with the state determine the consequence, so that Savage can represent acts as random variables on the space of states. We can use the Savage framework to give a semantics for the subjunctive conditionals that occur in the Gibbard-Harper account. The conditional: "If I were to do A, consequence C would follow," is true in a state just in case the act A maps that state onto consequence C. The set of states in which such a subjunctive conditional is true is the inverse image of the consequence C under the function A. Conversely, given the Gibbard-Harper account, one can reconstruct Savage-type states as appropriate bundles of subjunctive conditionals. These remarks fall somewhat short of showing that the theories are fully intertranslatable, but are sufficient for our purposes. The Stalnaker-Gibbard-Harper theory automatically inherits Good's theorem from its Savage counterpart.

What about the more general indeterministic forms of causal decision proposed by Lewis and myself? Here again we have one question rather than two, for a relation holds between our respective proposals analogous to that between Savage and Stalnaker-Gibbard-Harper. A state of the world, K, makes the subjunctive conditional "If I were to do A, CHANCE (C) = x," true just in case $\Pr[C|A \& K] = x$.

It will suffice then, to discuss one of the proposals. I choose my own. The expected utility of A therein:

$$\Sigma_{ij} \Pr(K_i) \Pr(C_j|A \& K_i) U(C_j \& A \& K_i),$$

can be thought of as an expansion of Savage's:

$$\Sigma_i \Pr(K_i) U(A \& K_i),$$

wherein Savage's $U(A \& K_i)$ is analyzed as:

$$\Sigma_j \Pr(C_j|A \& K_i) U(C_j \& A \& K_i).$$

Accordingly, Good's theorem extends to this case provided that this quantity is independent of each experimental result. A sufficient condition for this is that we have both the following:

(I) $U(C_j \& A \& K_i \& e_k) = U(C_j \& A \& K_i)$ for all i,j,k.
(II) $\Pr(C_j|A \& K_i \& e_k) = \Pr(C_j|A \& K_i)$ for all i,j,k.

(I) is one precise way of saying in this context that information is really free. (II) may be thought of as saying that all the experimental results do is give you information about the state of the world. They do not affect your belief about the conditional chance of a consequence on an act obtaining in a given state of the world. Given the intended interpretation where [K_i] is a sufficient partition for conditional chance of consequence on act, and thus state together with act determine chance of consequence, condition (II) will be fulfilled if the theory has been properly applied. Good's theorem holds for the most general form of causal decision theory.

VI. Deliberational Equilibrium and Rational Decision

The dominant rationality concept in Bayesian decision theory is maximization of expected utility. The dominant rationality concept in the theory of games is equilibrium. The equilibrium concept has been absent from most discussions of the philosophical foundations of decision theory because those discussions have neglected aspects of the *process of deliberation.*

In the simplest cases, the dynamics of deliberation may be trivial. One calculates expected utilities and maximizes. But there are more complex cases in which the very process of deliberation can generate new information relevant to the evaluation of the expected utilities at issue. In such cases, an act may have highest expected utility in terms of the probabilities available at the onset of deliberation, but look worse than other alternatives if we feed in the information that it is the act about to be chosen.

Such cases have moved toward center stage in the philosophical discussions of the respective merits of evidential and causal decision theory. To deal with them, apologists for both theories have suggested that some sort of equilibrium condition be added to the principle of maximum expected utility in the account of individual rational decision.

Richard Jeffrey and Ellery Eells explore the possibility that the class of equilibrium decisions under evidential decision theory may not include the anomalous cases that embarrass the theory. The idea is that the extra information generated by deliberation leading to a decision would screen off the spurious correlation between act and state of the world. And William Harper, responding to problems raised by Reed Richter, suggests that causal decision theory needs to be supplemented by an equilibrium requirement. Harper, following Jeffrey, calls this a requirement of "ratifiability."

The foregoing discussion of the value of knowledge in causal and evidential decision theory suggests the following two propositions:

(I) The addition of an equilibrium requirement is *inconsistent* with the expected utility principle in evidential decision theory.

(II) The equilibrium condition, in so far as it is legitimate, *follows* from the expected utility principle in causal decision theory.

I will discuss them in order.

If a decision maker finds himself about to make an equilibrium decision, there is no problem of consistency between the equilibrium condition and the expected utility hypothesis. But suppose that the act he has calculated to have maximum expected utility will not retain maximum expected utility if the information that he is about do it is used to update his probabilities. The "ratifiability" defense of evidential decision claims that the evidential decision theorist who has just selected "take one box" as the Bayes act is in this position; that were he to update and recalculate at the moment of truth he would regret not taking two boxes. Let us suppose for the sake of argument that a case can be made out for the dubious claim that the additional information will so render act and state independent. The question still remains as to whether the decision maker should update and recalculate. This is a question of the value of knowledge, and we have seen that even in the ideal case in which the evidential decision maker is offered a free knowledge of the state of nature the theory may lead him to reject that offer. That is, the evidential decision theorist may associate negative expected utility with a policy of moving to an equilibrium decision.

In causal decision theory, on the other hand, the general validity of Good's theorem suggests that in the absence of substansial processing costs, one should assimilate whatever information is generated by the process of deliberation. Deliberation then becomes a dynamic process of moving towards the apparent optimal act under conditions of informational feedback. In such a decision process, a decision maker cannot choose a nonequilibrium act. As he gets close to choosing the act, informational feedback will make it appear less than maximally attractive, and will point his deliberation in a different direction. In causal decision theory, a policy of selecting an equilibrium decision is almost a consequence of the expected utility principle.

Why almost? Well, in the first place there are real-world costs of computation that might be significant. But even if we idealize away these costs, there is a kind of fallacy of composition involved in assuming that one can justify a strategy of informational feedback by induction on the application of the Good theorem. Consider an artificial paradoxical situation in which there is always free information available, and the decision as to which act to perform could be postponed for any finite time. Under a strategy of always assimilating free information, the decision maker would never act! When the strategy as a whole is evaluated against "Choose the prior Bayes act," we have seen that the conditions of the theorem are violated. There is no choice of a posterior Bayes act subsequent to the deliberational process. Such a deliberational strategy is not Good thinking.

A correct Bayesian treatment of deliberation must evaluate deliberational strategies as wholes. The question of optimum deliberational strategies for problems of the kind we have been discussing is a large and complex question, which I will not address in any detail here. A few qualitative points can, however, be made on the basis of the foregoing discussion.

Suppose that you are confronted with two opaque boxes, A and B, and are to choose one and receive its contents. A mean demon, who you believe is a good predictor of your behavior conditional on either choice, has put money under the box she predicted you won't choose. Let's say $1,000 if it is A and & 1,100 if B, to make it interesting. Suppose you initially incline towards B, but when about to take B, you recalculate and find that A is more appealing; but when about to take B, you find A more appealing; etc. You find your deliberations oscillating, with convergence – if any – towards indecision rather than decision. You are suffering from Richter's complaint (Richter 1984). Richter's theory of its etiology is that there is something wrong with causal decision theory. Sharvy (1983) and Harper (1986) reply that you should consider mixed decisions, which do indeed include an equilibrium decision. Richter turns the screw. Randomization costs something. In fact, let's say that the reason that this is a mean demon is that she will boil you in oil if she catches you using a nondegenerate mixed strategy.

In this case I would say that the cause of Richter's complaint is not a defect in causal decision theory, but rather a misapplication of Good's theorem. The initial choice is not between "Decide now between A and B," and "Deliberate, generate free information, and then face essentially the same choice situation." Rather, if we take the story at face value, it is between the former and "Deliberate, choose A, B, or a mixed strategy; or fail to make any choice." If this is the only other sort of deliberational strategy open to you, and you assign any positive probability to deliberation's getting you into the kind of trouble described, causal decision theory can recommend "Decide now between A and B," in accord with Richter's intuition.

The sort of pathology to which Richter calls our attention is not simply the lack of an equilibrium decision, as is shown in the following example (Skyrms 1984, 83):

> You are to choose one of three shells [A_1, A_2, A_3], and will receive what is under it. No mixed acts are allowed. (If you attempt to randomize, even mentally, the attempt will be detected and you will be shot.) A very good predictor has predicted your choice. If he predicted A_1, he put 10 cents under shell one and nothing under the others. If he predicted A_2, he put $10 under shell two and $100 under shell three. If he predicted A_3, he put $20 under shell three and $200 under shell two.

Suppose you start deliberation with a very small probability of choosing A_1 and equal probabilities of choosing A_2 and A_3. If you deliberate continuously, moving towards maximum expected utility with informational feedback, you will suffer from Richter's complaint, hanging up between A_2 and A_3. Supposing that if you can't come to a decision, you get nothing, the expected utility of choosing the prior Bayes act is higher than that of the deliberational strategy that cannot lead to a decision.

The example was designed to show something else as well. Suppose that you also have another deliberational strategy under consideration: "Choose the equilibrium decision with highest prior expected utility" (Harper 1984). There is a unique equilibrium decision here, but the conditions for the Good theorem fail as before.

What can we say about the conditions under which the Good theorem *will* apply to a deliberational strategy? In the first place, the process as a whole must be cost free. If it is possible that the process does not lead to any decision, then there is an associated cost equal to the difference in payoff of the prior Bayes act and of no decision. For the process as a whole to be cost free, the expected costs of no decision must be zero. (This may happen because a deliberational strategy has prior probability zero of leading to no decision, even though it may possibly lead to no decision.) Otherwise the costs of deliberation by that strategy must be balanced against its benefits. (One way of guaranteeing a decision is to adopt a strategy with a time limit, such that if no pure act has been selected by the time limit, the mixed act corresponding to the decision maker's state of indecision about pure acts will be selected as a default.)

In the second place, the deliberational strategy as a whole must satisfy condition (M). Thus, for any fixed act, its initial expected utility must be equal to the initial expectation of its final expected utility. This can be thought of as a condition that the deliberational strategy be unbiased. One cannot, for instance, get away with a deliberational strategy designed to generate good news. And for deliberation to be nontrivial, one must be uncertain about where deliberation will lead. Since initial probability is the initial expectation of final probability, we have:

If $Pr_i[pr_f(p) = \alpha] = 1$, then $Pr_i(p) = \alpha$.

If you know where you're going, you're already there. Thus to the extent that deliberation generates information by computation, the results of computation must be initially uncertain.

Condition (M) can also give us some guidance regarding the internal structure of a deliberational strategy. Consider the deliberational strategy that computes expected utility; assigns probability one to the act with maximum expected utility (provided there is a unique one); revises probabilities of states of nature accordingly; and recomputes, etc. In the mean-demon case, this strategy oscillates between assigning probability one to box A and probability one to box B. From the point of view of automatic control, some damping in the system would be desirable. Condition (M) provides a less *ad hoc* justification. Applying it now to stages of deliberation we have:

If $Pr_i[_p] = 1$ then $Pr_i[pr_f(p) < 1] = 0$

or, by contraposition, *if you think that you may change your mind, you're not certain.* A strategy that expects informational feedback that may with some positive probability alter the Bayes act, and that proceeds stagewise in accordance with (M), will not leap immediately to the assignment of probability one to the prior

Bayes act, but will rather move some distance in the direction of attractive options. Very simple models of this kind show suprisingly nice properties of convergence to equilibrium decisions. (E.g., the Nash dynamics discussed in Skyrms [1986] "Deliberational Equilibria.")

Questions of deliberational dynamics become more interesting in game-theoretic contexts in which Bayesian players who know each other's initial starting points deliberate, while attempting to second-guess each other's deliberations. In such contexts, informational feedback may be more interesting than in examples like the mean demon, and individual deliberational equilibriums may be related to game theoretic equilibriums. Rational deliberation may lead Bayesian players from an initial position to a unique game-theoretic equilibrium solution in non-zero sum games with multiple equilibriums. In this way, deliberational dynamics is related to the "tracing procedure" of Harsanyi and Selten (Harsanyi 1975). Detailed discussion of these questions is beyond the scope of this paper, but perhaps enough has been said to show that considerations of the value of information must play a foundational role in the theory of deliberation.

VII. Condition M

The analysis of generalized Good thinking shows that condition M is of central importance. Fulfillment of condition M shows that you regard any impending belief change as a generalized learning experience. This applies even when the experience cannot be neatly summarized as being given a proposition in your probability space. Consider the example of Ulysses and the sirens. Prior to the encounter with the sirens, Ulysses has initial probabilities, Pr, about what his probabilities will be after hearing the siren song. The possible sensory siren inputs cannot be easily summarized as propositions in his probability space. Ulysses believes that the sirens have the power and predilection to cloud men's minds so that they cease to believe that the rocks are dangerous (R). Condition M fails:

$$Pr(R \mid pr_f(R) < .1) > .1,$$

and Ulysses believes that this sort of input can get him in deep trouble. Prudently, he makes arrangements (i) to prevent the input, and (ii) to prevent himself from acting effectively if (i) fails.

In the setting of the original Good theorem, condition M holds automatically. Since the learning experience consists of conditionalizing on the experimental result, e_k, final probability after the learning experience as a random variable in the initial probability space is identical to probability conditional on the partition of possible experimental results:

$$pr_f (\bullet) = Pr[\bullet \| \{e_k\}],$$

which is a point function that takes as its value the ordinary conditional probability $Pr[\bullet | e]$ for the member of the partition, e, that includes the point. But condition M holds by definition for probability conditional on a partition. Conversely, in the "dynamic probability" setting, with a finite number of possible pr_fs, we expand the initial small probability space by taking its product with space generated by taking the possible pr_fs as atoms. In the larger space, we have the partition of possible final probabilities (each of which, we are assuming, has positive prior probability). Condition M is necessary and sufficient for the final probability as a random variable to be equal to probability conditional on *this* partition. M is the principle that relates final probability to conditional probability.

M is a principle of dynamic *coherence*. An agent's degrees of belief are statically coherent at a time in the weak sense if it is impossible to make a "Dutch book" against him in a finite number of bets that he regards at that time as fair or favorable, such that he suffers a net loss in every possible outcome. An agent's degrees of belief are *statically coherent in the strong sense* if it is impossible to make a Dutch book against him using bets all of which he considers favorable (or, equivalently, assuming payoffs are real valued, using a countable number of bets that he considers fair or favorable). Static coherence in the strong sense requires the agent's degrees of belief to be a countably additive probability measure (Adams 1962), and we make this strong assumption in this paper. An agent is *dynamically coherent* (in strong and weak senses) if a Dutch book cannot be made against him by betting at different times, the bettor at the time of betting knowing no more than the agent does. (Goldstein 1983; Skyrms forthcoming a,b; Van Fraassen 1984).

An agent has probabilities now about her probabilities tomorrow. Consider the simple case in which there are only a finite number of probabilities that she may have tomorrow, each of which today has finite positive probability. We assume that tomorrow the agent will *know her own mind*: $Pr_f (pr_f = Pr_f) = 1$, so controversial features of higher order probability do not come into play. If we give the epistemic situation only this much structure, what can we say about dynamic coherence?

A necessary and sufficient condition for dynamic coherence is that the agent's initial probability, Pr_i, satisfy:

(M) $PR(q | pr_f = pr^*) = pr^*(q)$.

Necessity is established directly by describing a betting strategy that, by a finite number of bets today and tomorrow, can make a Dutch book against any agent who violates M (as in Goldstein 1983; or Van Fraassen 1984). Sufficiency is essential because the random variable pr_f can be construed as probability conditional on a partition. Any payoff function that the bettor can attain by betting at t_1 today and then tomorrow, if pr_f in some set S, can be attained by betting only at t_1 utilizing bets conditional on S (where S is a disjunction of a finite number of pr_2s). (Something similar can again be said with regard to more general versions of M. See Goldstein [1983]; Gaifman [1986]; Skyrms [forthcoming a]).

M is a condition for convergence.[3] Consider an infinite sequence of learning situations at time $t_1, t_2, \ldots$ and a corresponding sequence of probabilities $pr_1, pr_2, \ldots$ that indicate respectively the upshot of these learning situations. The situations might, for instance, be observations of outcomes of flips of a coin with unknown bias with updating by conditionalization. Prior to the sequence, we suppose that you have a big probability space on which $pr_1, pr_2, \ldots$ are random variables. Suppose also that you have the appropriate version of M to assure the expectation principle:

$$E\ [pr_{n+1} \parallel pr_1, \ldots pr_n] = pr_n.$$

(In our example, the random variable consisting of the expected value of probability of heads after observing $n+1$ tosses, conditional on the probabalities after observing the first n tosses, is identical to the probability after observing n tosses considered as a random variable.) Then the sequence of upcoming revised probabilities as random variables, $pr_1, pr_2, \ldots$ forms a nonnegative *martingale*. Then with probability one, the sequence $pr_1, pr_2, \ldots$ converges to a random variable pr_{ff} with:

$$E\ [pr_{ff}] = E\ [pr_1] = PR$$

The random variable, pr_{ff}, is final probability in the light of the whole learning sequence. In our example, pr_{ff} is a reasonable facsimile of "the true chance that the coin comes up heads." Your initial probability of heads, PR(H), is equal to your initial expectation of the true chance of heads, $E\ [pr_{ff}(H)]$. The example generalizes (see Dynkin 1978; Diaconis and Freedman 1981). Almost everything we know about convergence to a limiting relative frequency—the strong law of large numbers, de Finetti's theorem, and generalizations of de Finetti's theorem for various versions of partial exchangeability—are special cases of this martingale convergence argument.

Returning to the theme of this paper, what can we say about the value of knowledge in the long run? Suppose that learning experiences were really free in terms of time and opportunity costs, so that one could undertake an infinite number of learning experiences in a finite time, and then make a decision. By the martingale convergence theorem, the same argument used in section 4 for pr_f, goes through here for pr_{ff}, establishing in a general way the value of making a most informed decision.[4]

Notes

1. I would like to thank the Ramsey family and the Archives for Logical Positivism at the University of Pittsburgh for permission to reproduce this figure.

2. The general case is treated in my "On the Principle of Total Evidence with and without Observa-

tion and Sentences," in *Logic, Philosophy of Science and Epistemology*, Proceedings of the 11th International Wittgenstein Symposium (Holder-Pichler-Tempsky: Vienna, 1987).

3. The following argument assumes countable additivity.

4. Research partially supported by NSF grant SES-8605122.

References

Adams, E., and Rosenkrantz, R. 1980. Applying the Jeffrey Decision Model to Rational Betting and Information Acquisition. *Theory and Decision* 12: 1–20.

Adams, E. 1962. On Rational Betting Systems. *Archive fur Mathematische Logik und Grundlagenforschung* 6: 7–18; 112–28.

Armendt, B. 1980. Is there a Dutch Book theorem for Probability Kinematics? *Philosophy of Science* 47: 583–88.

——. 1986. Foundations of Causal Decision Theory. *Topoi* 1.

de Finetti, B. 1977. "Probabilities of Probabilities: A Real Problem or a Misunderstanding?" In *New Developments in the Applications of Bayesian Methods*, ed. Aykac and Brumat 1–10. Amsterdam: North-Holland.

Diaconis, P., and Freedman, D. 1981. "Partial Exchangeability and Sufficiency." In *Statistics: Applications and New Directions*, 205–36. Calcutta: Indian Statistical Institute.

Diaconis, P., and Zabell, S. 1982. Updating Subjective Probability. *Journal of the Americal Statistical Association* 77: 822–30.

Donkin, W. F. 1851. On Certain Questions Relating to the Theory of Probabilities. *Philosophical Magazine*.

Dynkin, E. B. 1978. Sufficient Statistics and Extreme Points. *Annals of Probability* 6: 705–30.

Eells, E., and Sober, E. 1986. Common Causes and Decision Theory. *Philosophy of Science* 53: 223–45.

Eells, E., 1982. *Rational Decision and Causality*. Cambridge: Cambridge University Press.

——. 1984. Metatickles and the Dynamics of Deliberation. *Theory and Decision* 17: 71–95.

——. 1985. Weirich on Decision Instability. *Australasian Journal of Philosophy* 63: 473–78.

——. 1984. Causal Decision Theory. In *PSA 1984* East Lansing: Philosophy of Science Assn.

Gaifman, H. 1988. "A Theory of Higher Order Probabilities." In *Causation, Chance and Credence*, ed. B. Skyrms and W. Harper. Dordrecht: Reidel.

Gibbard, A., and Harper, W. 1980. "Counterfactuals and Two Kinds of Expected Utility." In *IFS*, ed. Harper et al., 153–90. Dordrecht: Reidel.

Goldstein, M. 1983. The Prevision of a Prevision. *Journal of the American Statistical Association* 78: 817–19.

Good, I. J. 1950. *Probability and the Weighing of Evidence*. New York: Hafner.

——. 1967. On the Principle of Total Evidence. *British Journal for the Philosophy of Science* 17: 319–21.

——. 1974. A Little Learning Can Be Dangerous. *British Journal for the Philosophy of Science* 25: 340–42.

——. 1983. *Good Thinking: The Foundations of Probability and Its Applications*. Minneapolis: University Of Minnesota Press.

——. 1981. The Weight of Evidence Provided by Uncertain Testimony or from an Uncertain Event. *Journal of Statistical Computation and Simulation* 13: 56–60.

Graves, P. Forthcoming. "The Total Evidence Principle for Probability Kinematics." In *Philosophy of Science*.

Harper, W. 1986. Mixed Strategies and Ratifiability in Causal Decision Theory. *Erkenntnis* 24: 25–36.

———. 1984. Ratifiability and Causal Decision Theory: Comments on Eells and Seidenfeld. In *PSA 1984*. East Lansing: Philosophy of Science Association.

———. Forthcoming. "Causal Decision Theory and Game Theory." In *Causation in Decision, Belief Change and Statistics*, ed. Harper and Skyrms. Dordrecht: Reidel.

Harsanyi, J. C. 1967. Games with Incomplete Information Played by Bayesian Players, parts I,II,III. *Management Science* 14: 159–83, 320–34, 486–502.

———. 1973. Games with Randomly Disturbed Payoffs: A New Rationale for Mixed Strategy Equilibrium Points. *International Journal of Game Theory* 2: 1–23.

———. 1975. The Tracing Proceedure: A Bayesian Approach to Defining a Solution Concept for N-Person Non-cooperative Games. *International Journal of Game Theory* 4: 61–94.

Horwich, P. 1985. Decision Theory in the Light of Newcomb's Problem. *Philosophy of Science* 52: 431–50.

Jeffrey, R. 1965. *The Logic of Decision*. N.Y.: McGraw Hill; 2d rev. ed., Chicago: University of Chicago Press 1983).

———. 1968. "Probable Knowledge." In *The Problem of Inductive Logic*, ed. Lakatos. Amsterdam: North-Holland.

———. 1981. The Logic of Decision Defended. *Synthese* 48: 473–92.

Kyburg, H. 1980. Acts and Conditional Probabilities. *Theory and Decision* 12: 149–71.

———. Forthcoming. "Powers." In *Causation in Decision, Belief Change and Statistics*, ed. W. Harper and B. Skyrms. Dordrecht: Reidel.

Levi, I. 1975. Newcomb's Many Problems. *Theory and Decision* 6: 161–75.

———. 1982. A Note on Newcombmania. *The Journal of Philosophy* 79: 337–42.

———. 1983. The Wrong Box. *The Journal of Philosophy* 80: 534–42.

Lewis, D. 1979. Prisoner's Dilemma Is a Newcomb Problem. *Philosophy and Public Affairs* 8: 235–40.

———. 1980. Causal Decision Theory. *The Australasian Journal of Philosophy* 59: 5–30.

Lindley, D. V. 1965. *Introduction to Probability and Statistics*, pt. 2. Cambridge: Cambridge University Press.

Loeve, M. 1963. *Probability Theory*, 3d ed. New York: van Nostrand.

Marshak, J. 1975. Do Personal Probabilities of Probabilities have an Operational Meaning? *Theory and Decision* 6: 127–32.

Nash, J. 1951. Non-Cooperative Games. *Annals of Mathematics* 54: 286–95.

Nozick, R. 1969. "Newcomb's Problem and Two Principles of Choice." In *Essays in Honor of Carl G. Hempel*, ed. N. Rescher. Dordrecht: Reidel.

Raiffa, H., and Schlaiffer, R. 1961. *Applied Statistical Decision Theory* Boston: Harvard Graduate School of Business Administration.

Ramsey, F. P. 1931. *The Foundations of Mathematics and Other Essays*, ed. R. B. Braithwaite. New York: Harcourt Brace.

———. Unpublished. Manuscripts 005–20–01, 005–20–03, 003- 13–01 in the Archives for Scientific Philosophy in the Twentieth Century at the Hillman Library of the University of Pittsburgh.

Richter, R. 1984. Rationality Revisited. *The Australasian Journal of Philosophy* 62: 392–403.

———. 1986. Further Comments on Decision Instability. *The Australasian Journal of Philosophy* 64: 345–49.

Royden, H. I. 1968. *Real Analysis*, 2d ed. London: Macmillan.

Savage, L. J. 1954. *The Foundations of Statistics*. N.Y.: Wiley.

———. 1967. Difficulties in the Theory of Personal Probability. *Philosophy of Science* 34: 305–10.

Seidenfeld, T. 1984. Comments on Eells. In *PSA 1984*. East Lansing: Philosophy of Science Association.

Sharvy, R. 1983. Richter Destroyed. Circulated typescript.

Skyrms, B. 1980a. *Causal Necessity.* New Haven: Yale University Press.

——. 1980b. "Higher Order Degrees of Belief." In *Prospects for Pragmatism*, ed. D. H. Mellor. Cambridge: Cambridge University Press.

——. 1982. Causal Decision Theory. *The Journal of Philosophy* 79: 695–711.

——. 1986. Deliberational Equilibria. *Topoi 1.*

——. 1985. Ultimate and Proximate Consequences in Causal Decision Theory. *Philosophy of Science* 52: 608–11.

——. 1984. *Pragmatics and Empiricism.* New Haven: Yale University Press.

——. 1987a. "Coherence." In *Scientific Inquiry in Philosophical Perspective*, ed. N. Rescher, 225–42. Pittsburgh: University of Pittsburgh Press.

——. 1987b. "Dynamic Coherence." In *Foundations of Statistical Inference*, ed. I. B. MacNeill and G. J. Umphrey. Dordrecht: Reidel.

——. 1987c. Dynamic Coherence and Probability Kinematics. *Philosophy of Science* 54: 1–20.

——. 1987d. On the Principle of Total Evidence with and without Observation Sentences. In *Proceedings of the 11th International Wittgenstein Symposium*, 187–95. Vienna: Holder-Pichler-Tempsky.

——. 1987e. Updating, Supposing and MAXENT. *Theory and Decision* 22: 225–46.

——. 1988. "Conditional Chance." In *Probability and Causality: Essays in Honor of Wesley Salmon*, ed. J. Fetzer, 161–78. Dordrecht: Reidel.

——. Forthcoming a. "Deliberational Dynamics and the Foundations of Bayesian Game Theory." In *Epistemology* [Philosophical Perspectives vol. 2], ed. J. E. Tomberlin. Northridge, Calif.: Ridgeview.

——. Forthcoming b. Probability and Causation. *Journal of Econometrics.*

Sobel, J. H. 1983. Expected Utilities and Rational Actions and Choices. *Theoria* 49.

——. 1985. Circumstances and Dominance Arguments in a Causal Decision Theory. *Synthese* 52.

——. 1986. Notes on Decision Theory: Old Wine in New Bottles. *Australasian Journal of Philosophy* 64: 407–37.

Stalnaker, R. 1972. "Letter to David Lewis," In *Ifs* (1980), ed. Harper et al. Dordrecht: Reidel.

Teller, P. 1973. Conditionalization and Observation. *Synthese* 26: 218–58.

——. 1976. "Conditionalization, Observation, and Change of Preference." In *Foundations of Probability Theory, Statistical Inference, and Statistical Theories of Science*, ed. W. Harper and C. Hooker. Dordrecht: Reidel.

Uchii, S. 1973. Higher Order Probabilities and Coherence. *Philosophy of Science* 40: 373–81.

Van Fraassen, B. 1980. Rational Belief and Probability Kinematics. *Philosophy of Science* 47: 165–87.

——. 1984. Belief and the Will. *Journal of Philosophy* 81: 235–56.

Vickers, J. 1965. Some Remarks on Coherence and Subjective Probability. *Philosophy of Science* 32: 32–38.

Weirich, P. 1985. Decision Instability. *The Australasian Journal of Philosophy* 63: 465–72.

——. 1986. Decisions in Dynamic Settings. In *PSA 1986.* East Lansing: Philosophy of Science Assn.

Demystifying Underdetermination

Pure logic is not the only rule for our judgments; certain opinions which do not fall under the hammer of the principle of contradiction are in any case perfectly unreasonable.

(Pierre Duhem[1])

Introduction

This essay begins with some good sense from Pierre Duhem. The piece can be described as a defense of this particular Duhemian thesis against a rather more familiar doctrine to which Duhem's name has often been attached. To put it in a nutshell, I shall be seeking to show that the doctrine of underdetermination, and the assaults on methodology that have been mounted in its name, founder precisely because they suppose that the logically possible and the reasonable are coextensive. Specifically, they rest on the assumption that, unless we can show that a scientific hypothesis cannot possibly be reconciled with the evidence, then we have no epistemic grounds for faulting those who espouse that hypothesis. Stated so baldly, this appears to be an absurd claim. That in itself is hardly decisive, since many philosophical (and scientific) theses smack initially of the absurd. But, as I shall show below in some detail, the surface implausibility of this doctrine gives way on further analysis to the conviction that it is even more untoward and ill argued than it initially appears. And what compounds the crime is that precisely this thesis is presupposed by many of the fashionable epistemologies of science of the last quarter century. Before this complex indictment can be made plausible, however, there is a larger story that has to be told.

I am grateful to numerous friends for arguing through the ideas of this paper with me over the months that this piece has been in incubation. Among those who deserve more than honorable mention are M. Adeel, Jarrett Leplin, Debora Mayo, Alan Musgrave, Phil Quinn, Mary Tiles, Cassandra Pinnick, and, above all, Adolf Grünbaum, who has been jostling with me about Duhem and Quine for a quarter century.

There is abroad in the land a growing suspicion about the viability of scientific methodology. Polanyi, Wittgenstein, Feyerabend and a host of others have doubted, occasionally even denied, that science is or should be a rule-governed activity. Others, while granting that there are rules of the 'game' of science, doubt that those rules do much to delimit choice (e.g., Quine, Kuhn). Much of the present uneasiness about the viability of methodology and normative epistemology can be traced to a series of arguments arising out of what is usually called "the underdetermination of theories." Indeed, on the strength of one or another variant of the thesis of underdetermination, a motley coalition of philosophers and sociologists has drawn some dire morals for the epistemological enterprise.

Consider a few of the better-known examples: Quine has claimed that theories are so radically underdetermined by the data that a scientist can, if he wishes, hold on to *any* theory he likes, "come what may." Lakatos and Feyerabend have taken the underdetermination of theories to justify the claim that the only difference between empirically successful and empirically unsuccessful theories lay in the talents and resources of their respective advocates (i.e., with sufficient ingenuity, more or less *any* theory can be made to look methodologically respectable).[2] Boyd and Newton-Smith suggest that underdetermination poses several prima facie challenges to scientific realism.[3] Hesse and Bloor have claimed that underdetermination shows the *necessity* for bringing noncognitive, social factors into play in explaining the theory choices of scientists (on the grounds that methodological and evidential considerations alone are demonstrably insufficient to account for such choices).[4] H. M. Collins, and several of his fellow sociologists of knowledge, have asserted that underdetermination lends credence to the view that the world does little if anything to shape or constrain our beliefs about it.[5] Further afield, literary theorists like Derrida have utilized underdetermination as one part of the rationale for "deconstructionism" (in brief, the thesis that, since every text lends itself to a variety of interpretations and thus since texts underdetermine choice among those interpretations, texts have no determinant meaning).[6] This litany of invocations of underdeterminationist assumptions could be expanded almost indefinitely; but that is hardly called for, since it has become a familiar feature of contemporary intellectual discourse to endow underdetermination with a deep significance for our understanding of the limitations of methodology, and thus with broad ramifications for all our claims to knowledge—insofar as the latter are alleged to be grounded in trustworthy procedures of inquiry. In fact, underdetermination forms the central weapon in the relativistic assault on epistemology.

As my title suggests, I think that this issue has been overplayed. Sloppy formulations of the thesis of underdetermination have encouraged authors to use it— sometimes inadvertently, sometimes willfully—to support whatever relativist conclusions they fancy. Moreover, a failure to distinguish several distinct species of underdetermination—some probably viable, others decidedly not—has encouraged writers to lump together situations that ought to be sharply distinguished. Above all, inferences have been drawn from the fact of underdetermination that by no means follow from it. Because all that is so, we need to get as clear as we can about this slippery concept before we can decide whether underdetermination warrants the critiques of methodology that have been mounted in its name. That is the object of the next section of this paper. With those clarifications in hand, I will then turn in succeeding parts to assess some recent garden-variety claims about the methodological and epistemic significance of underdetermination.

Although this paper is one of a series whose larger target is epistemic relativism in general[7], my limited aim here is not to refute relativism in all its forms. It is rather to show that one important line of argument beloved of relativists, the argument from underdetermination, will not sustain the global conclusions that they claim to derive from it.

Vintage Versions of Underdetermination

Humean Underdetermination. Although claims about underdetermination have been made for almost every aspect of science, those that interest philosophers most have to do specifically with claims about the underdetermination of *theories*. I shall use the term "theory" merely to refer to any set of *universal statements* that purport to describe the natural world.[8] Moreover, so as not to make the underdeterminationists' case any harder to make out than it already is, I shall—for purposes of this essay- suppose, with them, that single theories by themselves make no directly testable assertions. More or less everyone, relativist or nonrelativist, agrees that "theories are underdetermined" in some sense or other; but the seeming agreement about that formula disguises a dangerously wide variety of different meanings.

Our first step in trying to make some sense of the huge literature on underdetermination comes with the realization that there are two quite distinct families of theses, both of which are passed off as "the" thesis of underdetermination. Within each of these "families," there are still further differentiating features. The generic and specific differences between these versions, as we shall see shortly, are not minor or esoteric. They assert different things; they presuppose different things; the arguments that lead to and from them are quite different. Nonetheless each has been characterized, and often, as "*the* doctrine of underdetermination."

The first of the two generic types of underdetermination is what I shall call, for obvious reasons, deductive or *Humean underdetermination* (HUD). It amounts to one variant or other of the following claim:

HUD   For any finite body of evidence, there are indefinitely many mutually contrary theories, each of which logically entails that evidence.

The arguments for HUD are sufficiently familiar and sufficiently trivial that they need no rehearsal here. HUD shows that the fallacy of affirming the consequent is indeed a deductive fallacy (like so many other interesting patterns of inference in science); that the method of hypothesis is not logically probative; that successfully "saving the phenomena" is not a robust warrant for detachment or belief. I have no quarrels with either HUD or with the familiar arguments that can be marshaled for it. But when duly considered, HUD turns out to be an extraordinarily *weak* thesis about scientific inference, one that will scarcely sustain any of the grandiose claims that have been made on behalf of underdetermination.

Specifically, HUD is weak in two key respects: First, it addresses itself only to the role of *deductive logic* in scientific inference; it is wholly silent about whether the rules of a broader ampliative logic underdetermine theory choice. Secondly, HUD provides no motivation for the claim that *all* theories are reconcilable with any given body of evidence; it asserts rather that indefinitely many theories are so. Put differently, even if our doxastic policies were so lax that they permitted us to accept as rational any belief that logically entailed the evidence, HUD would not sanction the claim (which we might call the "*thesis of cognitive egalitarianism*") that all rival theories are thereby equally belief-worthy or equally rational to accept.

Despite these crucial and sometimes overlooked limitations of its scope, HUD still has some important lessons for us. For instance, HUD makes clear that theories cannot be "deduced from the phenomena" (in the literal, non-Newtonian sense of that phrase). It thus establishes that the resources of deductive logic are insufficient, no matter how extensive the evidence, to enable one to determine for certain that any theory is true. But for anyone comfortable with the nowadays familiar mixture of (a) fallibilism about knowledge and (b) the belief that ampliative inference depends on modes of argument that go beyond deductive logic, none of that is either very surprising or very troubling.

As already noted, HUD manifestly does *not* establish that all theories are equally good or equally well supported, or that falsifications are inconclusive or that any theory can be held on to, come what may. Nor, finally, does it suggest, let alone entail, that the methodological enterprise is hopelessly flawed because methodological rules radically underdetermine theory selection. Indeed, consistently with HUD, one could hold (although I shall not) that the ampliative rules of scientific method fully determine theory choice. HUD says nothing whatever about whether *ampliative* rules of theory appraisal do or do not determine theory choice uniquely. What HUD teaches, and all that it licenses, is that if one is prepared to accept only those theories that can be proven to be true, then one is going to have a drastically limited doxastic repertoire.

Mindful of the some of the dire consequences (enumerated above) that several authors have drawn from the thesis of underdetermination, one is inclined to invoke minimal charity by saying that Humean underdetermination must not be quite what they have in mind. And I think we have independent evidence that they do not. I have dwelt on this weak form of underdetermination to start with because, as I shall try to show below, it is the only *general* form of underdetermination that has been incontrovertibly established. Typically, however, advocates of underdetermination have a much stronger thesis in mind. Interestingly, when attacked, they often fall back on the truism of HUD; a safe strategy since HUD is unexceptionable. They generally fail to point out that HUD will support none of the conclusions that they wish to draw from underdetermination. By failing to distinguish between HUD and stronger (and more controversial) forms of underdetermination, advocates of undifferentiated underdetermination thus piggyback their stronger claims on this weaker one. But more of that below.

The Quinean Reformulations of Underdetermination.[9] Like most philosophers, Quine of course accepts the soundness of HUD. But where HUD was silent on the key question of ampliative underdetermination, Quine (along with several other philosophers) was quick to take up the slack. In particular, Quine has propounded two distinct doctrines, both of which have direct bearing on the issues before us. The first, and weaker, of these doctrines I shall call *the nonuniqueness thesis*. It holds that: *for any theory, T, and any given body of evidence supporting T, there is at least one rival (i.e. contrary) to T that is as well supported as T.*[10] In his more ambitious (and more influential) moments, Quine is committed to a much stronger position, which I call *the egalitarian thesis*. It insists that: *every theory is as well supported by the evidence as any of its rivals.*[11] Quine nowhere explicitly expresses the egalitarian thesis in precisely this form. But it will be the burden of the following analysis to show that Quine's numerous pronouncements on the retainability of theories, in the face of virtually any evidence, presuppose the egalitarian thesis, and make no sense without it. What follows is not meant to be an exegesis of Quine's intentions; it is meant, rather, as an exploration of whether Quine's position on this issue will sustain the broad implications that many writers (sometimes including Quine himself) draw from it.

What distinguishes both the nonuniqueness thesis and the egalitarian thesis from HUD is that they concern ampliative rather than deductive underdetermination; that is, they centrally involve the notion of "empirical support," which is after all the central focus of ampliative inference. In this section and the first part of the next, I shall focus on Quine's discussion of these two forms of ampliative underdetermination (especially the egalitarian thesis), and explore some of their implications. The egalitarian thesis is sufficiently extreme—not to say epistemically pernicious—that I want to take some time showing that some versions of Quine's holism are indeed committed to it. I shall thus examine its status in coniderable detail before turning in later sections to look at some other prominent ccounts of ampliative underdetermination.

Everyone knows that Quine, in his "Two Dogmas of Empiricism," maintained that:

(0)   one may hold onto any theory whatever in the face of any evidence whatever.[12]

Crucial here is the sense of "may" involved in this extraordinary claim. If taken as asserting that human beings are psychologically capable of retaining beliefs in the face of overwhelming evidence against them, then it is a wholly uninteresting truism, borne out by every chapter in the saga of human folly. But if Quine's claim is to have any bite, or any philosophical interest, it must be glossed along roughly the following lines:

(1)   It is rational to hold onto any theory whatever in the face of any evidence whatever.

I suggest this gloss because I suppose that Quine means to be telling us something about scientific rationality; and it is clear that (0), construed descriptively, has *no* implications for normative epistemology. Combined with Quine's counterpart claim that one is also free to jettison any theory one is minded to, (1) appears to assert the *equirationality* of all rival theoretical systems. Now, what grounds does Quine have for asserting (1)? One might expect that he could establish the plausibility of (1) only in virtue of examining the relevant rules of rational theory choice and showing, if it could be shown, that those rules were always so ambiguous that, confronted with any pair of theories and any body of evidence, they could never yield a decision procedure for making a choice. Such a proof, if forthcoming, would immediately undercut virtually every theory of empirical or scientific rationality. But Quine *nowhere,* neither in "Two Dogmas . . . " nor elsewhere, engages in a general examination of ampliative rules of theory choice.

His specific aim in propounding (0) or (1) is often said to be to exhibit the ambiguity of falsification or of *modus tollens.* The usual reading of Quine here is that he has shown the impotence of negative instances to disprove a theory, just as Hume had earlier showed the impotence of positive instances to prove a theory. Indeed, it is this gloss that establishes the parallel between Quine's form of the thesis of underdetermination and HUD. Between them, they seem to lay to rest any prospect for a purely deductive logic of scientific inference.

But what is the status of (1)? I have already said that Quine nowhere engages in an exhaustive examination of various rules of rational theory choice with a view to showing them impotent to make a choice between all pairs of theories. Instead, he is content to examine a *single* rule of theory choice, what we might call the Popperian gambit. That rule says, in effect, "reject theories that have (known) falsifying instances." Quine's strategy is to show that this particular rule radically underdetermines theory choice. I intend to spend the bulk of this section examining Quine's case for the claim that this particular rule underdetermines theory choice. But the reader should bear in mind that even if Quine were successful in his dissection of this particular rule (which he is not), that would still leave unsettled the question whether other ampliative rules of detachment suffer a similar fate.

How does he go about exhibiting the underdeterminative character of falsification? Well, Quine's explicit arguments for (1) in "Two Dogmas . . . " are decidedly curious. Confronted, for instance, with an apparent refutation of a claim that "there are brick houses on Elm Street," we can—he says—change the meaning of the terms so that (say) "Elm Street" now refers to Oak Street, which adventitiously happens to have brick houses on it, thereby avoiding the force of the apparent refutation. Now this is surely a Pickwickian sense of "holding onto a theory come what may," since what we are holding onto here is not what the theory asserted, but the (redefined) string of words constituting the theory.[13] Alternatively, says Quine, we can always change the laws of logic if need be. We might, one supposes, abandon *modus tollens,* thus enabling us to maintain a theory in the face of evidence that, under a former logical regime, was falsifying of it; or we could jettison *modus ponens* and thereby preclude the possibility that the theory we are concerned to save is "implicated" in any schema of inference leading to the awkward prediction. If one is loathe to abandon such useful logical devices (and Quine is), other resources are open to us. We could, says Quine, dismiss the threatening evidence "by pleading hallucination."[14]

But are there no constraints on when it is reasonable to abandon selected rules of logic or when to label evidence specious (because the result of hallucination) or when to redefine the terms of our theories? Of course, it is (for all I know) humanly possible to resort to any of these stratagems, as a descriptivist reading of (0) might suggest. But nothing Quine has said thus far gives us any grounds to believe, as (1) asserts, that it will ever, let alone *always,* be rational to do so. Yet his version of the thesis of underdetermination, if he means it to have any implications for normative epistemology, requires him to hold that it is rational to use some such devices.[15] Hence he would appear to be committed to the view that epistemic rationality gives us no grounds for avoiding such maneuvers. (On Quine's view, the only considerations that we could possibly invoke to block such stratagems have to do with pragmatic, not epistemic, rationality.[16]) Thus far, the argument for ampliative underdetermination seems made of pretty trifling stuff.

But there is a fourth, and decidedly nontrivial, stratagem that Quine envisages for showing how our Popperian principle underdetermines theory choice. This is the one that has received virtually all the exegetical attention; quite rightly too, since Quine's arguments on the other three are transparently question begging because they fail to establish the rationality of holding onto any theory in the face of any evidence. Specifically, Quine proposes that a threatened statement or theory can always be immunized from the threat of the recalcitrant evidence by making suitable adjustments in our auxiliary theories. It is here that the familiar "Duhem-Quine thesis" comes to the fore. What confronts experience in any test, according to both Quine and Duhem, is an entire theoretical structure (later dubbed by Quine "a web of belief") consisting inter alia of a variety of theories. Predictions, they claim, can never be derived from single theories but only from collectives consisting of multiple theories, statements of initial and boundary conditions, assumptions about instrumentation, and the like. Since (they claim) it is whole systems and whole systems alone that make predictions, when those predictions go awry it is theory complexes, not individual theories, that are indicted via *modus tollens*. But, so the argument continues, we cannot via *modus tollens* deduce the falsity of any component of a complex from the falsity of the complex as a whole. Quine put it this way:

> But the failure [of a prediction] falsifies only a block of theory as a whole, a conjunction of many statements. The failure shows that one or more of those statements is false, but it does not show which.[17]

Systems, complexes or "webs" apparently turn out to be unambiguously falsifiable on Quine's view; but the choice between individual theories or statements making up these systems is, in his view, radically underdetermined.

Obviously, this approach is rather more interesting than Quine's other techniques for saving threatened theories, for here we need not abandon logic, redefine the terms in our theories in patently ad hoc fashion, nor plead hallucinations. The thesis of underdetermination in this particular guise, which I shall call Quinean underdetermination (QUD), can be formulated as follows:

QUD   Any theory can be reconciled with any recalcitrant evidence by making suitable adjustments in our other assumptions about nature.

Before we comment on the credentials of QUD, we need to further disambiguate it. We especially need to focus on the troublesome phrase "can be reconciled with." On a weak interpretation, this would be glossed as "can be made logically compatible with the formerly recalcitrant evidence." I shall call this the "*compatibilist version of QUD.*' On a stronger interpretation, it might be glossed as "can be made to function significantly in a complex that entails" the previously threatening evidence. Let us call this the "*entailment version of QUD.*" To repeat, the compatibilist version says that any theory can be made *logically compatible* with any formerly threatening evidential report; the entailment interpretation insists further that any theory can be made to function essentially in a *logical derivation* of the erstwhile refuting instance.

The compatibilist version of QUD can be trivially proven. All we need do, given any web of belief and a suspect theory that is part of it, is to remove (*without replacement*) any of those ancillary statements within the web needed to derive the recalcitrant prediction from the theory. Of course, we may well lose enormous explanatory power thereby, and the web may lose much of its pragmatic utility thereby, but there is nothing in deductive logic that would preclude any of that.

The entailment version of QUD, by contrast, insists that there is always a set of auxiliary assumptions that can replace others formerly present, and that will allow the *derivation,* not of the wrongly predicted result, but of precisely what we have observed. As Grünbaum, Quinn, Laudan and others have shown,[18] neither Quine nor anyone else has ever produced a general existence proof concerning the availability either in principle or in practice of suitable (i.e., nontrivial) theory-saving auxiliaries. Hence the entailment version of QUD is without apparent warrant. For a time (circa 1962), Quine himself conceded as much.[19] That is by now a familiar result. But what I think needs much greater emphasis than it has received is the fact that, *even if nontrivial auxiliaries existed that would satisfy the demands of the entailment version of QUD, no one has ever shown that it would be rational to prefer a web that included them and the threatened theory to a rival web that dispensed with the theory in question.* Indeed, as I shall show in detail, what undermines *both* versions of QUD is that neither logical *compatibility* with the evidence nor logical *derivability* of the evidence is sufficient to establish that a theory exhibiting such empirical compatibility and derivability is rationally acceptable.

It will prove helpful to distinguish four different positive relations in which a theory (or the system in which a theory is embedded) can stand to the evidence. Specifically, a theory (or larger system of which it is a part) may:

- be logically compatible with the evidence;
- logically entail the evidence;
- explain the evidence;
- be empirically supported by the evidence.

Arguably, none of these relations reduces to any of the others; despite that, Quine's analysis runs all four together. But what is especially important for our purposes is the realization that *satisfaction of either the compatibility relation or the entailment relation fails to establish either an explanatory relation or a relation of empirical support.* For instance, theories may entail statements that they nonetheless do not explain; self-entailment being the most obvious example. Equally, theories may entail evidence statements, yet not be empirically supported by them (e.g., if the theory was generated by the algorithmic manipulation of the "evidence" in question).

So, when QUD tells us that any theory can be "reconciled" with any bit of recalcitrant evidence, we are going to have to attend with some care to what that reconciliation consists in. Is Quine claiming, for instance, that any theory can — by suitable modifications elsewhere — continue to function as part of an *explanation* of a formerly recalcitrant fact? Or is he claiming, even more ambitiously, that any formerly recalcitrant instance for a theory can be transformed into a *confirming instance* for it?

As we have seen, the only form of QUD that has been firmly established is compatibilist Quinean underdetermination (an interpretation that says a theory can always be rendered logically compatible with any evidence, provided we are prepared to give up enough of our other beliefs); so I shall begin my discussion there. Saving a prized, but threatened, theory by abandoning the auxiliary assumptions once needed to link it with recalcitrant evidence clearly comes at a price. Assuming that we give up those beliefs without replacement (and recall that this is the only case that has been made plausible), we not only abandon an ability to say anything whatever about the phenomena that produced the recalcitrant experience; we also now give up the ability to explain all the other things which those now-rejected auxiliaries enabled us to give an account of—with no guarantee whatever that we can find alternatives to them that will match their explanatory scope.

But further and deeper troubles lurk for Quine just around the corner. For it is not just explanatory scope that is lost; it is also *evidential support*. Many of those phenomena that our web of belief could once give an account of (and which presumably provided part of the good reasons for accepting the web with its constituent theories) are now beyond the resources of the web to explain and predict. That is another way of saying that the revised web, stripped of those statements formerly linking the theory in question with the mistaken prediction, now has substantially less empirical support than it once did; assuming, of course, that the jettisoned statements formerly functioned to do more work for us than just producing the discredited prediction.[20] Which clearly takes things from bad to worse. For now Quine's claim about the salvageability of a threatened theory turns out to make sense just in case the only criterion of theory appraisal is logical compatibility with observation. If we are concerned with issues like explanatory scope or empirical support, Quine's QUD in its compatibilist version cuts no ice whatsoever.

Clearly, what is wrong with QUD, and why it fails to capture the spirit of (1), is that it has dropped out any reference to the *rationality* of theory choices, and specifically theory rejections. It doubtless is possible for us to jettison a whole load of auxiliaries in order to save a threatened theory (where "save" now means specifically "to make it logically compatible with the evidence"), but Quine nowhere establishes the reasonableness or the rationality of doing so. And if it is plausible, as I believe it is, to hold that scientists are aiming (among other things) at producing theories with broad explanatory scope and impressive empirical credentials, then it has to be said that Quine has given us no arguments to suppose that any theory we like can be doctored up so as to win high marks on those scores.

This point underscores the fact that too many of the discussions of underdetermination in the last quarter century have proceeded in an evaluative vacuum. They imagine that if a course of action is logically possible, then one need not attend to the question of its rationality. But if QUD is to carry any epistemic force, it needs to be formulated in terms of the rationality of preserving threatened theories. One might therefore suggest the following substitute for QUD, (which was itself a clarification of (1)):

(2)   any theory can be rationally retained in the face of any recalcitrant evidence.

Absent strong arguments for (2) or its functional equivalents, Quinean holism, the Duhem-Quine thesis and the (non-Humean) forms of underdetermination appear to pose no threat in principle for an account of scientific methodology or rationality. The key question is whether Quine, or any of the other influential advocates of the methodological significance of underdetermination, have such arguments to make.

Before we attempt to answer that question, a bit more clarification is called for, since the notion of retainment, let alone rational retainment, is still less than transparent. I propose that we understand that phrase to mean something along these lines: to say that a theory can be rationally retained is to say that reasons can be given for holding that theory, or the system of which it is a part, as true (or empirically adequate) that are (preferably stronger than but) as least as strong as the reasons that can be given for holding as true (or empirically adequate) any of its *known* rivals. Some would wish to give this phrase a more demanding gloss; they would want to insist that a theory can be rationally held only if we can show that the reasons in its behalf are stronger than those for all its *possible* rivals, both extant and those yet-to-be-conceived. That stronger gloss, which I shall resist subscribing to, would have the effect of making it even harder for Quine to establish (2) than my weaker interpretation does. Because I believe that theory choice is generally a matter of comparative choice among extant alternatives, I see no reason why we should saddle Quine and his followers with having to defend (2) on its logically stronger construal. More to the point, if I can show that the arguments on behalf of the weaker construal fail, that indeed the weaker construal is false, it follows that its stronger counterpart fails as well, since the stronger entails the weaker. I therefore propose emending (2) as follows:

(2*)   any theory can be shown to be as well supported by any evidence as any of its known rivals.

Quine never formulates this thesis as such, but I have tried to show that defending a thesis of this sort is incumbent on anyone who holds, as Quine does, that any theory can be held true, come what may. Duly considered, (2*) is quite a remarkable thesis, entailing as it does that all the known contraries to every known theory are equally well supported. Moreover, (2*) is our old friend, the egalitarian thesis. If correct, (2*) entails (for instance) that the flat-earth hypothesis is as sound as the oblate-spheroid hypothesis[21]; that it is as reasonable to believe in fairies at the bottom of my garden as not. But, for all its counter-intuitiveness, this is precisely the doctrine to which authors like Quine, Kuhn, and Hesse are committed.[22] (In saying that Quine is committed to this position, I do not mean that he would avow it if put to him directly; I doubt that very much. My claim rather is (a), that Quine's argument in "Two Dogmas . . . " commits him to such a thesis, and (b), that those strong relativists who look to Quine as having espoused and established the eqalitarian thesis are exactly half right. I prefer to leave it to Quine exegetes to decide whether the positions of the *later* Quine allow him to be exonerated of the charge that his more recent writing run afoul of the same problem.)

One looks in vain in "Two Dogmas . . . " for even the whiff of an argument that would make the egalitarian thesis plausible. As we have seen, Quine's only marginally relevant points there are his suppositions (1) that any theory can be made logically compatible with any evidence (statement) and (2) that any theory can function in a network of statements that will entail any particular evidence statement.[23] But what serious epistemologist has ever held either (a) that bare logical compatibility with the evidence constituted adequate reason to accept a scientific theory,[24] or (b) that logical entailment of the evidence by a theory constituted adequate grounds for accepting a theory? One might guess otherwise. One might imagine that some brash hypothetico-deductivist would say that any theory that logically entailed the known evidence was acceptable. If one conjoins this doctrine with Quine's claim (albeit one that Quine has never made out) that every theory can be made to logically entail any evidence, then one has the makings of the egalitarian thesis. But such musings cut little ice, since no serious twentieth-century methodologist has ever espoused, without crucial qualifications, logical compatibility with the evidence or logical derivability of the evidence as a sufficient condition for detachment of a theory.[25]

Consider some familiar theories of evidence to see that this is so. Within Popper's epistemology, two theories , T_1 and T_2, that thus far have the same positive instances, e, may nonetheless be differentially supported by e. For instance, if T_1 predicted e before e was determined to be true, whereas T_2 is produced after e is known, then e (according to Popper) constitutes a good test of T_1 but no test of T_2. Bayesians too insist that rival (but nonequivalent) theories sharing the same known positive instances are not necessarily equally well confirmed by those instances. Indeed, if two theories begin with different prior probabilities, then their posterior probabilities must be different, *given the same positive instances*.[26] But that is just to say that even if two theories enjoy precisely the same set of known confirming instances, *it does not follow that they should be regarded as equally well confirmed by those instances*. All of which is to say that showing that rival theories enjoy the same "empirical support"—in any sense of that term countenanced by (2*)—requires more than that those rivals are compatible with, or capable of entailing, the same "supporting" evidence. (2*) turns out centrally to be a claim in the theory of evidence and, since Quine does not address the evidence relation in "Two Dogmas . . . ," one will not find further clarification of this issue there.[27]

Of course, "Two Dogmas . . . " was not Quine's last effort to grapple with these issues. Some of these themes recur prominently in *Word and Object,* and it is worth examining some of Quine's arguments about underdetermination to be found there. In that work, Quine explicitly if briefly addresses the question, already implicit in "Two Dogmas . . . ," whether ampliative rules of theory choice underdetermine theory choice.[28] Quine begins his discussion there by making the relatively mild claim that scientific methodology, along with any imaginable body of evidence, *might possibly* underdetermine theory choice. As he wrote:

> *conceivably* the truths about molecules are only partially determined by any ideal organon of scientific method plus all the truths that can be said in common sense terms about ordinary things.[29]

Literally, the remark in this passage in unexceptionable. Since we do not yet know what the final "organon of scientific method" will look like, it surely is "conceivable" that the truth status of claims about molecular structure might be underdetermined by such an organon. Three sentences later, however, this claim about the conceivability of ampliative underdetermination becomes a more ambitious assertion about the *likelihood* of such underdetermination:

> The incompleteness of determination of molecular behavior by the behavior of ordinary things . . . remains true even if we include all past, present and future irritations of all the far-flung surfaces of mankind, and probably *even if we throw in [i.e., take for granted] an in fact achieved organon of scientific method besides.*[30]

As it stands, and as it remains in Quine's text, this is no argument at all, but a bare assertion. But it is one to which Quine returns still later:

> we have no reason to suppose that man's surface irritations even unto eternity admit of any systematization that is scientifically better or simpler than all possible others. It seems *likelier*, if only on account of symmetries or dualities, that countless alternative theories would be tied for first place.[31]

Quite how Quine thinks he can justify this claim of "likelihood" for ampliative underdetermination is left opaque. Neither here nor elsewhere does he show that *any* specific ampliative rules of scientific method[32] actually underdetermine theory choice—let alone that the rules of a "final methodology" will similarly do so. Instead, on the strength of the notorious ambiguities of simplicity (and by some hand-waving assertions that other principles of method may "plausibly be subsumed under the demand for simplicity"[33]—a claim that is anything but plausible), Quine asserts "in principle," that there is "probably" no theory that can uniquely satisfy the "canons of any ideal organon of scientific method."[34] In sum, Quine fails to show that theory choice is ampliatively underdetermined even by *existing* codifications of scientific methodology (all of which go considerably beyond the principle of simplicity), let alone by all possible such codifications.[35]

More important for our purposes, even if Quine were right that no ideal organon of methodology could ever pick out any theory as uniquely satisfying its demands, we should note—in the version of underdetermination contained in the last passage from Quine—how drastically he has apparently weakened his claims from those of "Two Dogmas . . . ." That essay, you recall, had espoused the egalitarian thesis that *any* theory can be reconciled with any evidence. We noted how much stronger that thesis was than the nonuniqueness thesis to the effect that there will always be some rival theories reconcilable with any finite body of evidence. But in *Word and Object,* as the passages I have cited vividly illustrate, *Quine is no longer arguing that any theory can be reconciled with any evidence;*[36] he is maintaining rather that, no matter what our evidence and no matter what our rules of appraisal, there will always remain the possibility (or the likelihood) that the choice will not be uniquely determined. But that is simply to say that there will (probably) always be at least one contrary to any given theory that fits the data equally well—a far cry from the claim, associated with QUD and (2*), that *all* the contraries to a given theory will fit the data equally well. In a sense, therefore, Quine appears in *Word and Object* to have abandoned the egalitarian thesis for the nonuniqueness thesis, since the latter asserts not the epistemic equality of all theories but only the epistemic equality of certain theories.[37] That surmise aside, it is fair to say that *Word and Object* does nothing to further the case for Quine's egalitarian view that "any theory can be held true come what may."

Some terminological codification might be useful before we proceed, since we have reached a natural breaking point in the argument. As we have seen, one can distinguish between (a) *descriptive* (0) and (b) normative (1, 2, 2*) forms of underdetermination, depending upon whether one is making a claim about what people are capable of doing or what the rules of scientific rationality allow.[38] One can also distinguish between (c) *deductive* and (d) *ampliative* underdetermination, depending upon whether it is the rules of deductive logic (HUD) or of a broadly inductive logic or theory of rationality that are alleged to underdetermine choice (QUD). Further, we can distinguish between the claims that theories can be reconciled with recalcitrant evidence via establishing (e) *compatibility* between the two or (f) a one-way *entailment* between the theory and the recalcitrant evidence or (g) equivalence of support between rival theories. Finally, one can distinguish between (h) the doctrine that choice is underdetermined between at least one of the contraries of a theory and that theory (*nonuniqueness*) and (i) the doctrine that theory choice is underdetermined between every contrary of a theory and that theory ("cognitive *egalitarianism*").

Using this terminology, we can summarize such conclusions as we have reached to this point: In "Two Dogmas . . . ," Quine propounded a thesis of normative, ampliative, egalitarian underdetermination. Whether we construe that thesis in its compatibilist or entailment versions, it is clear that Quine has said nothing that makes plausible the idea that every prima facie refuted theory can be embedded in a rationally acceptable (i.e., empirically well-supported) network of beliefs. Moreover, "Two Dogmas . . . " developed an argument for under-determination for only one rationality principle among many, what I have been calling the Popperian gambit. This left completely untouched the question whether other rules of theory choice suffered from the same defects that Quine thought Popper's did. Perhaps with a view to remedying that deficiency, Quine argued – or, rather, alleged without argument – in *Word and Object* that *any* codi-fication of scientific method would underdetermine theory choice. Unfortunately, *Word and Object* nowhere delivers on its claim about underdetermination.

But suppose, just for a moment, that Quine had been able to show what he claimed in *Word and Object,* to wit, the nonuniqueness thesis. At best, that result would establish that for any well-confirmed theory, there is in principle at least one other theory that will be equally well-confirmed by the same evidence. That is an interesting thesis to be sure, and possibly a true one, although Quine has given us no reason to think so. (Shortly, we shall examine arguments of other authors that seem to provide some ammunition for this doctrine.) But even if true, the nonuniqueness thesis will not sustain the critiques of methodology that have been mounted in the name of underdetermination. Those critiques are all based, implicitly or explicitly, on the strong, egalitarian reading of underdetermination. They amount to saying that the project of developing a methodology of science is a waste of time since, no matter what rules of evidence we eventually produce, those rules will do nothing to delimit choice between rival theories. The charge that methodology is toothless pivots essentially on the viability of QUD in its am-pliative, egalitarian version. Nonuniqueness versions of the thesis of ampliative underdetermination at best establish that methodology will not allow us to pick out a theory as uniquely true, no matter how strong its evidential support. (*Word and Object's* weak ampliative thesis of underdetermination, even if sound, would provide no grounds for espousing the strong underdeterminationist thesis implied by the "any theory can be held come what may" dogma.[39])

Theory choice may or may not be ampliatively underdetermined in the sense of the nonuniqueness thesis; that is an open question. But however that issue is resolved, that form of underdetermination poses no challenge to the methodologi-cal enterprise. What would be threatening to, indeed debilitating for, the methodological enterprise is if QUD in its egalitarian version were once estab-lished. Even though Quine offers no persuasive arguments in favor of normative, egalitarian, ampliative underdetermination, there are several other philosophers who appear to have taken up the cudgels on behalf of precisely such a doctrine. It is time I turned to their arguments.

Ampliative Underdetermination

With this preliminary spade work behind us, we are now in a position to see that the central question about underdetermination, at least so far as the philosophy of science is concerned, is the issue of ampliative underdetermination. Moreover, as we have seen, the threat to the epistemological project comes, not from the nonuniqueness version of underdetermination, but from the egalitarian version. (That version states that any theory can be embedded in a system that will be as strongly supported by the evidence as any rival is supported by the same evidence.) The question is whether anyone has stronger arguments than Quine's for the methodological underdetermination of theory choice. Two plausible contenders for that title are Nelson Goodman and Thomas Kuhn. I shall deal briefly with them in turn.

Goodman's *Fact, Fiction and Forecast* is notorious for posing a particularly vivid form of ampliative underdetermination, in the form of the grue/green, and related, paradoxes of induction. Goodman is concerned there to deliver what Quine had elsewhere merely promised, namely, a proof that the inductive rules of scientific method underdetermine theory choice in the face of any conceivable evidence. The general structure of Goodman's argument is too familiar to need any summary here. But it is important to characterize carefully what Goodman's result shows. I shall do so utilizing terminology we have already been working with. Goodman shows that one specific rule of ampliative inference (actually a whole family of rules bearing structural similarities to the straight rule of induction) suffers from this defect: Given any pair (or n-tuple) of properties that have previously always occurred together in our experience, it is possible to construct an indefinitely large variety of contrary theories, all of which are compatible with the inductive rule: "If, for a large body of instances, the ratio of the successful instances of a hypothesis is very high compared to its failures, then assume that the hypothesis will continue to enjoy high success in the future." All these contraries will (along with suitable initial conditions) entail all the relevant past observations of the pairings of the properties in question. Thus, in one of Goodman's best-known examples, the straight rule will not yield an algorithm for choosing between "All emeralds are green" and "All emeralds are grue"; it awards them equally good marks.

There is some monumental question begging going on in Goodman's setting up of his examples. He supposes without argument that—since the contrary inductive extrapolations all have the same positive instances (to date)—the inductive logician must assume that the extrapolations from each of these hypotheses are all rendered equally likely by those instances. Yet we have already had occasion to remark that "possessing the same positive instances" and "being equally well confirmed" boil down to the same thing only in the logician's never-never land. (It was Whewell, Peirce and Popper who taught us all that theories sharing the same positive instances need not be regarded as equally well tested or equally belief- worthy.) But Goodman does have a point when he directs our attention to the fact that the straight rule of induction, as often stated, offers no grounds for distinguishing between the kind of empirical support enjoyed by the green hypothesis and that garnered by the grue hypothesis.

Goodman himself believes, of course, that this paradox of induction can be overcome by an account of the entrenchment of predicates. Regardless whether one accepts Goodman's approach to that issue, it should be said that strictly he does not hold that theory choice is underdetermined; on his view, such ampliative underdetermination obtains only if we limit our organon of scientific methodology to some version of the straight rule of induction.

But, for purposes of this paper, we can ignore the finer nuances of Goodman's argument since, even if a theory of entrenchment offered no way out of the paradox, and even if the slide from "possessing the same positive instances" to "being equally well confirmed" was greased by some plausible arguments, Goodman's arguments can provide scant comfort to the relativist's general repudiation of methodology. Recall that the relativist is committed, as we have seen, to arguing an egalitarian version of the thesis of ampliative underdetermination, i.e., he must show that all rival theories are equally well supported by any conceivable evidence. But there is nothing whatever in Goodman's analysis – even if we grant *all* its controversial premises – that could possibly sustain such an egalitarian conclusion. Goodman's argument, after all, does not even claim to show apropos of the straight rule that it will provide support for any and every hypothesis; his concern, rather, is to show that there will always be a family of contrary hypotheses between which it will provide no grounds for rational choice. The difference is crucial. If I propound the hypothesis that "All emeralds are red" and if my evidence base happens to be that all previously examined emeralds are green, then the straight rule is unambiguous in its insistence that my hypothesis be rejected. The alleged inability of the straight rule to distinguish between green- and grue-style hypotheses provides no ammunition for the claim that such a rule can make no epistemic distinctions whatever between rival hypotheses. If we are confronted with a choice between (say) the hypotheses that all emeralds are red and that all are green, then the straight rule gives us entirely unambiguous advice concerning which is better supported by the relevant evidence. Goodmanian underdetermination is thus of the nonuniqueness sort. When one combines that with a recognition that Goodman has examined but one among a wide variety of ampliative principles that arguably play a role in scientific decision making, it becomes clear that no global conclusions whatever can be drawn from Goodman's analysis concerning the general inability of the rules of scientific methodology for strongly delimiting theory choice.

But we do not have to look very far afield to find someone who does propound a strong (viz., egalitarian) thesis of ampliative underdetermination, one which, if sound, would imply that the rules of methodology were never adequate to enable one to choose between any rival theories, regardless of the relevant evidence. I refer, of course, to Thomas Kuhn's assertion in *The Essential Tension* to the effect that the shared rules and standards of the scientific community *always* underdetermine theory choice.[40] Kuhn there argues that science is guided by the use of several methods (or, as he prefers to call them, "standards"). These include the demand for empirical adequacy, consistency, simplicity, and the like. What Kuhn says about these standards is quite remarkable. He is not making the point that the later Quine and Goodman made about the methods of science; namely, that for any theory picked out by those methods, there will be indefinitely many contraries to it that are equally compatible with the standards. On the contrary, Kuhn is explicitly pushing the same line that the early Quine was implicitly committed to, viz., that the methods of science are inadequate ever to indicate that any theory is better than any rival, regardless of the available evidence. In the language of this essay, it is the egalitarian form of underdetermination that Kuhn is here proposing.

Kuhn, of course, does not use that language, but a brief rehearsal of Kuhn's general scheme will show that egalitarian underdetermination is one of its central underpinnings. Kuhn believes that there are divergent paradigms within the scientific community. Each paradigm comes to be associated with a particular set of practices and beliefs. Once a theory has been accepted within an ongoing scientific practice, Kuhn tells us, there is nothing that the shared standards of science can do to dislodge it. If paradigms do change, and Kuhn certainly believes that they do, this must be the result of "individual" and "subjective" decisions by individual researchers, not because there is anything about the methods or standards scientists share that ever requires the abandonment of those paradigms and their associated theories. In a different vein, Kuhn tells us that a paradigm always looks good by its own standards and weak by the standards of its rivals and that there never comes a point at which adherence to an old paradigm or resistance to a new one ever becomes "unscientific."[41] In effect, then, Kuhn is offering a paraphrase of the early Quine, but giving it a Wittgensteinean twist: "once a theory/paradigm has been established within a practice, it can be held on to, come what may." The shared standards of the scientific community are allegedly impotent ever to force the abandonment of a paradigm, and the specific standards associated with any paradigm will always give it the nod.

If this seems extreme, I should let Kuhn speak for himself. "*Every* individual choice between competing theories," he tells us, "depends on a mixture of objective and subjective factors, or of shared and individual criteria."[42] It is, in Kuhn's view, no accident that individual or subjective criteria are used alongside the objective or shared criteria, for the latter "are not by themselves sufficient to determine the decisions of individual scientists."[43] Each individual scientist "must complete the objective criteria [with 'subjective considerations'] before any computations can be done."[44] Kuhn is saying here that the shared methods or standards of scientific research are *always* insufficient to justify the choice of one theory over another.[45] That could only be so if (2*) or one of its functional equivalents were true of those shared methods.

What arguments does Kuhn muster for this egalitarian claim? Well, he asserts that all the standards that scientists use are ambiguous and that "individuals may legitimately differ about their application to concrete cases."[46] "Simplicity, scope, fruitfulness and even accuracy can be judged differently . . . by different people."[47] He is surely right about some of this. Notoriously, one man's simplicity is another's complexity; one may think a new approach fruitful, while a second may see it as sterile. But such fuzziness of conception is precisely why most methodologists have avoided falling back on these hazy notions for talking about the empirical warrant for theories. Consider a different set of standards, one arguably more familiar to philosophers of science:

- prefer theories that are internally consistent;
- prefer theories that correctly make some predictions that are surprising given our background assumptions;
- prefer theories that have been tested against a diverse range of kinds of phenomena to those that have been tested only against very similar sorts of phenomena.

Even standards such as these have some fuzziness around the edges, but can anyone believe that, confronted with *any* pair of theories and *any* body of evidence, these standards are so rough-hewn that they could be used indifferently to justify choosing either element of the pair? Do we really believe that Aristotle's physics correctly made the sorts of surprising predictions that Newton's physics did? Is there any doubt that Cartesian optics, with its dual insistence on the instantaneous propagation of light and that light traveled faster in denser media than in rarer ones, violated the canon of internal consistency?

Like the early Quine, Kuhn's wholesale holism commits him to the view that, consistently with the shared canons of rational acceptance, any theory or paradigm can be preserved in the face of any evidence. As it turns out, however, Kuhn no more has plausible arguments for this position than Quine had. In each case, the idea that the choice between changing or retaining a theory/paradigm is ultimately and always a matter of personal preference turns out to be an unargued dogma. In each case, if one takes away that dogma, much of the surrounding edifice collapses.

Of course, none of what I have said should be taken to deny that all forms of underdetermination are bogus. They manifestly are not. Indeed, there are several types of situations in which theory choice is indeed underdetermined by the relevant evidence and rules. Consider a few:

a) We can show that for some rules, and for certain theory pairs, theory choice is underdetermined for certain sorts of evidence. Consider the well-known case of the choice between the astronomical systems of Ptolemy and Copernicus. If the only sort of evidence available to us involves reports of line-of-sight positions of planetary position, and if our methodological rule is something like "Save the phenomena," then it is easy to prove that any line-of-sight observation that supports Copernican astronomy also supports Ptolemy's.[48] (It is crucial to add, of course, that if we consider other forms of evidence besides line-of-sight planetary position, this choice is not strongly underdetermined.)

b) We can show that for some rules and for some local situations, theory choice is underdetermined, regardless of the sorts of evidence available. Suppose our only rule of appraisal says, "Accept that theory with the largest set of confirming instances," and that we are confronted with two rival theories that have the same known confirming instances. Under these special circumstances, the choice is indeterminate.[49]

What is the significance of such limited forms of ampliative underdetermination as these? They represent interesting cases to be sure, but none of them—taken either singly or in combination—establishes the soundness of strong ampliative underdetermination as a general doctrine. Absent sound arguments for global egalitarian underdetermination (i.e., afflicting every theory on every body of evidence), the recent dismissals of scientific methodology turn out to be nothing more than hollow, anti-intellectual sloganeering.

I have thus far been concerned to show that the case for strong ampliative underdetermination has not been convincingly made out. But we can more directly challenge it by showing its falsity in specific concrete cases. To show that it is ill conceived (as opposed to merely unproved), we need to exhibit a methodological rule, or set of rules, a body of evidence, and a local theory choice context in which the rules and the evidence would *unambiguously* determine the theory preference. At the formal level it is of course child's play to produce a trivial rule that will unambiguously choose between a pair of theories. (Consider the rule: "Always prefer the later theory.") But, unlike the underdeterminationists,[50] I would prefer real examples, so as not to take refuge behind contrived cases.

The history of science presents us with a plethora of such cases. But I shall refer to only one example in detail, since that is all that is required to make the case. It involves the testing of the Newtonian celestial mechanics by measurements of the "bulging" of the earth.[51] The Newtonian theory predicted that the rotation of the earth on its axis would cause a radical protrusion along the equator and a constriction at the poles—such that the earth's actual shape would be that of an oblate spheroid, rather than (as natural philosophers from Aristotle through Descartes had maintained) that of a uniform sphere or a sphere elongated along the polar axis. By the early eighteenth century, there were well-established geodesic techniques for ascertaining the shape and size of the earth (to which all parties agreed). These techniques involved the collection of precise measurements of distance from selected portions of the earth's surface. (To put it oversimply, these techniques generally involved comparing measurements of chordal segments of the earth's polar and equatorial circumferences.[52]) Advocates of the two major cosmogonies of the day, the Cartesians and the Newtonians, looked to such measurements as providing decisive evidence for choosing between the systems of Descartes and Newton.[53] At great expense, the Paris Académie des Sciences organized a series of elaborate expeditions to Peru and Lapland to collect the appropriate data. The evidence was assembled by scientists generally sympathetic to the Cartesian/Cassini hypothesis. Nonetheless, it was *their* interpretation, as well as everyone else's, that the evidence indicated that the diameter of the earth at its equator was significantly larger than along its polar axis. This result, in turn, was regarded as decisive evidence showing the superiority of Newtonian over Cartesian celestial mechanics. The operative methodological rule in the situation seems to have been something like this:

> when two rival theories, T_1 and T_2, make conflicting predictions that can be tested in a manner that presupposes neither T_1 nor T_2, then one should accept whichever theory makes the correct prediction and reject its rival.

(I shall call this rule R_1.) We need not concern ourselves here with whether R_1 is methodologically sound. The only issue is whether it underdetermines a choice between these rival cosmogonies. It clearly does not. Everyone in the case in hand agreed that the measuring techniques were uncontroversial; everyone agreed that Descartes's cosmogony required an earth that did not bulge at the equator and that Newtonian cosmogony required an oblately spheroidal earth.

Had scientists been prepared to make Quine-like maneuvers, abandoning (say) *modus ponens*, they obviously could have held on to Cartesian physics "come what may." But that is beside the point, for if one suspends the rules of inference, then there are obviously no inferences to be made. What those who hold that underdetermination undermines methodology must show is that methodological rules, even when scrupulously adhered to, fail to sustain the drawing of any clear preferences. As this historical case makes clear, the rule cited and the relevant evidence required a choice in favor of Newtonian mechanics.

Let me not be misunderstood. I am not claiming that Newtonian mechanics was "proved" by the experiments of the Académie des Sciences, still less that Cartesian mechanics was "refuted" by those experiments. Nor would I suggest for a moment that the rule in question (R_1) excluded all possible rivals to Newtonian mechanics. What is being claimed, rather, is that this case involves a certain plausible rule of theory preference that, when applied to a specific body of evidence and a specific theory choice situation, yielded (in conjunction with familiar rules of deductive logic and of evidential assessment) *unambiguous* advice to the effect that one theory of the pair under consideration should be *rejected*. That complex of rules and evidence *determined* the choice between the two systems of mechanics, for anyone who accepted the rule(s) in question.

Underdetermination and the "Sociologizing of Epistemology"

If (as we saw in the first section) some scholars have been too quick in drawing ampliative morals from QUD, others have seen in such Duhem-Quine-style underdetermination a rationale for the claim that science is, at least in large measure, the result of social processes of "negotiation" and the pursuit of personal interest and prestige. Specifically, writers like Hesse and Bloor have argued that, because theories are deductively underdetermined (HUD), it is reasonable to expect that the adoption by scientists of various ampliative criteria of theory evaluation is the result of various social, "extra-scientific" forces acting on them. Such arguments are as misleading as they are commonplace.[54]

The most serious mistake they make is that of supposing that *any* of the normative forms of underdetermination (whether deductive or ampliative, weak or strong) entails anything whatever about what *causes* scientists to adopt the theories or the ampliative rules that they do. Consider, for instance, Hesse's treatment of underdetermination in her recent *Revolutions and Reconstructions in the Philosophy of Science*. She there argues that, since Quine has shown that theories are deductively underdetermined by the data, it follows that theory choice must be based, at least in part, on certain "non-logical," "extra-empirical" criteria for what counts as a good theory.[55] Quine himself would probably agree with that much. But Hesse then goes on to say that:

> it is only a short step from this philosophy of science to the suggestion that adoption of such [non-logical, extra-empirical] criteria, that can be seen to be different for different groups and at different periods, should be explicable by social rather than logical factors.

The thesis being propounded by these writers is that since the rules of deductive logic by themselves underdetermine theory choice, it is only natural to believe that the choice of ampliative criteria of theory evaluation (with which a scientist supplements the rules of deductive logic) are to be explained by "social rather than logical factors." It is not very clear from Hesse's discussion precisely what counts as a "social factor"; but she evidently seems to think—for her argument presupposes—that everything is either deductive logic or sociology. To the extent that a scientist's beliefs go beyond what is deductively justified, Hesse seems to insist, to that degree is it an artifact of the scientist's social environment. (Once again, we find ourselves running up against the belief—against which Duhem inveighs in the opening quotation—that formal logic exhausts the realm of the "rational.")

Hesse's contrast, of course, is doubly bogus. On the one side, it presupposes that there is nothing social about the laws of logic. But since those laws are formulated in a language made by humans and are themselves human artifacts fashioned to enable us to find our way around the world, one could hold that the laws of logic are at least in part the result of social factors. But if one holds, with Hesse, that the laws of formal logic are not the result of social factors, then what possible grounds can one have for holding that the practices that constitute ampliative logic or methodology are apt to be primarily sociological in character?

What Hesse wants to do, of course, is to use the fact of logical underdetermination (HUD) as an argument for taking a sociological approach to explaining the growth of scientific knowledge. There may or may not be good arguments for such an approach. But, as I have been at some pains to show in this essay, the underdetermination of theory choice by deductive logic is not among them.

There is another striking feature of her treatment of these issues. I refer to the fact that Hesse thinks that a semantic thesis about the relations between sets of propositions (and such is the character of the thesis of deductive underdetermination) might sustain *any* causal claim whatever about the factors that lead scientists to adopt the theoretical beliefs they do. Surely, whatever the causes of a scientist's acceptance of a particular (ampliative) criterion of theory evaluation may be (whether sociological or otherwise), the thesis of deductive underdetermination entails nothing whatever about the character of those causes. The Duhem-Quine thesis is, in all of its many versions, a thesis about the logical relations between certain statements; it is not about, nor does it directly entail anything about, the causal interconnections going on in the heads of scientists who believe those statements. Short of a proof that the causal linkages between propositional attitudes mirror the formal logical relations between propositions, theses about logical underdetermination and about causal underdetermination would appear to be wholly distinct from one another. Whether theories are deductively determined by the data, or radically underdetermined by that data; in neither case does *anything* follow concerning the contingent processes whereby scientists are caused to utilize extralogical criteria for theory evaluation.

The point is that normative matters of logic and methodology need to be sharply distinguished from empirical questions about the causes of scientific belief. None of the various forms of normative underdetermination that we have discussed in this essay entails anything whatever about the causal factors responsible for scientists adopting the beliefs that they do. Confusion of the idiom of good reasons and the idiom of causal production of beliefs can only make our task of understanding either of them more difficult.[56] And there is certainly no good reason to think (with Hesse and Bloor) that, because theories are deductively underdetermined, the adoption by scientists of ampliative criteria 'should be explicable by social rather than logical factors.' It may be true, of course, that a sociological account can be given for why scientists believe what they do; but the viability of that program has nothing to do with normative underdetermination. The slide from normative to causal underdetermination is every bit as egregious as the slide (discussed earlier) from deductive to ampliative underdetermination. The wonder is that some authors (e.g., Hesse) make the one mistake as readily as the other.

David Bloor, a follower of Hesse in these matters, produces an interesting variant on the argument from underdetermination. He correctly notes two facts about the history of science: sometimes a group of scientists changes its "system of belief," even though there is "no change whatsoever in their evidential basis."[57] "Conversely," says Bloor, "systems of belief can be and have been held stable in the face of rapidly changing and highly problematic inputs from experience."[58] Both claims are surely right; scientists do not necessarily require new evidence to change their theoretical commitments, nor does new evidence—even prima facie refuting evidence—always cause them to change their theories. But the conclusion that Bloor draws from these two commonplaces about belief change and belief maintenance in science comes as quite a surprise. For he thinks these facts show that *reasonable scientists are free to believe what they like, independently of the evidence.* Just as Quine had earlier asserted that scientists can hold any doctrine immune from refutation or, alternatively, they can abandon any deeply entrenched belief, so does Bloor hold that there is virtually no connection between beliefs and evidence. He writes: "So [sic] the stability of a system of belief [including science] is the prerogative of its users."[59] Here would seem to be underdetermination with a vengeance! But once the confident rhetoric is stripped away, this emerges—like the parallel Quinean holism on which it is modeled—as a clumsy non sequitur. The fact that scientists sometimes give up a theory in the absence of anomalies to it, or sometimes hold on to a theory in the face of prima facie anomalies for it, provides no license whatever for the claim that scientists can rationally hold on to any system of belief they like, just so long as they choose to do so.

Why do I say that Bloor's examples about scientific belief fail to sustain the general morals he draws from them? Quite simply because his argument confuses necessary with sufficient conditions. Let us accept without challenge the desiderata Bloor invokes: scientists sometimes change their mind in the absence of evidence that would seem to force them to, and scientists sometimes hang on to theories even when those theories are confronted by (what might appear to be) disquieting new evidence. What the first case shows, and all that it shows, is that the theoretical preferences of scientists are influenced by factors other than purely empirical ones. But that can scarcely come as a surprise to anyone. For instance, even the most ardent empiricists grant that considerations of simplicity, economy and coherence play a role in theory appraisal. Hence, a scientist who changes his mind in the absence of new evidence *may* simply be guided in his preferences by those of his standards that concern the nonempirical features of theory. Bloor's second case shows that new evidence is not necessarily sufficient to cause scientists to change their minds even when that evidence is prima facie damaging to their beliefs. Well, to a generation of philosophers of science raised to believe that theories proceed in a sea of anomalies, this is not exactly news either.

What is novel is Bloor's suggestion that one can derive from the conjunction of these home truths the thesis that scientists—quite independent of the evidence—can reasonably decide when to change their beliefs and when not to, irrespective of what they are coming to learn about the world. But note where the argument goes astray: it claims that because certain types of evidence are neither necessary nor sufficient to occasion changes of belief, it follows that no evidence can ever compel a rational scientist to change his beliefs. This is exactly akin to saying that, because surgery is not always necessary to cure gall stones, nor always sufficient to cure them, it follows that surgery is never the appropriate treatment of choice for gall stones. In the same way, Bloor argues that because beliefs sometimes change reasonably in the absence of new evidence and sometimes do not change in the face of new evidence, it follows that we are always rationally free to let our social interests shape our beliefs.

Conclusion

We can draw together the strands of this essay by stating a range of conclusions that seem to flow from the analysis:

- The fact that a theory is deductively underdetermined (relative to certain evidence) does *not* warrant the claim that it is ampliatively underdetermined (relative to the same evidence).
- Even if we can show in principle the nonuniqueness of a certain theory with respect to certain rules and evidence (i.e., even if theory choice is weakly underdetermined by those rules), it does not follow that that theory cannot be rationally judged to be better than its extant rivals (viz., that the choice is strongly underdetermined).
- The *normative* underdetermination of a theory (given certain rules and evidence) does not entail that a scientist's belief in that theory is causally underdetermined by the same rules and evidence, and vice versa.
- The fact that *certain* ampliative rules or standards (e.g., simplicity) may strongly underdetermine theory choice does not warrant the blanket (Quinean/Kuhnian) claim that all rules similarly underdetermine theory choice.

None of this involves a denial (a) that theory choice is always deductively underdetermined (HUD) or (b) that the nonuniqueness thesis may be correct. But one may grant all that and still conclude from the foregoing that no one has yet shown that established forms of underdetermination do anything to undermine scientific methodology as a venture, in either its normative or its descriptive aspect. The relativist critique of epistemology and methodology, insofar as it is based on arguments from underdetermination, has produced much heat but no light whatever.

Appendix

In the main body of the paper, I have (for ease of exposition) ignored the more *holistic* features of Quine's treatment of underdetermination. Thus, I have spoken about single theories (a) having confirming instances, (b) entailing observation statements, and (c) enjoying given degrees of evidential support. Most of Quine's self-styled advocates engage in similar simplifications. Quine himself, however, at least in most of his moods, denies that single theories exhibit (a), (b), or (c). It is, on his view, only *whole systems* of theories that link up to experience. So if this critique of Quine's treatment of underdetermination is to have the force required, I need to recast it so that a thoroughgoing holist can see its force.

The reformulation of my argument in holistic terms could proceed along the following lines. The nested or systemic version of the nonuniqueness thesis would insist that: *For any theory, T, embedded in a system, S, and any body of evidence, e, there will be at least one other system, S'* (containing a rival to T), *such that S' is as well supported by e as S is.* The stronger, nested egalitarian thesis would read: *For any theory, T, embedded in a system, S, and any body of evidence, e, there will be systems, S_1, S_2, . . . , S_n, each containing a different rival to T, such that each is as well supported by e as S.*

Both these doctrines suffer from the defects already noted afflicting their nonholistic counterparts. Specifically, Quine has not shown that, for any arbitrarily selected rival theories, T_1 and T_2, there are respective nestings for them, S_1 and S_2, that will enjoy equivalent degrees of empirical support. Quine can, with some degree of plausibility, claim that it will be possible to find systemic embeddings for T_1 and T_2 such that S_1 and S_2 will be logically compatible with all the relevant evidence. And it is even remotely possible, I suppose, that he could show that there were nestings for T_1 and T_2 such that S_1 and S_2 respectively entailed all the relevant evidence. But as we have seen, such a claim is a far cry from establishing that S_1 and S_2 exhibit equal degrees of empirical support. Thus, Quine's epistemic egalitarianism is as suspect in its holistic versions as in its atomistic counterpart.

Notes

1. Pierre Duhem, *Aim and Structure of Physical Theory*, 217.
2. Lakatos once put the point this way:

A brilliant school of scholars (backed by a rich society to finance a few well-planned tests) might succeed in pushing any fantastic programme [however "absurd"] ahead, or, alternatively, if so inclined, in overthrowing any arbitrarily chosen pillar of "established knowledge." (In I. Lakatos and A. Musgrave, eds., *Criticism and the Growth of Knowledge* (Cambridge: Cambridge University Press, 1970), 187–88.)

3. See especially R. Boyd, "Realism, Underdetermination, and a Causal Theory of Evidence," *Nous*, 7 (1973): 1–12. W. Newton-Smith goes so far as to entertain (if later to reject) the hypothesis that "given that there can be cases of the underdetermination of theory by data, realism . . . has to be rejected." ("The Underdetermination of Theories by Data," in N. R. Hilpinen, ed., *Rationality in Science* (Dordrecht: Reidel, 1980), 105.) (Compare John Worrall, "Scientific Realism and Scientific Change," *The Philosophical Quarterly*, 32 (1982): 210–31.)

4. See chap. 2 of Hesse's *Revolutions and Reconstructions in the Philosophy of Science* (Notre Dame: Notre Dame Press, 1980) and D. Bloor's *Knowledge and Social Imagery* (London: Routledge, 1976) and "The Strengths [sic] of the Strong Programme," *Philosophy of the Social Sciences*, 11 (1981): 199–214.

5. See H. Collins's essays in the special number of *Social Studies of Science*, 11 (1981). Among Collins's many fatuous *obiter dicta*, my favorites are these:

• "the natural world in no way constrains what is believed to be" (*ibid.*, 54); and

• "the natural world has a small or nonexistent role in the construction of scientific knowledge" (3). Collins's capacity for hyperbole is equaled only by his tolerance for inconsistency, since (as I have shown in "Collins's Blend of Relativism and Empiricism," *Social Studies of Science*, 12 [1982], 131–33) he attempts to argue for these conclusions by the use of empirical evidence! Lest it be supposed that Collins's position is idiosyncratic, bear in mind that the self-styled "arch-rationalist," Imre Lakatos, could also write in a similar vein, apropos of underdetermination, that:

The direction of science is determined primarily by human creative imagination and not by the universe of facts which surrounds us. Creative imagination is likely to find corroborating novel evidence even for the most "absurd" programme, if the search has sufficient drive (Lakatos, *Philosophical Papers* [Cambridge: Cambridge University Press, 1978], vol. 1, 99.)

6. For a discussion of many of the relevant literary texts, see A. Nehamas, "The Postulated Author," *Critical Inquiry*, 8 (1981): 133–49.

7. See, for instance, my "The Pseudo-Science of Science?," *Philosophy of the Social Sciences,* 11 (1981): 173–98; "More on Bloor," *Philosophy of the Social Sciences,* 12 (1982): 71–74; "Kuhn's Critique of Methodology," in J. Pitt, ed., *Change and Progress in Modern Science* (Dordrecht: Reidel, 1985), 283–300; "Explaining the Success of Science: Beyond Epistemic Realism and Relativism," G. Gutting *et. al.*, eds., *Science and Reality: Recent Work in the Philosophy of Science* (Notre Dame: Notre Dame Press, 1984), 83–105; "Are All Theories Equally Good?" in R. Nola ed., *Relativism and Realism in Science* (Dordrecht: Reidel), 117–39; "Cognitive Relativism," in R. Egidi, ed., *La Svolte Relativistica* (Rome: Franco Angeli, 1988) 203–24; "Relativism, Naturalism and Reticulation," *Synthese*, 71 (1987), 114–39; "Methodology: Its Prospects," *PSA-86*, vol. 2, P. Machamer, ed., forthcoming; and "For Method (and Against Feyerabend)," in J. Brown, ed., *Festschrift for Robert Butts*, forthcoming.

8. There are, of course, more interesting conceptions of "theory" than this minimal one; but I do not want to beg any questions by imposing a foreign conception of theory on those authors whose work I shall be discussing.

9. Quine has voiced a preference that the view I am attributing to him should be called "the holist thesis," rather than a "thesis of underdetermination." (See especially his "On Empirically Equivalent Systems of the World," *Erkenntnis*, 9 (1975): 313–28.) I am reluctant to accept his terminological recommendation here, both because Quine's holism is often (and rightly) seen as belonging to the family of underdetermination arguments, and because it has become customary to use the term underdetermination to refer to Quine's holist position. I shall be preserving the spirit of Quine's recommendation, however, by insisting that we distinguish between what I call "nonuniqueness" (which is very close to what Quine himself calls "underdetermination") and egalitarianism (which represents one version of Quinean holism). (For a definition of these terms, see below.)

10. It is important to be clear that Quine's nonuniqueness thesis is *not* simply a restatement of

HUD, despite certain surface similarities. HUD is entirely a *logico-semantic* thesis about deductive relationships; it says nothing whatever about issues of empirical support. The nonuniqueness thesis, by contrast, is an *epistemic* thesis.

11. Obviously, the egalitarian thesis entails the nonuniqueness thesis, but not conversely.

12. Quine specifically put it this way: "Any statement can be held true come what may, if we make drastic enough adjustments elsewhere in the system [of belief]." ("Two Dogmas of Empiricism," in S. Harding, ed., *Can Theories Be Refuted?* (Dordrecht: Reidel, 1976), 6. I am quoting from the version of Quine's paper in the Harding volume since I will be citing a number of other works included there.)

13. Grünbaum, in his *Philosophical Problems of Space and Time*, 2d ed., (Dordrecht: Reidel 1974), 590–610, has pointed to a number of much more sophisticated, but equally trivial, ways of reconciling an apparently refuted theory with recalcitrant evidence.

14. Quine in Harding (see note 12 above), 60. In a much later, back-tracking essay ("On Empirically Equivalent Systems of the World," *Erkenntnis*, 9 (1975): 313–28), Quine seeks to distance himself from the proposal, implied in "Two Dogmas . . . ," that it is always (rationally) possible to reject 'observation reports'. Specifically, he says that QUD "would be wrong if understood as imposing an equal status on all the statements in a scientific theory and thus denying the strong presumption in favor of the observation statements. It is this [latter] bias which makes science empirical" (*ibid.*, p. 314).

15. In fact, of course, Quine thinks that we generally do (should?) not use such stratagems. But his only argument for avoiding such tricks, at least in "Two Dogmas of Empiricism," is that they make our theories more complex and our belief systems less efficient. On Quine's view, neither of those considerations carries any epistemic freight.

16. Quine, in Harding (see note 12 above), 63. I am not alone in finding Quine's notion of pragmatic rationality to be epistemically sterile. Lakatos, for instance, remarks of Quine's "pragmatic rationality": "I find it irrational to call this 'rational' " (Lakatos, *Philosophical Papers* (Cambridge: Cambridge University Press, 1978), vol. 1, 97n).

17. W. Quine, *Ontological Relativity and Other Essays* (New York: Columbia University Press, 1969), 79.

18. See especially A. Grünbaum, *Philosophical Problems of Space and Time* (Dordrecht: Reidel 1974), 585–92 and Larry Laudan, "Grünbaum on 'the Duhemian Argument'," in S. Harding, *Can Theories Be Refuted?* (Dordrecht: Reidel, 1975).

19. In a letter to Grünbaum, published in Harding (see note 12 above), p. 132, Quine granted that "the Duhem-Quine thesis" (a key part of Quine's holism and thus of QUD) "is untenable if taken nontrivially." Quine even goes so far as to say that the thesis is not "an interesting thesis as such." He claims that all he used it for was to motivate his claim that meaning comes in large units, rather than sentence-by-sentence. But just to the extent that Quine's QUD is untenable on any nontrivial reading, then so is his epistemic claim that any theory can rationally be held true come what may. Interestingly, as late as 1975, and despite his concession that the D-Q thesis is untenable in its nontrivial version, Quine was still defending his holistic account of theory testing (see below in text).

20. And if they did not, the web would itself be highly suspect on other epistemic grounds.

21. Or, more strictly, that there is a network of statements that includes the flat-earth hypothesis and that is as well confirmed as any network of statements including the oblate-spheroid hypothesis.

22. Since I have already discussed Quine's views on these matters, and will treat Kuhn's in the next section, I will limit my illustration here to a brief treatment of Hesse's extrapolations from the underdetermination thesis. The example comes from Mary Hesse's recent discussion of underdetermination in her *Revolutions and Reconstructions in the Philosophy of Science*. She writes:

> Quine points out that scientific theories are never logically determined by data, and that there are consequently [sic] always in principle alternative theories that fit the data more or less adequately. (See note 4 above, 32–33)

Hesse appears to be arguing that, because theories are deductively underdetermined, it follows that numerous theories will always fit the data "more or less adequately." But this conclusion follows not at all from Quine's arguments, since the notion of "adequacy of fit" between a theory and the data is an epistemic and methodological notion, not a logical or syntactic one. I take it that the claim that a theory fits a given body of data "more or less *adequately*" is meant to be, among other things, an indication that the data lend a certain degree of support to the theory that they "fit." As we have already seen, there may be numerous rival theories that fit the data (say in the sense of entailing them); yet that implies nothing about equivalent degrees of support enjoyed by those rival theories. It would do so only if we subscribed to some theory of evidential support that held that "fitting the data" was merely a matter of entailing it, or approximately entailing it (assuming counterfactually that this latter expression is coherent). Indeed, it is generally true that *no* available theories exactly entail the available data; so sophisticated inductive-statistical theories must be brought to bear to determine which fits the data best. We have seen that Quine's discussion of underdetermination leaves altogether open the question whether there are always multiple theories that "fit the data" equally well, when that phrase is acknowledged as having extra-syntactic import. If one is to establish that numerous alternative theories "fit the data more or less adequately," then one must give arguments for such ampliative underdetermination that goes well beyond HUD and any plausible version of QUD.

23. I remind the reader again that neither Quine nor anyone else has successfully established the cogency of the entailment version of QUD, let alone the explanatory or empirical support versions thereof.

24. If it did, then we should have to say that patently nonempirical hypotheses like "The Absolute is pure becoming" had substantial evidence in their favor.

25. In his initial formulation of the qualitative theory of confirmation, Hempel toyed with the idea of running together the entailment relation and the evidential relation; but he went on firmly to reject it, not least for the numerous paradoxes it exhibits.

26. Consider, for sake of simplicity, the case where two theories each entail a true evidence statement, e. The posterior probability of each theory is a function of the ratio of the prior probability of the theory to the prior probability of e. Hence if the two theories began with different priors, they must end up with different posterior probabilities, *even though supported by precisely the same evidence*.

27. It is generally curious that Quine, who has had such a decisive impact on contemporary epistemology, scarcely ever—in "Two Dogmas . . . " or elsewhere—discussed the rules of ampliative inference. So far as I can see, Quine generally believed that ampliative inference consisted wholly of hypothetico-deduction and a simplicity postulate!

28. As we shall eventually see, the kind of underdetermination advocated in *Word and Object* has no bearing whatever on (2*) or QUD.

29. W. V. Quine, *Word and Object* (Cambridge: Cambridge University Press, 1960), 22, my italics.

30. *Ibid.*, p. 22, my italics. There is, of course, this difference between these two passages: The first says that commonsense talk of objects may conceivably underdetermine theory preferences, whereas the second passage is arguing for the probability that sensations underdetermine theory choice. In neither case does Quine give us an argument.

31. *Ibid.*, 23, my italics.

32. Except a vague version of the principle of simplicity.

33. *Ibid.*, 21.

34. *Ibid.*, 22–23.

35. In some of Quine's more recent writings (see especially his "On Empirically Equivalent Systems of the World," *Erkenntnis*, 9 (1975): 313–28), he has tended to soften the force of underdetermination in a variety of ways. As he now puts it, "The more closely we examine the thesis [of underdetermination], the less we seem to be able to claim for it as a 'theoretical thesis' " (*ibid.*, 326).

He does, however, still want to insist that "it retains significance in terms of what is practically feasible" (*ibid.*). Roughly speaking, Quine's distinction between theoretical and practical underdetermination corresponds to the situations we would be in if we had all the available evidence (theoretical underdetermination) and if we had only the sort of evidence we now possess (practical underdetermination). If the considerations that I have offered earlier are right, the thesis of practical Quinean underdetermination is as precarious as the thesis of theoretical underdetermination.

36. Quine does not repudiate the egalitarian thesis in *Word and Object*; it simply does not figure here.

37. In some of Quine's later gyrations (esp. his "On Empirically Equivalent Systems of the World") he appears to waver about the soundness of the nonuniqueness thesis, saying that he does not know whether it is true. However, he still holds on there to the egalitarian thesis, maintaining that it is "plausible" and "less beset with obscurities" than HUD (*ibid.*, 313). He even seems to think that nonuniqueness depends argumentatively on the egalitarian thesis, or at least, as he puts it, that "the holism thesis [egalitarianism] lends credence to the underdetermination theses [nonuniqueness]." (*ibid.*) This is rather like saying that the hypothesis that there are fairies at the bottom of my garden lends credence to the hypothesis that something is eating my carrots.

38. E.g., the difference between Quine's (0) and (1).

39. Quine's repeated failures to turn any of his assertions about normative underdetermination into plausible arguments may explain why, since the mid-1970s, he has been distancing himself from virtually all the strong readings of his early writings on this topic. Thus, in his 1975 paper on the topic, he offers what he calls "my latest tempered version" of the thesis of underdetermination. It amounts to a variant of nonuniqueness thesis. ("The thesis of underdetermination . . . asserts that our system of the world is bound to have empirically equivalent alternatives . . . " *ibid.*, 327.) Significantly, Quine is now not even sure whether he believes this thesis: "This, for me, is [now] an open question" (*ibid.*).

40. What follows is a condensation of a much longer argument, which can be found, with appropriate documentation, in my "Kuhn's Critique of Methodology" (see note 7 above).

41. Apropos the resistance to the introduction of a new paradigm, Kuhn claims that the historian "will not find a point at which resistance becomes illogical or unscientific" (*The Structure of Scientific Revolutions*, Chicago: University of Chicago Press, 1962, 159).

42. Kuhn, *The Essential Tension* (Chicago: University of Chicago press, 1970), 325. My italics.

43. *Ibid.* My italics.

44. *Ibid.*, 329. My italics.

45. In *Structure of Scientific Revolutions*, Kuhn had maintained that the refusal to accept a theory or paradigm "is not a violation of scientific standards" (159).

46. Kuhn, *The Essential Tension*, 322.

47. *Ibid.*, 322.

48. See, for instance, Derek Price, "Contra-Copernicus," in M. Clagett, ed., *Criticial Problems in the History of Science* (Madison, 1959), 197–218.

49. A similar remark can be made about several of Popper's rules about theory choice. Thus, Miller and Tièchy have shown that Popper's rule "accept the theory with greater verisimilitude" underdetermines choice between incomplete theories; and Grünbaum has shown that Popper's rule "prefer the theory with a higher degree of falsifiability" underdetermines choice between mutually incompatible theories.

50. Recall Quine's claim that we can hang on to any statement we like by changing the meaning of its terms.

51. See, for instance, I. Todhunter, *History of the Theories of Attraction and the Figure of the Earth* (New York: Dover, 1962).

52. Typically, astronomical measurements of angles subtended at meridian by stipulated stars were used to determine geodetic distances.

53. In fact, the actual choice during the 1730s, when these measurements were carried out, was between a Cassini-emended version of Cartesian cosmogony (which predicted an *oblong* form for the earth) and Newtonian cosmology (which required an *oblate* shape).

54. Indeed, most of so-called radical sociology of knowledge rests on just such confusions about what does and does not follow from underdetermination.

55. M. Hesse (see note 4 above), 33.

56. This is not to say, of course, that there are no contexts in which it is reasonable to speak of reasons as causes of beliefs and actions. But it is to stress that logical relations among statements cannot unproblematically be read off as causal linkages between propositional attitudes.

57. Bloor, "Reply to Buchdahl," *Studies in History and Philosophy of Science*, 13 (1982): 306.

58. *Ibid.*

59. *Ibid.* In his milder moments, Bloor attempts to play down the radicalness of his position by suggesting (in my language) that it is the nonuniqueness version of underdetermination rather than the egalitarian version that he is committed to. Thus, he says at one point that "I am not saying that any alleged law would work in any circumstances" ("Durkheim and Mauss Revisited," *Studies in History and Philosophy of Science*, 13 [1982]: 273). But if indeed Bloor believes that the stability of a system of belief *is* the prerogative of its users, then it seems he *must* hold that any "alleged law" could be made to work in any conceivable circumstances; otherwise, there would be some systems of belief that it was not at the prerogative of the holder to decide whether to hang on to.

Dubbing and Redubbing: The Vulnerability of Rigid Designation

The issues I shall consider in this paper are central to my current research, and their centrality is a source of present difficulty. In the book on which I am at work, these issues arise only after lengthy prior discussion has led to conclusions I must here present as premises. Limited illustration and evidence for those premises will follow, but only in the later portions of this paper, where they are put to work.

What I am presupposing will be suggested by the following claim: To understand some body of past scientific belief, the historian must acquire a lexicon that here and there differs systematically from the one current in his or her own day. Only by using that older lexicon can historians accurately render certain of the statements that are basic to the science under scrutiny. Those statements are not accessible by means of a translation that uses the current lexicon, not even if it is expanded by the addition of selected terms from its predecessor.

I shall elaborate that claim in the first of the three sections of this paper and illustrate it in the second by an extended analysis of some interrelated terms from the vocabulary of Newtonian mechanics. The final section will apply what I have said to some standard assertions about meaning and/or reference made by proponents of causal theory. Scientific development, I shall then suggest, has from time to time involved sets of scientific terms in systematically interrelated acts of redubbing. Only for the periods between those acts, I shall argue, does dubbing result in rigid designation. That sketch supplies the route on which I now embark.[1]

This paper is a somewhat reduced and considerably revised version of a draft, "Possible Worlds in History of Science," prepared for discussion at a Nobel Symposium held in August 1986. That fuller version, also much revised, has since been published in: Sture Allén, *Possible Worlds in Humanities, Arts, and Sciences* (Berlin: 1988). The abridged form was prepared for the annual Chapel Hill Philosophy Colloquium in October 1986. Both versions, in their published forms, owe much to the cogent criticism and advice of Barbara Partee, as well as to that of my colleagues Ned Block, Sylvain Bromberger, Dick Cartwright, Jim Higginbotham, Paul Horwich, and Judy Thomson.

*

A historian reading an out-of-date scientific text characteristically encounters passages that make no sense. That is an experience I have had repeatedly whether my subject was an Aristotle, a Newton, a Volta, a Bohr, or a Planck.[2] It has been standard to ignore such passages or to dismiss them as the products of error, ignorance, or superstition, and that response is occasionally appropriate. More often, however, sympathetic contemplation of the troublesome passages results in a different diagnosis. The apparent textual anomalies are artifacts, products of misreading.

For lack of an alternative, the historian has been understanding words and phrases in the text as he or she would if they had occurred in contemporary discourse. Through much of the text that way of reading proceeds without difficulty; most terms in the historian's vocabulary are still used as they were by the author of the text. But some sets of interrelated terms are not, and it is failure to isolate those terms and to discover how they were used that has permitted the passages in question to seem anomalous. Apparent anomaly is thus ordinarily evidence of the need for local adjustment of the lexicon, and it often provides clues to the nature of that adjustment as well. An important clue to problems in reading Aristotle's physics is provided by the discovery that the term translated 'motion' in his text refers not simply to change of position but to all changes characterized by two end points. Similar difficulties in reading Planck's early papers begin to dissolve with the discovery that, for Planck before 1907, 'the energy element $h\nu$' referred, not to a physically indivisible atom of energy (later to be called 'the energy quantum'), but to a mental subdivision of the energy continuum, any point on which could be physically occupied.

These examples all turn out to involve more than mere changes in the use of terms, thus illustrating what I had in mind years ago when speaking of the "incommensurability" of successive scientific theories.[3] In its original mathematical use 'incommensurability' meant "no common measure," for example of the hypotenuse and side of an isosceles right triangle. Applied to a pair of theories in the same historical line, the term meant that there was no common language into which both could be fully translated.[4] Some statements constitutive of the older theory could not be stated in any language adequate to express its successor and vice versa.

Incommensurability thus equals untranslatability, but what incommensurability bars is not quite the activity of professional translators. Rather, it is a quasi-mechanical activity governed in full by a manual that specifies, as a function of context, which string in one language may, *salva veritate*, be substituted for a given string in the other. Translation of that sort is Quinean, and the point at which I aim will be suggested by the remark that most or all of Quine's arguments for the indeterminacy of translation can with equal force be directed to an opposite conclusion: Instead of there being an infinite number of translations compatible with all normal dispositions to speech behavior, there are often none at all.

With that much, Quine might very nearly agree. His arguments require that a choice be made, but they do not dictate its outcome. In his view, one must either entirely abandon traditional notions of meaning, of intension, or else one must give up the assumption that language is, or could be, universal, that anything expressible in one language, or by using one lexicon, can be expressed also in any other. His own conclusion—that meaning must be abandoned—follows only because he takes universality for granted, and this paper will suggest that there is no sufficient basis for doing so. To possess a lexicon, a structured vocabulary, is to have access to the varied set of worlds which that lexicon can be used to describe. Different lexicons—those of different cultures or different historical periods, for example—give access to different sets of possible worlds, largely but never entirely overlapping. Though a lexicon may be enriched to yield access to worlds previously accessible only with another, the result is peculiar, a point to be elaborated below. In order that the "enriched" lexicon continue to serve some essential functions, the terms added during enrichment must be rigidly segregated and reserved for a special purpose.[5]

What has made the assumption of universal translatability so nearly inescapable is, I believe, its deceptive similarity to a quite different one, in this case an assumption that I share: Anything that can be said in one language can, with sufficient imagination and effort, be *understood* by a speaker of another. What is prerequisite to such understanding, however, is not translation but language learning. Quine's radical translator is, in fact, a language learner. If he succeeds, which I think no principle bars, he will become bilingual. But that does not ensure that he or anyone else will be able to translate from his newly acquired language to the one with which he was raised. Though learnability could in principle imply translatability, the thesis that it does so needs to be argued. Much philosophical discussion instead takes it for granted. Quine's *Word and Object* provides a notably explicit case in point.[6]

I am suggesting, in short, that the problems of translating a scientific text, whether into a foreign tongue or into a later version of the language in which it was written, are far more like those of translating literature than has generally been supposed. In both cases the translator repeatedly encounters sentences that can be rendered in several alternative ways, none of which captures them completely. Difficult decisions must then be made about which aspects of the original it is most important to preserve. Different translators may differ, and the same translator may make different choices in different places, even though the term involved is in neither language ambiguous. Such choices are governed by standards of responsibility, but they are not determined by them. In these matters there is no such thing as being merely right or wrong. The preservation of truth values when translating scientific prose is as delicate a task as the preservation of resonance and emotional tone in the translation of literature. Neither can be fully achieved; even responsible approximation requires the greatest tact and taste. In the scientific case, these generalizations apply, not only to passages that make explicit use of theory, but also and more significantly to those their authors took to be merely descriptive.

Unlike many people who share my generally structuralist leanings, I am not attempting to erase or even to reduce the gap generally thought to separate literal from figurative use of language. On the contrary, I cannot imagine a theory of figurative use—a theory, for example, of metaphor and other tropes—that did not presuppose a theory of literal meanings. Nor, to turn from theory to practice, can I imagine how words could be employed effectively in tropes like metaphor except within a community whose members have previously assimilated their literal use.[7] My point is simply that the literal and the figurative uses of terms are alike in their dependence on preestablished associations between words.

That remark provides entrée to a theory of meaning, but only two aspects of that theory are centrally relevant to the arguments that follow, and I must here restrict myself to them. First, knowing what a word means is knowing how to use it for communication with other members of the language community within which it is current. But that ability does not imply that one knows something that attaches to the word by itself—its meaning, say, or its semantic markers. With occasional exceptions, words do not have meanings individually, but only through their associations with other words within a semantic field. If the use of an individual term changes, then the use of the terms associated with it normally changes as well.

The second aspect of my developing view of meaning is both less standard and more consequential. Two people may use a set of interrelated terms in the same way but employ different sets (in principle, totally disjunct sets) of field coordinates in doing so. Examples will be found in the next section of this paper; meanwhile, the following metaphor may prove suggestive. The United States can be mapped in many different coordinate systems. Individuals with different maps will specify the location of, say, Chicago by means of a different pair of coordinates. But all will nevertheless locate the same city, provided that the maps are scaled to preserve the relative distances between the items mapped. The metric that accompanies each of the various sets of coordinates must, that is, be chosen to preserve the structural geometrical relations within the mapped area.[8]

*   *

I have so far dealt in general assertions, omitting both illustration and defense. The argument that begins to supply them will proceed in two stages. The first examines part of the lexicon of Newtonian mechanics, especially the interrelated terms 'force', 'mass', and 'weight'. It asks what one need and need not know to be a member of the community that uses these terms, and indicates how possession of such knowledge constrains what members of the community can express. The second stage examines implications of these constraints for discussions of scientific development, especially for the application to it of causal theory.

The vocabulary in which the phenomena of a field like mechanics are described and explained is itself a historical product, developed over time, and repeatedly transmitted, in its then-current state, from one generation to its successor. In the case of Newtonian mechanics, the required cluster of terms has been stable for some time, and transmission techniques are relatively standard. Examining them will suggest characteristics of what the student acquires in the course of becoming a licensed practitioner of the field.[9]

Before the exposure to Newtonian terminology can usefully begin, other significant portions of the lexicon must be in place. Students must, for example, already have a vocabulary adequate to refer to physical objects and to their locations in space and time. Onto this they must have grafted a mathematical vocabulary rich enough to permit the quantitative description of trajectories and the analysis of velocities and accelerations of bodies moving along them.[10] Also, at least implicitly, they must command a notion of extensive magnitude, a quantity whose value for the whole of a body is the sum of its values for the body's parts. Quantity of matter provides a standard example. These terms can all be acquired without resort to Newtonian theory, and the student must control them before that theory can be learned. The other lexical items required by that theory — most notably 'force', 'mass', and 'weight' in their Newtonian senses — can only be acquired together with the theory itself.

Five aspects of the way in which these Newtonian terms are learned require particular illustration and emphasis. First, as already indicated, learning cannot begin until a considerable antecedent vocabulary is in place. Second, in the process through which the new terms are acquired, definition plays a negligible role. Rather than being defined, these terms are introduced by exposure to examples of their use, examples provided by someone who already belongs to the speech community in which they are current. That exposure often includes actual exhibits, for example in the student laboratory, of one or more exemplary situations to which the terms in question are applied by someone who already knows how to use them. The exhibits need not be actual, however. The exemplary situations may instead be introduced by a description conducted primarily in terms drawn from the antecedentally available vocabulary, but in which the terms to be learned also appear here and there. The two processes are for the most part interchangeable, and most students encounter them both, in some mix or other. Both include an indispensable ostensive or stipulative element: terms are taught through the exhibit, direct or by description, of situations to which they apply.[11] The learning that results from such a process is not, however, about words alone but equally about the world in which they function. When I use the phrase 'stipulative descriptions' in what follows, the stipulations I have in mind will be simultaneously and inseparably about both the substance and the vocabulary of science, about both the world and the language.

A third significant aspect of the learning process is that exposure to a single exemplary situation seldom or never supplies enough information to permit the student to use a new term. Several examples of varied sorts are required, often accompanied by examples of apparently similar situations to which the term in question does not apply. The terms to be learned, furthermore, are seldom applied to these situations in isolation, but are instead embedded in whole sentences or statements, among which are some usually referred to as laws of nature.

Fourth, among the statements involved in learning one previously unknown term are some that include other new terms as well, terms that must be acquired together with the first. The learning process thus interrelates a set of new terms, giving structure to the lexicon that contains them. Finally, though there is usually considerable overlap between the situations to which individual language learners are exposed (and even more between the accompanying statements), individuals can in principle communicate fully even though they acquired the terms with which they do so along very different routes. To the extent that the process I am describing supplies individuals with anything resembling a definition, it is not a definition that need be shared by other members of the speech community.

For illustrations, consider first the term 'force'. The situations that exemplify a force's presence are of varied sorts. They include, for example, muscular exertion, a stretched string or spring, a body possessed of weight (note the occurrence of another of the terms to be learned), or, finally, certain sorts of motion. The last is particularly important and presents particular difficulties to the student. As Newtonians use 'force', not all motions signify the presence of its referent, and examples that display the distinction between forced and force-free motions are therefore required. Their assimilation, furthermore, demands the suppression of a highly developed pre-Newtonian intuition. For children and Aristotelians, the standard example of a forced motion is the hurled projectile. Force-free motion is for them exemplified by the falling stone, the spinning top, or the rotating flywheel. For the Newtonian, all of these are cases of forced motion. The only example of a Newtonian force-free motion is motion in a straight line at constant speed, and that can be exhibited directly only in interplanetary space. Teachers nevertheless try. (I still remember the contrived lecture-demonstration—a block of ice sliding on a sheet of glass—that helped me undo prior intuitions and acquire the Newtonian concept of 'force'.) But for most students the main path to this key aspect of the use of the term is provided by Newton's first law of motion: In the absence of an external force applied to it, a body moves continuously at constant speed in a straight line. It exhibits, by description, the motions that require no force.[12]

More will need to be said about 'force', but let me first look briefly at its two

Newtonian companions, 'weight' and 'mass'. The first refers to a particular sort of force, the one that causes a physical body to press on its supports while at rest or to fall when unsupported. In this still-qualitative form the term 'weight' is available prior to Newtonian 'force' and is used during the latter's acquisition. 'Mass' is usually introduced as equivalent to 'quantity of matter', where matter is the substrate underlying physical bodies, the stuff of which quantity is conserved as the qualities of material bodies change. Any feature that, like weight, picks out a physical body, is an index also of the presence of matter and of mass. As in the case of 'weight' and unlike the case of 'force', the qualitative features by which one picks out the referents of 'mass' are identical with those of pre-Newtonian usage.

But the Newtonian use of all three terms is quantitative, and the Newtonian form of quantification alters both their individual uses and the interrelationships between them.[13] Only the unit measures may be established by convention; the scales must be chosen so that weight and mass are extensive quantities and so that forces can be added vectorially. (Contrast the case of temperature, in which both unit and scale can be chosen by convention.) Once again, the learning process requires the juxtaposition of statements involving the terms to be learned with situations drawn directly or indirectly from nature.

Begin with the quantification of 'force'. Students acquire the full quantitative concept by learning to measure forces with a spring balance or some other elastic device. Such instruments had appeared nowhere in scientific theory or practice before Newton's time, when they took over the conceptual role previously played by the pan balance. But they have since been central, for reasons that are conceptual rather than pragmatic. The use of a spring balance to exhibit the proper measure of force requires, however, recourse to two statements ordinarily described as laws of nature. One of these is Newton's third law, which states, for example, that the force exerted by a weight on a spring is equal and opposite to the force exerted by the spring on the weight. The other is Hooke's law, which states that the force exerted by a stretched spring is proportional to the spring's displacement. Like Newton's first law, these are first encountered during language learning, where they are juxtaposed with examples of situations to which they apply. Such juxtapositions play a double role, simultaneously stipulating how the word 'force' is to be used and how the world populated by forces behaves.

Turn now to the quantification of the terms 'mass' and 'weight'. It illustrates with special clarity a key aspect of the lexical acquisition process, one that has not yet been considered. To this point, my discussion of Newtonian terminology has probably suggested that, once the required antecedent vocabulary is in place, students learn the terms that remain by exposure to some single specifiable set of examples of their use. Those particular examples may well have seemed to provide necessary conditions for the acquisition of those terms. In practice, however, cases of that sort are very rare. Usually there are alternate sets of examples that will serve for the acquisition of the same term or terms. And, though it usually makes no difference to which set of these examples an individual has, in fact, been exposed, there are special circumstances in which the differences between sets prove very important.

In the case of 'mass' and 'weight', one of these alternate sets is standard. It is able to supply the missing elements of both vocabulary and theory together, and it probably therefore enters the lexical acquisition process for all students. But logically other examples would have done as well, and for most students some of them also play a role. Begin with the standard route, which first quantifies 'mass' in the guise of what today is called 'inertial mass'. Students are presented with Newton's second law — force equals mass times acceleration — as a description of the way moving bodies actually behave; but the description makes essential use of the still incompletely established term 'mass'. That term and the second law are thus acquired together, and the law can thereafter be used to supply the missing measure: the mass of a body is proportional to its acceleration under the influence of a known force. For purposes of concept acquisition, centripetal-force apparatus provides a particularly effective way to make the measurement.

Once mass and the second law have been added to the Newtonian lexicon in this way, the law of gravity can be introduced as an empirical regularity. Newtonian theory is applied to observation of the heavens and the attractions manifest there are compared to those between the Earth and bodies resting on it. The mutual attraction between bodies is thus shown to be proportional to the product of their masses, an empirical regularity that can be used to introduce the still missing aspects of the Newtonian term 'weight'. 'Weight' is now seen to denote a relational property, one that depends on the presence of two or more bodies. It can therefore, unlike mass, differ from one location to another, at the surface of the Earth and of the Moon, for example. That difference is captured only by the spring balance, not by the previously standard pan balance, which yields the same reading at all locations. What the pan balance measures is mass, a quantity that depends only on the body and on the choice of a unit measure.

Because it establishes both the second law and the use of 'mass', the sequence just sketched provides the most direct route to many applications of Newtonian theory.[14] That is why it plays so central a role in introducing the theory's vocabulary. But it is not, as previously indicated, required for that purpose, and, in any case, it rarely functions alone. Let me now consider a second route along which the use of 'mass' and 'weight' can be established. It starts from the same point as the first, by quantifying the notion of force with the aid of a spring balance. Next, 'mass' is introduced in the guise of what is today labeled 'gravitational mass'. A stipulative description of the way the world is provides students with the notion of gravity as a universal force of attraction between pairs of material bodies, its magnitude proportional to the mass of each. With the missing aspects of 'mass' thus supplied, weight can be explained as a relational property, the force resulting from gravitational attraction.

That is a second way to establish the use of the Newtonian terms 'mass' and 'weight'. With them in hand Newton's second law, the still missing component of Newtonian theory, can be introduced as empirical, a consequence simply of observation. For that purpose, centripetal force apparatus is again appropriate, no longer to measure mass, as it did on the first route, but now rather to determine the relation between applied force and the acceleration of a mass previously measured by gravitational means. The two routes thus differ in what must be stipulated about nature in order to learn Newtonian terms, what can be left instead for empirical discovery. On the first route, the second law enters stipulatively, the law of gravitation empirically. On the second, their epistemic status is reversed. In each case one, but only one, of the laws is, so to speak, built into the lexicon. I do not quite want to call such laws analytic, for experience with nature was essential to their initial formulation. Yet they do have something of the necessity that the label 'analytic' implies. Perhaps 'synthetic a priori' comes closer.

There are, of course, still other ways in which the quantitative elements of 'mass' and 'weight' can be acquired. For example, Hooke's law having been introduced together with 'force', the spring balance can be stipulated as the measure of weight, and mass can be measured, again by stipulation, in terms of the vibration period of a weight at the end of a spring. In practice, several of these applications of Newtonian theory usually enter into the process of acquiring Newtonian language, information about the lexicon, and information about the world being distributed in an indivisible mix among them. Under those circumstances, one or another of the examples introduced during lexical acquisition can, when occasion requires, be adjusted or replaced in the light of new observations. Other examples will maintain the lexicon stable, keeping in place a set of quasi necessities equivalent to those initially induced by language learning.

Clearly, however, only a certain number of examples may be altered piecemeal in this way. If too many require adjustment, then it is no longer individual laws or generalizations that are at stake, but the very vocabulary in which they are stated. A threat to that vocabulary is, however, a threat also to the theory or laws essential to its acquisition and use. Could Newtonian mechanics withstand revision of the second law, of the third law, of Hooke's law, or the law of gravity? Could it withstand the revision of any two of these, of three, or of all four? These are not questions that individually have yes or no answers. Rather, like Wittgenstein's "Could one play chess without the queen?", they suggest the strains placed on a lexicon confronted by questions that its designer, whether God or cognitive evolution, did not anticipate its being required to answer.[15] What should one have said when confronted by an egg-laying creature that suckles its young? Is it a mammal or is it not? These are the circumstances in which, as Austin put it, "*We don't know what to say. Words literally fail us.*"[16] Such circumstances, if they endure for long, call forth a locally different lexicon, one that permits an answer, but to a slightly altered question: "Yes, the creature is a mammal" (but to be a mammal is not what it was before). The new lexicon opens new possibilities, ones that could not have been stipulated by the use of the old.

To clarify what I have in mind, let me suppose that there are only two ways in which use of the terms 'mass' and 'weight' can be acquired: one that stipulates the second law and finds the law of gravity empirically; another that stipulates the law of gravity and discovers the second law empirically. Suppose further that the two routes are exclusive; students traverse one or the other so that the necessities of the lexicon and the contingencies of experiment are kept separate on each. Clearly, these two routes are very different, but the differences will not ordinarily interfere with full communication among those who use the terms. All will pick out the same objects and situations as the referents of the terms they share, and all will agree about the laws and other generalizations governing these objects and situations. All are thus fully participants in a single speech community. What individual speakers may differ about is the epistemic status of generalizations that community members share, and such differences are not usually important. Indeed, in *ordinary* scientific discourse, they do not emerge at all. While the world behaves in anticipated ways—the ones for which the lexicon evolved—these differences between individual speakers are of little or no consequence.

But change of circumstance may make such differences consequential. Imagine that a discrepancy is discovered between Newtonian theory and observation, for example celestial observations of the motion of the lunar perigee. Scientists who had learned Newtonian 'mass' and 'weight' along the first of my two lexical-acquisition routes would be free to consider altering the law of gravity as a way to remove the anomaly. On the other hand, they would be bound by language to preserve the second law. On the other hand, scientists who had acquired 'mass' and 'weight' along my second route would be free to suggest altering the second law but would be bound by language to preserve the law of gravity. A difference in the language-learning route, one that had had no effect while the world behaved as anticipated, would become the source of a difference of opinion when anomalies were found.

Now suppose that neither the revisions that preserved the second law nor those that preserved the law of gravity proved effective in eliminating anomaly. The next step would be an attempt at revisions that altered both laws together, and those revisions the lexicon will not, in its present form, permit.[17] Such attempts are often successful nonetheless, but they require recourse to such devices as metaphorical extension, devices that alter the meanings of lexical items themselves. After such revision, say the transition to an Einsteinian vocabulary, one can write down strings of symbols that *look like* revised versions of the second law and the law of gravity. But the resemblance is deceptive, because some symbols in the new strings attach to nature differently from the corresponding symbols in the old, thus distinguishing between situations which, in the antecedently available vocabulary, were the same.[18] They are the symbols for terms whose acquisition involved laws that have changed form with the change of theory: the differences between the old laws and the new are reflected by the terms acquired with them. Each of the resulting lexicons then gives access to its own set of possible worlds, and the two sets are disjoint. Translations involving terms introduced with the altered laws are impossible.

The impossibility of translation does not, of course, bar users of one lexicon from learning the other. And having done so, they can join the two together, enriching their initial lexicon by adding to it sets of terms from the new one they have acquired. That is, as I have argued elsewhere, what historians do both for themselves and for their readers.[19] But the sense of enrichment involved is peculiar. It is like the enrichment that gives philosophers an alternative set of terms for describing emeralds: not 'blue', 'green', and the traditional roster of color terms, but 'grue', 'bleen', and the names of the other occupants of the corresponding spectrum.[20] One set of terms is projectible, supports induction, the other not. One set of terms is available for descriptions of the world, the other is reserved for the special purposes of the philosopher, the historian, the writer of certain sorts of fiction. However useful these lexical add-ons may be, the integrity of language requires that they be segregated, reserved for distinct sorts of discourse. If communication is to succeed, then all participants in discourse must know at all times which set of terms is being used.

Students of literature have long taken for granted that metaphor and its companion devices (those that alter the interrelations among words) provide entrée to new worlds and make translation impossible by doing so. Similar characteristics have been widely attributed to the language of political life and, by some, to the entire range of the human sciences. But the natural sciences, dealing objectively with the real world (as they do), are generally held to be immune. Their truths (and falsities) are thought to transcend the ravages of temporal, cultural, and linguistic change. I am suggesting, of course, that they cannot do so. Neither the descriptive nor the theoretical language of a natural science provides the bedrock such transcendence would require. I shall not in this paper even attempt to deal with the philosophical problems consequent upon that point of view.[21] Rather, I shall attempt to reinforce them by examining a technique that claims to set them aside.

*   *   *

Change of lexicon, I have been arguing, can change the meaning of some group or groups of interrelated terms. Problems about the possibility of truth-preserving translations result, and efforts to avoid them have in recent years taken a characteristic form. Truth values, these efforts emphasize, depend only on reference, not on meaning or mode of use. Discussions of truth value need not, therefore, invoke meaning at all, at least not in a sense so nearly traditional as the one I appear to have invoked here.[22]

Among such efforts, the most influential is the causal theory of reference, and many of the advances achieved with its aid are likely to prove permanent. But causal theory, which invokes an original act of baptism or dubbing as an essential determinant of reference, is intrinsically historical, and its expositors resort repeatedly to putative examples from scientific development. These examples, I believe, quite regularly fail in ways that are both consequential and illuminating. To see what is involved, I conclude this paper by examining two well-known examples developed by Hilary Putnam, the philosopher who has most explicitly applied causal theory to history.[23]

Excluding proper names, I doubt that there is any set of terms for which causal theory works precisely; but it comes very close to doing so for terms like 'gold', and the plausibility of its application to natural-kind terms depends on the existence of such cases. Terms that behave like 'gold' ordinarily refer to naturally occurring, widely distributed, functionally significant, and easily recognized substances. Such terms occur in the languages of most or all cultures, retain their original use over time, and refer throughout to the same sorts of samples. There is little problem about translating them, for they occupy closely equivalent positions in all lexicons. 'Gold' is among the closest approximations we have to an item in a neutral, mind-independent observation vocabulary.

When a term is of this sort, modern science can often be used not only to specify the common essence of its referents but actually to single them out. Modern theory, for example, identifies gold as the substance with atomic number 79, and licenses specialists to identify it by the application of such techniques as x-ray spectroscopy. Neither the theory nor the instrument was available seventy-five years ago, but it is nevertheless reasonable to suggest, as Putnam does, that the referents of 'gold' are and have always been the same as the referents of 'substance with atomic number 79'. Exceptions to that equation are few, and they result primarily from the ever-increasing refinement of our ability to detect impurities and forgeries. For the causal theorist, therefore, 'having atomic number 79', is *the* essential property of gold—the single property such that, if gold in fact does have it, then it has it necessarily. Other properties—yellowness and ductility, for example—are superficial and correspondingly contingent. Kripke suggests that gold might even be blue, its apparent yellowness resulting from an optical illusion.[24] Though individuals may, in fact, use color and other superficial characteristics when picking out samples of gold, that practice tells nothing essential about the referents of the term.

'Gold' presents a relatively special case, however, and what is special about it obscures essential limitations on the conclusions it will support. More representative is Putnam's most developed example, 'water', and the problems that arise with it are still more severe in the case of such other widely discussed terms as 'heat' and 'electricity'.[25] For water the discussion divides in two parts. In the first, which is the more familiar, Putnam imagines a possible world containing Twin Earth, a planet just like our own except that the stuff called 'water' by Twin Earthians is not H_2O but a different liquid with a very long and complicated chemical formula abbreviated XYZ. "Indistinguishable from water at normal temperature and pressure," XYZ is the stuff that on Twin Earth quenches thirst, rains from the skies, and fills oceans and lakes, much as water does here. If a spaceship from Earth ever visits Twin Earth, Putnam writes:

> then, the supposition at first will be that 'water' has the same meaning on Earth and Twin Earth. This supposition will be corrected when it is discovered that 'water' on Twin Earth is XYZ, and the Earthian spaceship will report somewhat as follows: "On Twin Earth the word 'water' means XYZ." (See n. 23.)

As in the case of gold, superficial qualities like quenching thirst or raining from the skies have no role in determining to what substance the term 'water' properly refers.

Two aspects of Putnam's fable require special notice. First, the fact that Twin Earthians call XYZ by the name 'water' (the same symbol that Earthians use for the stuff that lies in lakes, quenches thirst, etc.) is an irrelevancy. The difficulties presented by this story will emerge more clearly if the visitors from Earth use their own language throughout. Second, and presently central, whatever the visitors call the stuff that lies in Twin Earthian lakes, the report they send home should not be about language but about chemistry. It must take some form like: "Back to the drawing board! Something is badly wrong with chemical theory."

The terms 'XYZ' and 'H_2O' are drawn from modern chemical theory, and that theory is incompatible with the existence of a substance with properties very nearly the same as water but described by an elaborate chemical formula. Such a substance would, among other things, be too heavy to evaporate at normal terrestrial temperatures. Its discovery would present the same problems as the simultaneous violation of Newton's second law and the law of gravity described in the last section. It would, that is, demonstrate the presence of fundamental errors in the chemical theory that gives meanings to compound names like 'H_2O' and the unabbreviated form of 'XYZ'. Within the lexicon of modern chemistry, a world containing both our Earth and Putnam's Twin Earth is lexically possible, but the composite statement that describes it is necessarily false. Only with a differently structured lexicon, one shaped to describe a very different sort of world, could one, without contradiction, describe the behavior of 'XYZ' at all, and in that lexicon 'H_2O' might no longer refer to what we call 'water'.

So much for the first part of Putnam's argument. In the second he applies it more concretely to the referential history of 'water', suggesting that "we roll the time back to 1750," and continuing:

At that time chemistry was not developed on either Earth or Twin Earth. The typical Earthian speaker of English did not know water consisted of hydrogen and oxygen, and the typical Twin Earthian speaker of English did not know 'water' consisted of XYZ. . . . Yet the extension of the term 'water' was just as much H_2O on Earth in 1750 as in 1950; and the extension of the term 'water' was just as much XYZ on Twin Earth in 1750 as in 1950. (See n. 23.)

In journeys through time, as in those through space, Putnam suggests, it is chemical formula, not superficial qualities, that determines whether a given substance is water.

For present purposes, attention can be restricted to Earthian history, and on Earth Putnam's argument for 'water' is the same as it was for 'gold'. The extension of 'water' is determined by the original sample together with the relation sameness-of-kind. That sample dates from before 1750, and the nature of its members has been stable. So has the relation sameness-of-kind, though *explanations* of what it is for two bodies to be of the same kind have varied widely. What matters, however, is not explanations but what gets picked out, and identifying samples of H_2O is, according to causal theory, the best means yet found to pick out samples of the same kind as the original set. Give-or-take a few discrepancies at the margins, discrepancies due to refinement of technique or perhaps to change of interest, 'H_2O' refers to the same samples that 'water' referred to in either 1750 or 1950. Apparently causal theory has rendered the referents of 'water' immune to changes in the concept of water, the theory of water, and the way samples of water are picked out. The parallel between causal theory's treatment of 'gold' and of 'water' seems complete.

But in the case of water, difficulties arise. 'H_2O' picks out samples not only of water but also of ice and steam. H_2O can exist in all three states of aggregation – solid, liquid, and gaseous – and it is therefore not the same as water, at least not as picked out by the term 'water' in 1750. The difference in items referred to is, furthermore, by no means marginal, like that due to impurities for example. Whole categories of substance are involved, and their involvement is by no means accidental. In 1750 the primary differences between the species recognized by chemists were still more or less those between what are now called the states of aggregation. Water, in particular, was an elementary body of which liquidity was an essential property. For some chemists the term 'water' referred to the generic liquid, and it had done so for many more only a few generations before. Not until the 1780s, in an episode long known as the "Chemical Revolution," was the taxonomy of chemistry transformed so that a chemical species might exist in all three states of aggregation. Thereafter, the distinction between solids, liquids, and gases became physical, not chemical. The discovery that *liquid* water was a compound of two *gaseous* substances, hydrogen and oxygen, was an integral part of that larger transformation and could not have been made without it.

This is not to suggest that modern science is incapable of picking out the stuff that people in 1750 (and most people still) label 'water'. That term refers to *liquid* H_2O. It should be described not simply as H_2O but as close-packed H_2O particles in rapid relative motion. Marginal differences again aside, samples answering that compound description are the ones picked out in 1750 and before by the term 'water'. But this modern description leads to a new network of difficulties, difficulties that may ultimately threaten the concept of natural kinds and that meanwhile must bar the automatic application of causal theory to them.

Causal theory was initially developed with notable success for application to proper names. Its transfer from them to natural-kind terms was facilitated — perhaps made possible — by the fact that natural kinds, like single individual creatures, are denoted by short and apparently arbitrary names, names coextensive with those of the corresponding kind's single essential property. Our examples have been 'gold' paired with 'having atomic number 79', and 'water' paired with 'being H_2O'. The latter member of each pair names a property, of course, as the name coupled with it does not. But so long as only a single essential property is required by each natural kind that difference is inconsequential. When two non-coextensive names are required, however — 'H_2O' and 'liquidity' in the case of water — then each name, if used alone, picks out a larger class than the pair does when conjoined, and the fact that they name properties becomes central. For if two properties are required, why not three or four? Are we not back to the standard set of problems that casual theory was intended to resolve: which properties are essential, which accidental; which properties belong to a kind by definition, which are only contingent? Has the transition to a developed scientific vocabulary really helped at all?

I think it has not. The lexicon required to label attributes like being-H_2O or being-close-packed-particles-in-rapid-relative-motion is rich and systematic. No one can use any of the terms that it contains without being able to use a great many. And given that vocabulary, the problems of choosing essential properties arise again, except that the properties involved can no longer be dismissed as superficial. Is deuterium hydrogen, for example, and is heavy water really water? And what may one say about a sample of close-packed particles of H_2O in rapid relative motion at the critical point — under the conditions of temperature and pressure, that is, at which the liquid, solid, and gaseous states are indistinguishable? Is it really water? The use of theoretical rather than superficial properties offers great advantages, of course. There are fewer of the former; the relations between them are more systematic; and they permit both richer and more precise discriminations. But they come no closer to being essential or necessary properties than the superficial ones they appear to supplant.

The inverse argument proves even more significant. The so-called superficial properties are no less necessary than their apparently essential successors. To say that water is liquid H_2O is to locate it within an elaborate lexical and theoretical system. Given that system, as one must be in order to use the label, one can in principle predict the superficial properties of water (just as one could those of XYZ), compute its boiling and freezing points, the optical wavelengths that it will transmit, and so on.[26] If water is liquid H_2O, then these properties are necessary to it. If they were not realized in practice, that would be a reason to doubt that water really was H_2O.

This last argument applies also to the case of gold, in which causal theory apparently succeeded. 'Atomic number' is a term from the lexicon of atomic-molecular theory. Like 'force' and 'mass', it must be learned together with other terms deployed in that theory, and the theory itself must play a role in the acquisition process. When the process is complete, one can replace the label 'gold' with 'atomic number 79', but one can then also replace the label 'hydrogen' with 'atomic number 1', 'oxygen' with 'atomic number 8', and so on to a total well over a hundred. And one can do something more important as well. Invoking such other theoretical properties as electronic charge and mass, one can in principle, and to a considerable extent in fact, predict the superficial qualities—density, color, ductility, conductivity, and so on—that samples of the corresponding substance will possess at normal temperatures. Those properties are no more accidental than having-atomic-number-79. That color is a superficial property does not make it a contingent one. Furthermore, in a comparison of superficial and theoretical qualities, the former have a double priority. If the theory that posits the relevant theoretical properties could not predict these superficial qualities or some of them, there would be no reason to take it seriously. If gold were blue for a normal observer under normal conditions of illumination, its atomic number would not be 79. In addition, superficial properties are the ones called upon in those difficult cases of discrimination characteristically raised by new theories. Is deuterium really hydrogen, for example? Are viruses alive?[27]

What remains special about 'gold' is simply that, unlike 'water', only one of the underlying properties recognized by modern science—having atomic number 79—need be called upon to pick out members of the sample to which the term has continued through history to refer.[28] 'Gold' is not the only term that possesses or closely approximates this characteristic. So do many of the basic-level referring terms used in everyday speech, including the everyday use of the term 'water'. But not all everyday terms are of this sort. 'Planet' and 'star' now categorize the world of celestial objects differently from the way they did before Copernicus, and the differences are not well described by phrases like "marginal adjustment" or "zeroing in." Similar transitions have characterized the historical development of virtually all the referring terms of the sciences, including the most elementary: 'force', 'species', 'heat', 'element', 'temperature', and so on. It is terms like these that have provided this paper's primary concern, and a three-item summary of what it has had to say about them can now bring it quickly to a close.

Dealing with such Newtonian terms as 'mass', 'force', and 'weight', the second section of this paper emphasized that they are learned in use. The student first encounters them in authoritative statements about the world by someone who already knows how to use them and who illustrates their correct use by juxtaposing statements that contain them with exemplary situations to which those statements do or do not apply. Two parallels to that position are implicit in the third section's discussion of causal theory. First, the role played by actual examples in anchoring terms of the lexicon to the world recurs in causal theory's emphasis on an original act of dubbing, which supplies examples canonical for later use. Second, the emphasis on language learning—on the way the lexicon is transmitted from one generation to the next—is duplicated in causal theory's emphasis upon the chain linking later users of a term to the canonical sample. These are important parallels, especially because their recognition makes it possible to isolate the key respect in which the position taken in this paper diverges from that of causal theory. Dubbing is here seen as a process that recurs again and again through history. The putative "original sample" may mark the beginning of the chain, if any beginning is needed, but there is nothing privileged about its membership. The sets of canonical examples used in transmitting the lexicon change in the course of time, and not all the changes can properly be viewed as mere adjustments.

Many of them, of course, are small adjustments, for example those due to refinements in the process of purifying gold. But others are both systematic and wide reaching. Changes in the sample required to transmit a term like 'force' cannot be made in isolation, but require simultaneous changes in the samples used to introduce such terms as 'mass' and 'weight'. Moving circular motions from the category of force-free to that of forced motions required shifting uniform linear motions from forced to force-free. Simultaneously, the exhibit of a new instrument, the spring balance, was required to introduce the Newtonian term 'force', and examples that introduced the corresponding term 'weight' were required to exhibit not simply the body whose weight was in question, but one or more other bodies in gravitational interaction with it. Similarly, the development of the chemical lexicon in which H_2O is embedded required an almost total readjustment in the samples used to introduce the basic chemical kinds. No workable form of chemistry could have survived the change that placed liquid water in the same category as ice and steam but continued elsewhere to regard the divisions between states of aggregation as chemically fundamental. In short, dubbing and the procedures that accompany it ordinarily do more than place the dubbed object together with other members of its kind. They also locate it with respect to other kinds, placing it not simply within a taxonomic category but within a taxonomic system. Only while that system endures do the names of the kinds it categorizes designate rigidly.

A lexicon that embodies such interrelationships between terms necessarily also embodies knowledge of the world those terms can be used to describe, and that knowledge may be placed at risk. As time goes on and new demands are placed on the lexicon, conditions may be encountered that defy description. The Newtonian lexicon could not, without internal contradiction, describe a world in which both Newton's second law and the law of gravity were violated. Prior to the eighteenth century, the lexicon of chemistry could not provide a coherent description of a world in which a sample of liquid could change its state of aggregation without simultaneously changing its chemical kind. But these changes, both in mechanics and in chemistry, nevertheless came about, together with the change of lexicon they required. And the new lexicon, once consolidated, displayed the same sorts of limitations as its predecessor. It could not, that is, be used to to provide coherent descriptions of some aspects of the world described by its predecessor. Here and there the old and new lexicons embodied differently structured, nonhomologous taxonomies, and statements involving terms from the regions where the two differed were not translatable between them.

Those untranslatable statements are the ones from which this paper began, the ones that make no sense to the historian who encounters them in an out-of-date text, and which lead him or her to abandon translation in favor of language learning. Returning to them closes a circle. Though well aware that challenging problems are located within it, I shall not start another lap here and now.

Notes

1. Throughout this paper I shall continue to speak of the lexicon, of terms, and of statements. My concern, however, is actually with conceptual or intensional categories more generally, e.g., with those that may reasonably be attributed to animals or to the perceptual system.

2. For Newton, see my "Newton's '31st Query' and the Degradation of Gold," *Isis* 42 (1951), 296–98. For Bohr, see John L. Heilbron and Thomas S. Kuhn, "The Genesis of the Bohr Atom," *Historical Studies in the Physical Sciences*, 1 (1969), 211–90, where the nonsense passages that gave rise to the project are quoted on p. 271. For an introduction to the other examples mentioned, see my "What are Scientific Revolutions?" in L. J. Daston, M. Heidelberger, and L. Kruger, eds., *The Probabilistic Revolution*, vol. 1, *Ideas in History* (Cambridge: MIT Press, 1987), 7–22.

3. For a fuller and more nuanced discussion of this point and those that follow see my "Commensurability, Comparability, Communicability," in P. D. Asquith and T. Nickles, eds., *PSA 1982*, vol. 2 (East Lansing: Philosophy of Science Association, 1983), 669–88.

4. My original discussion described nonlinguistic as well as linguistic forms of incommensurability. That I now take to have been an overextension resulting from my failure to recognize how large a part of the apparently nonlinguistic component was acquired with language during the learning process. The acquisition during language learning of what I once took to be incommensurability with respect to instrumentation is, for example, illustrated by the discussion of the spring balance in the next section of this paper.

5. To speak of different lexicons as giving access to different sets of possible worlds is not simply to add one more kind of accessibility relation to those that currently generate different kinds of modal

necessity. There is no type of necessity corresponding to lexical accessibility. Excepting purely analytic statements or combinations of statements, for which see below, no statement framable in a given lexicon is necessary simply because it can be accessed in that lexicon. More generally, lexical accessibility seems to cut across the standard set of accessibility relations. Perhaps all possible-world analyses of modalities and semantics should be relativized to the lexicons with which the appropriate set of possible worlds might have been stipulated or described.

6. W. V. O. Quine, *Word and Object* (New York: John Wiley & Sons; Cambridge: MIT Press; 1960), 47, 70f.

7. See my "Metaphor in Science," in Andrew Ortony, ed., *Metaphor and Thought* (Cambridge: Cambridge University Press, 1979), 409-19.

8. Some preliminary indications of what these cryptic remarks intend are supplied in my "Commensurability, Comparability, Communicability" (See note 3 above).

9. I discuss the *transmission* of a lexicon because it is a source of clues to what the individual's possession of a lexicon entails. Nothing, however, depends upon the lexicon's being acquired by transmission. The consequences would be the same if, for example, it were a consequence of genetics or else implanted by a skilled neurosurgeon. I shall, for example, shortly emphasize that transmitting a lexicon requires repeated recourse to concrete examples. Implanting the same lexicon surgically would, I am suggesting, have involved implanting the memory traces left by exposure to such examples.

10. In practice, the techniques for describing velocities and accelerations along trajectories are usually learned in the same courses that introduce the terms to which I turn next. But the first set can be acquired without the second, whereas the second cannot be acquired without the first.

11. The terms "ostension" and "ostensive" have two different uses, which for present purposes need to be distinguished. In one, these terms imply that *nothing but* the exhibit of a word's referent is needed to learn or to define it. In the other, they imply only that *some* exhibit is required during the acquisition process. I shall, of course, be using the second sense of the terms. The propriety of extending them to cases in which description in an antecedent vocabulary replaces an actual exhibit depends on recognizing that description does not supply a string of words equivalent to the statements containing the words to be learned. Rather it enables students to visualize the situation and apply to the visualization the same mental processes (whatever they may be) that would otherwise have been applied to the situation as perceived.

12. Newton's first law is a logical consequence of his second, and Newton's reason for stating them separately has long been a puzzle. The answer may well lie in pedagogic strategy. If Newton had permitted the second law to subsume the first, his readers would have had to sort out his use of 'force' and of 'mass' together, an intrinsically difficult task further complicated by the fact that the terms had previously been different not only in their individual use but in their interrelation. Separating them to the extent possible displayed the nature of the required changes more clearly.

13. Though my analysis diverges from theirs, many of the considerations that follow (as well as a few of those introduced above) were suggested by contemplation of the techniques developed by J. D. Sneed and Wolfgang Stegmüller for formalizing physical theories, especially by their manner of introducing theoretical terms. Note also that these remarks suggest a route to the solution of a central problem of their approach, how to distinguish the core of a theory from its expansions. For this problem see my paper, "Theory Change as Structure Change: Comments on the Sneed Formalism," *Erkenntnis*, 10 (1976), 179-99.

14. All applications of Newtonian theory depend on understanding 'mass', but for many of them 'weight' is dispensable.

15. Twenty-five years ago the quotation was a standard part of what I now discover was a merely oral tradition. Though clearly "Wittgensteinian," it is not to be found in any of Wittgenstein's published writings. I preserve it here because of its recurrent role in my own philosophical development

and because I've found no published substitute that so clearly prohibits the response that the question might be answerable if only there were more information.

16. J. L. Austin, "Other Minds," in *Collected Philosophical Papers* (Oxford: Oxford University Press, 1961), 44–84. The quoted passage occurs on p. 56, and the italics are Austin's. For examples from literature of situations in which words fail us, see James Boyd White, *When Words Lose their Meaning*, (Chicago: University of Chicago Press, 1984). I have compared an example from the sciences witih one from developmental psychology in "A Function for Thought Experiments," reprinted in *The Essential Tension* (Chicago: University of Chicago Press, 1977), 240–65.

17. At this point I will seem to be reintroducing the previously banished notion of analyticity, and perhaps I am. Using the Newtonian lexicon, the statement "Newton's second law and the law of gravity are both false" is itself false. Furthermore, it is false by virtue of the meanings of the Newtonian terms 'force' and 'mass'. But it is not—unlike the statement "Some bachelors are married"—false by virtue of the *definitions* of those terms. The meanings of 'force' and 'mass' are not embodiable in definitions, but rather in their relation to the world. The necessity to which I here appeal is not so much analytic as synthetic a priori.

18. In fact, for the Newton-to-Einstein transition, the most significant lexical change is in the antecedent kinematic vocabulary for space and time, and it moves from there upward into the vocabulary of mechanics.

19. See the reference cited in note 3.

20. Like the Newtonian terms I have been examining, the terms in any color vocabulary form an interrelated set. One cannot alter one of them without making corresponding alterations in a number of the others as well. Note, however, that the parallel I am drawing is incomplete. Because their differences do not affect the structure of the color vocabulary itself, one can translate between the projectable 'blue'/'green' vocabulary and the unprojectable vocabulary containing 'bleen'/'grue'.

21. Despite my critics, I do not think that the position developed here leads to relativism, but the threats to realism are real and require much discussion, which I expect to provide in another place. These problems have already emerged repeatedly in this paper: in transitions between object language and metalanguage, for example, or in constant substitution of talk about how the world used to be, for talk about how people thought it was. They will emerge again below in my implied refusal to suppose, with Putnam, that the need to make drastic changes in the set of objects to which a term once referred indicates that it did not, in fact, refer at all.

22. Views that, like mine, depend on talking about the way words are actually used, the situations in which they apply, are regularly charged with invoking a "verification theory of meaning," not currently a respectable thing to do. But in my case at least that charge does not hold. Verification theories attribute meanings to individual sentences and through them to the individual terms those sentences contain. Each term has a meaning determined by the way in which sentences containing it are verified. I have been suggesting, however, that with occasional exceptions terms do not *individually* have meanings at all. More important, the view sketched above insists that people may use the same lexicon, refer to the same items with it, and yet pick out those items in different ways. Reference is a function of the shared structure of the lexicon, but not of the varied feature spaces within which individuals represent that structure. There is, however, a second charge, closely related to verificationism, of which I am guilty. Those who maintain the independence of reference and meaning also maintain that metaphysics is independent of epistemology. No view like mine (in the respects presently at issue there are a number) is compatible with so stringent a separation. The distinction between metaphysics and epistemology can be drawn only from within a position that involves both.

23. "The Meaning of Meaning," in *Mind, Language and Reality* (Cambridge: Cambridge University Press, 1975), 215–71, especially 223ff. All the quotations that follow are from the latter pages. Putnam has, I believe, now abandoned significant components of the essentialist viewpoint that underlies this paper, moving from it to a view ("internal realism") with significant parallels to my own. But

few philosophers have followed him: both the examples and the viewpoint discussed below are still very much alive.

24. *Naming and Necessity* (Cambridge: Harvard University Press, 1980), 118.

25. The force of Putnam's discussion depends in part upon an equivocation that needs to be eliminated. As used in everyday life or by the laity, 'water' has through history behaved much like 'gold'. But that is not the case within the community of scientists and philosophers to which Putnam's argument needs to be applied.

26. Laypeople can, of course, say that water is H_2O without controlling the fuller lexicon or the theory that it supports. But their ability to communicate by doing so depends upon the presence of experts in their society. The laity must be able to identify the experts and say something of the nature of the relevant expertise. And the experts must, in turn, command the lexicon, the theory, and the computations.

27. At issue, of course, is where to draw the boundary lines that delimit the referents of 'water', 'living thing', and so on, a problem that arises from and seems to threaten the notion of natural kinds. That notion is closely modeled on the concept of a biological species, and discussions of causal theory repeatedly invoke the relation between a particular gene type and a corresponding species (often tigers) to illustrate the relation said to hold between a natural kind and its essence between H_2O and water, for example, or between atomic number 79 and gold. But even individuals who are unproblematically members of the same species have differently constituted sets of genes. Which sets are compatible with membership in that species is a subject of continuing debate, both in principle and practice, and the subject of the argument is always which superficial *properties* (e.g., the ability to interbreed) the members of a species must share.

28. Even for gold this generalization is not *altogether* correct. As mentioned above, scientific progress does result in marginal adjustments of the original samples of gold by virtue of "our increased ability to detect impurities" (see n. 23). But what it is for gold to be pure is determined in part by theory. If gold is the substance with atomic number 79, then even a single atom with a different atomic number constitutes an impurity. But if gold is, as it was in antiquity, a metal that ripens naturally in the earth, changing gradually from lead through iron and silver to gold in the process, then there is no single form of matter that is gold *tout court*. When the ancients applied the term "gold" to samples from which we might withhold it, they were not always simply mistaken.

Scientific Revolutions and Scientific Rationality: The Case of the "Elderly Holdout"

1. Introduction: The Rationality of Scientific Change and the "Elderly Holdout"

Many parents feel that there is no longer enough emphasis at elementary school on the "three rs". No one could complain that recent philosophy of science has failed to emphasise *its* "three rs": revolutions, rationality, and realism. In this paper I risk being a pain in the rs by returning to the well-worn topic of the rationality of revolutionary scientific change (a topic that in turn raises the problem of scientific realism). I concentrate on what appears to be a particularly sharp challenge to the claim that the development of science has been a rational affair, and try to use that challenge to clarify some aspects of the claim.

The (well-known) challenge stems from Kuhn's *The Structure of Scientific Revolutions* and concerns fundamental theory change. Kuhn's view, remember, was that an individual scientist's switch to a new paradigm is a "conversion experience" that "cannot be forced": reason—in the form of the "objective factors" of traditional philosophy of science (empirical accuracy, simplicity, and the like)—certainly plays a role but it *never dictates* the switch to the new paradigm.[1]

Consequently, on Kuhn's view, it is never actually irrational to *resist* the switch. This is true even of those "elderly holdouts" who continue to resist after pretty well all their colleagues have switched allegiance. There is, claims Kuhn, *no* "point at which resistance becomes illogical or unscientific." An elderly holdout, like Priestley "holding out" against Lavoisier's oxygen theory, may infuriate

This paper was originally written for the NEH Institute on Philosophy of Science at the University of Minnesota, and an early version was delivered there in April 1986. I should like to thank Philip Kitcher for the invitation to contribute to the Institute, and him and Wade Savage for their hospitality. Early versions were also delivered in Pittsburgh and in Edinburgh: I thank Adolf Grünbaum, Nick Rescher, David Bloor, and Steve Shapin for helpful, critical comments. I have also benefited from criticisms of previous drafts of the paper by my colleagues John Watkins and Elie Zahar.

his colleagues by his stubbornness, but they have no right to brand him irrational or unscientific:

> Lifelong resistance, particularly from those whose productive careers have committed them to an older tradition . . . is not a violation of scientific standards. (1962, 151)

But if Priestley did not violate scientific standards in resisting the oxygen theory of combustion, are we not forced to say that modern day creationists, for example, equally do not "violate scientific standards" in resisting Darwinism? Indeed, if reason never dictates a preference for a new theory (even once the revolutionary dust has largely settled) *are there any* scientific standards to violate? Worries like these have frequently been expressed by Kuhn's critics. And Kuhn himself has almost equally frequently, but generally unsuccessfully, attempted to lay such worries to rest — claiming that, when properly understood, his views on this particular matter are rather less challenging to philosophical orthodoxy than might meet the eye.

The present paper attempts to clarify this confused situation via a case study of one elderly holdout. The historical details will, I hope, supply *both* a test of Kuhn's views *and* the means of clarifying at any rate some aspects of the general claim that the development of science has been a predominantly rational affair.

2. Sir David Brewster and the Wave Theory of Light

Obviously a case study of a holdout requires a scientist to do the resisting and a scientific revolution to resist. The scientific revolution I have chosen is the early nineteenth-century revolution in optics, which saw the wave theory of light emerge triumphant over its previously entrenched Newtonian emissionist rival. Among the handful of significant holdouts against this revolution two stand out: Jean-Baptiste Biot and David Brewster. For various reasons (not least because he published the more explicit accounts of his reasons for holding out) I have chosen Brewster.

First, his credentials. Brewster was certainly no peripheral or negligible figure. He was the discoverer of a great many of the properties of polarized light, especially elliptically polarized light; he discovered "Brewster's law," relating the polarizing angle and refractive index of transparent substances; he discovered a whole new class of doubly refracting crystals, the "biaxal crystals"; he discovered that ordinary unirefringent matter can be made birefringent by the application of mechanical pressure; and he discovered the hitherto unknown general phenomenon of selective absorption.

So Brewster was a significant scientist, and he was certainly some sort of holdout. In 1831 another knight of the realm, Sir George Biddel Airy, published a

Mathematical Tract on the Undulatory Theory of Light, which began with the words:

> The Undulatory Theory of Optics is presented to the reader as having the same claims to his attention as the Theory of Gravitation; namely that it is certainly true. . . . (Airy 1831, vii)

In that same year, Brewster presented a "Report on the Present State of Physical Optics" to the British Association for the Advancement of Science, in which he asserted that the undulatory theory is "still burthened with difficulties, and cannot claim our implicit assent," (Brewster 1833a, 318). Two years later he reported:

> I have not yet ventured to kneel at the new shrine [that is, the shrine of the wave theory] and I must acknowledge myself subject to the national weakness which urges me to venerate, and even to support, the falling temple in which Newton once worshipped. (1833b, 361)

Although Brewster's "official" published position was usually one of (alleged) theoretical *neutrality*, there can be no doubt that he retained to the end of his life a soft spot for the Newtonian theory; that is, for the *ancien régime* in this revolution.[2]

The main features, as I see them, of Brewster's attitude towards the wave theory and its Newtonian rival are the following.

(2a) Brewster's Acceptance That the Wave Theory Was Empirically More Successful

Brewster time and again expressed great admiration for the theory and fully acknowledged that it had enjoyed unparalleled explanatory and predictive success:

> I have long been an admirer of the singular power of this theory to explain some of the most perplexing phenomena of optics; and the recent discoveries of Professor Airy, Mr Hamilton and Mr Lloyd afford the finest examples of its influence in predicting new phenomena. (1833b, 360)[3]

Elsewhere (1833a, 318) he talked of the "theory of undulations, with all its power and all its beauty."

(2b) Brewster's Belief That the Wave Theory, Despite its Empirical Success, Could Not Be True

Despite its empirical success, Brewster quite explicitly held that the wave theory was, when considered as a "physical theory," false. He produced two main arguments in this connection.

The first was of a general methodological kind: namely that the fact that the wave theory explained and predicted a whole range of phenomena did not establish it as a physical truth. Instead:

> Twenty theories . . . may all enjoy the merit of accounting for a certain class of facts provided they have all contrived to interweave some common principle to which these facts are actually related. (1833b, 360)

Brewster admitted that the wave theory's predictive success did indeed mean that:

> it must contain amongst its assumptions (though as a physical theory it may still be false) some principle which is inherent in . . . the real producing cause of the phenomena of light. . . . (Brewster 1838, 306)

But he went on to make his own opinion clear that, as a physical theory – as, that is, a fully realistically interpreted claim about the universe – the wave theory is indeed false. In particular, a full, realistic interpretation committed the wave theory to "an ether, invisible, intangible, imponderable, inseparable from all bodies, and extending from our own eye to the remotest verge of the starry heavens" (ibid.). This was always too much – or perhaps too little – for Brewster to swallow.

Whether or not Brewster would have subscribed to the strong view – that a theory may be totally empirically adequate and yet still not "physically" true – is not clear. It is not clear precisely because his second argument was that the wave theory's explanatory and predictive success, though impressive, was definitely limited. Indeed on Brewster's view there were two important areas in which the wave theory failed, and failed badly.

These were dispersion and selective absorption. Since similar methodological conclusions are drawn from the two cases, I shall just concentrate on the second phenomenon, which Brewster in fact discovered. Brewster found that if sunlight is passed through certain gases and then dispersed in a prism, its spectrum is marked by a whole series of sharp, dark absorption lines. For example, he noted more than a thousand such lines in the spectrum of sunlight that had passed through "nitrous acid gas." This implies that such media absorb particular elements of the solar spectrum (or at any rate very narrow ranges of such elements), while transmitting other elements "infinitesimally close" to the absorbed ones. This was not a question of the refutation of a particular version of the wave theory – instead it marked a difficulty for the whole approach. This was because, whatever the details, the *general* story that the wave theory was forced to tell concerning this phenomenon was, as Brewster emphasized, so farfetched. Let's concentrate for sake of illustration on just one dark line in the spectrum of light transmitted through "oxalate of chromium and potash." The wave theory is forced to say that the ether within that gas:

> freely undulates to a red ray whose index of refraction, in flint glass, is 1.6272, and also to another red ray whose index is 1.6274; while . . . its ether will

not undulate at all to a red ray of intermediate refrangibility whose index is 1.6273! (1833b, 362)

In other words, whatever detailed account may eventually be given, a tiny difference in the length of a wave must be supposed to produce a black-and-white change from free passage through the ether within the gas to no passage at all. Brewster pointed out (1833a, 321) that:

> There is no fact analogous to this in the phenomena of sound, and I can form no conception of a simple elastic medium so modified by the particles of the body which contains it, as to make such an extraordinary selection of the undulations which it stops or transmits. . . .

He quite reasonably concluded that the phenomenon "presents a formidable difficulty to the undulatory theory." As we shall see, he also felt that this was an area where the emission theory scored over the wave theory, since it was easy to produce at any rate outline suggestions for how selective absorption might be accommodated within the emission approach.

(2c) Brewster's Disagreement with the Wave Theorists Over the Way Forward

So far as the friends of the wave theory were concerned, Brewster's claims, considered simply as claims about the *present* version of the wave theory, were perfectly reasonable and indeed completely uncontroversial. Both Airy and Baden Powell, each of whom responded directly to Brewster, freely acknowledged that the wave theory *as it stood* had no adequate explanation either for dispersion or for selective absorption. Airy and Powell both insisted that the main question, and their main disagreement with Brewster, was about the appropriate *response* to those admitted difficulties. In other words, the main disagreement was *heuristic*.

Brewster does *seem* to have believed that, despite all the difficulties that had mounted against it, there was life left in the Newtonian emissionist theory. He echoed Herschel's sentiment expressed some ten years earlier that, were sufficient talent and energy invested in the emission theory, it might yet turn the tables of scientific superiority on its undulatory rival. Correspondingly, Brewster felt that the near monopoly of theoretical talent that the wave theory had attracted by the 1830s was unhealthy. He believed that the experimental difficulties he had pointed to were sufficient to justify a less committed, more theory-neutral approach than the one that then dominated optics in both Britain and France.

Airy and Powell both argued that, on the contrary, the only reasonable response to the empirical difficulties that beset the wave theory was renewed commitment to it, aimed at solving those difficulties within the general approach. They each argued as follows. First, the wave theory was already overwhelmingly superior in terms of empirical support. Airy and Powell both accepted that there were phenomena (including those that Brewster had highlighted) that the wave theory could not (or could not yet) explain. However they insisted that there were lots of phenomena that the wave theory explained and/or predicted, but that the emission theory could not deal with adequately at all; *and* that there was no phenomenon that the Newtonian theory could adequately explain (let alone predict) that the wave theory could not.[4]

Secondly, while the corpuscular theory had run into difficulty after difficulty and failed successfully to overturn any of them, the wave approach had, on the contrary, already demonstrated the capacity to be modified in the light of experimental difficulties in a scientifically fruitful way. Airy in particular stressed the example of Fresnel's switch from longitudinal to transverse waves.

What had happened in that case was, in outline, roughly as follows. Wave theorists before Fresnel had all assumed that the ether is an extremely rare and subtle fluid—how else could the planets move so freely through it? It is a theorem of mechanics that fluids transmit only *longitudinal* (pressure) waves. (Longitudinal waves are ones in which the particles of the medium oscillate in the *same direction* as the overall transmission of the wave through the medium; an example being a sound wave in air.) Fresnel's own initial theory was indeed that light is a longitudinal wave. However, he and Arago then established experimentally that if, say, the two beams emerging from the two slits in the double-slit experiment are polarized at right angles to one another (by passage through suitably oriented crystal plates), then the interference fringes disappear. It seemed that light beams polarized in mutually orthogonal planes *fail* to interfere (or, rather, fail to produce interference fringes). Neither Fresnel nor any other wave theorist had, at this stage, any theory of the polarization of light. But, so long as the light waves were assumed longitudinal, the precise account of what happened when light is polarized could make no difference. Assuming that the wave theory is *at all* correct, the longitudinal assumption alone means that the disturbances in the two coherent and near-parallel beams (the slits are, remember, very close together) must themselves be near parallel and hence must alternately interfere constructively and destructively for different path differences. The Fresnel-Arago experiment, therefore, put the wave theory into deep trouble. Fresnel took a still deeper breath and switched to the *transverse* wave theory: to the theory that the ether particles oscillate at right angles to the direction of the propagation of light. This yields an easy theoretical account of the process of polarization: the disturbance in an unpolarized beam has components in *all* planes through the direction of propagation; polarization (linear or plane polarization, that is) consists in restricting the disturbance to one such plane. This explained the apparent "sidedness" of polarized beams, and also explained the Fresnel-Arago results. The oscillations in beams that are polarized orthogonally are assumed themselves to be orthogonal. Hence, although the two sets of oscillations certainly interfere or superpose—to produce (in general) elliptically polarised light—they operate at right angles rather than along the same line, and hence can never *destructively* interfere so as to produce fringes. Although it straightforwardly dealt with this difficulty over polarized light, the switch to the transverse theory certainly required a deep breath. This was because elastic media can transmit such waves only if they exhibit resistance to sheer, that is, only if they are *solids*. But how could the planets move completely freely through an *elastic solid* ether?[5]

But whatever the conceptual difficulties, Fresnel's new transverse theory scored stunning empirical successes. Not least when Hamilton showed in 1830 that the transverse theory entails the hitherto unsuspected phenomena of internal and external conical refraction—predictions that were confirmed by Lloyd in 1833. Airy's point about this theoretical shift was that the phenomena of polarized light discovered by Fresnel and Arago were major difficulties for the original version of the wave theory, no less major than the difficulties now cited by Brewster against the new version of the theory. But, in the earlier case, rather than give up the whole theory, Fresnel had modified it and had in this way produced a theory whose empirical virtues Brewster himself now rightly applauded. On the other hand, Airy told Brewster:

> Had Fresnel proceeded as you (apparently) would wish us to proceed, the undulatory theory would not now have existed. (1833, 423)

And so the major predictive success would have been missed.

The wave theory had already shown the ability to turn major difficulties into major successes. Its only well-articulated rival seemed hopeless: the emission theory had simply stumbled from one difficulty to the next, without producing anything remotely resembling the predictive success of the wave theory. Airy and Powell both acknowledged that important empirical problems *did* face the wave theory in the 1830s, but given the overall methodological situation, the only reasonable reaction to these problems seemed to be to work on the wave theory in the attempt to eliminate them. As Baden Powell put it:

> no sound philosopher would for a moment think of abandoning so hopeful a track, and none but the most ignorant or perverse would find in the obstacles which beset the wave theory anything but the most powerful stimulus to pursue it. (Powell 1841, iii)

3. Did Brewster Blunder?

So, these are the main elements of Brewster's view of the wave-particle rivalry in optics in the 1830s, compared to the views of the contemporary proponents of the wave theory. Was Brewster's position "irrational"? The question, along indeed with the whole issue of the rationality of theory change in science, stands in desperate need of clarification. In this section I shall try to bring the general issue into focus through an analysis of Kuhn's influential views on theory change. This analysis will, in turn, sharpen the questions that need to be asked about Brewster's particular views.

(3a) Reason and Theory Choice

Kuhn's account of paradigm change in *The Structure of Scientific Revolutions*, and in particular his claim that resistance of the new paradigm is never irrational, led critics to accuse him of holding that the decision to adopt a new paradigm is never "based on good reasons of any kind, factual or otherwise,"[6] and hence of making fundamental theoretical changes in science "a matter for mob psychology."[7] Kuhn has directly confronted such criticisms in his article "Objectivity, Value Judgement and Theory Choice."[8] He explains that he has never denied that "reason," in the form of the "objective factors" from the philosopher of science's "traditional list" (including such factors as empirical accuracy and scope, consistency, simplicity, and fruitfulness) plays a crucially important role in theory change:

> I agree entirely with the traditional view that [these objective factors] play a vital role when scientists must choose between an established theory and an upstart competitor . . . [T]hey provide the *shared* basis for theory choice. (322)

However, these objective factors supply no "algorithm for theory choice." At any rate when the choice is still a live one in science, they never *dictate* the choice of one of the rival theories. This is for two main reasons. First, single factors often turn out to be ambiguous when applied to the theories *as they stood at the time when the choice was being made*. It is often assumed, for example, that the Copernican heliocentric theory was empirically more accurate (that is, had a better detailed fit with the empirical data) than the Ptolemaic theory. This *eventually* became true, but only as a result of the work of Galileo, Kepler and others — who had clearly then already "chosen" the Copernican view for different reasons (if for any *reasons* at all). Secondly, even if the single factors in the list of objective virtues each point in a definite direction, it is by no means always the *same* direction. Again taking the Copernican revolution as example, and again taking the rival theories as they then stood (say in 1543), while *simplicity* (in a certain special sense) favored the Copernican theory, *consistency* (with other accepted theories) unambiguously favored the Ptolemaic theory.

It follows, therefore, claims Kuhn, that the objective factors must always be supplemented by "subjective" (or, rather, individual or idiosyncratic) factors in order to deliver an unambiguous preference:

My point is, then, that every individual choice between competing theories depends on a mixture of objective and subjective factors, or of shared and individual criteria. (325)

This account of theory choice, according to Kuhn, diverges comparatively little from that "currently received" in the philosophy of science (321). He accepts, on his part, that the "traditional" objective factors play a vital role. While the philosophers have, on their part, abandoned the idea of an entirely objective algorithm for theory choice, or, at any rate, have relegated it to a practically unattainable ideal, they have further accepted that, as a *matter of fact*, subjective (or idiosyncratic) factors have played a role in the choices actually made by scientists. The gibes about "mob pyschology," therefore, "manifest total misunderstanding" (321): properly understood, he and the philosophers of science are in broad agreement. In particular, he does not deny (and never has denied) that those who switch to a new paradigm in a scientific revolution have *good reasons* for doing so. The only thing is that those who stick with the old paradigm do so for "good reasons" too: "there are always at least some good reasons for each possible choice" (328).

Is this rather cozy view of no real conflict correct? One major problem is that current philosophy of science is altogether less monolithic than Kuhn seems to assume (philosophy of science is "pre-paradigmatic"). There is no single received view in philosophy of science on theory change. At least two broad traditions need to be differentiated. One — the subjectivist tradition — is represented by personalist-Bayesianism. According to this view a person is rational if he assigns degrees of belief to the theories available to him in such a way that these degrees of belief obey the probability calculus and the principle of conditionalization.[9] The latter requires that the rational agent's *posterior* degree of belief in a theory T, that is, his degree of belief in view of the actual evidence e that has accumulated at some later stage, should be measured by the conditional probability $p(T,e)$. The value of this latter quantity is, of course, dependent on the *prior* probability that the agent assigns to T, that is, roughly speaking, his degree of belief in T ahead of any (new) empirical evidence. On the personalist view, this prior probability is a purely subjective matter. This means that there is indeed no real clash between this view and Kuhn's account as just presented. It is in fact easy to give a personalist-Bayesian reconstruction of the theory choices of pretty well any scientist by making suitable assumptions about his distribution of priors. Did Fresnel switch to the wave theory while Brewster resisted it and stuck with the corpuscular theory? Well then, Brewster clearly gave the wave theory a very low prior probability — a much lower one than did Fresnel. Did Einstein resist the quantum theory while acknowledging the strength of the evidence in its favor at a time when Bohr was already fully persuaded of the theory by the same evidence? Well then, Bohr clearly assigned a higher prior probability to the quantum theory than did Einstein. The personalist-Bayesian can show that in the limit, as evidence accumulates, the degrees of belief of all rational scientists in certain circumstances converge on the same values, irrespective of their personal prior probabilities. But of course in real situations we are never at the limit, and resistance by a scientist to any theory in any evidential situation can be explained as rational by ascribing to him a sufficiently low *prior* degree of belief in the theory.

It is, however, precisely this aspect of personalist-Bayesianism that is found objectionable by the defenders of a second major tradition in philosophy of science. The objectivists, as they might be called, see personalist-Bayesians as in effect abandoning the whole idea that scientific change is a rational affair.[10] On the objectivist view, some scientific judgments about the relative merits of competing theories *are* dictated by objective factors (although not necessarily those on Kuhn's list). On this view, if some particular scientist, because of subjective or idiosyncratic considerations, fails to concur in such a judgment, then the scientist does indeed "violate scientific standards." Those who criticized Kuhn for making scientific change an irrational affair clearly belong to this *second* tradition: and it is by no means clear that Kuhn's subsequent clarification of his views does anything to reduce the gulf between him and them and hence to lead to a withdrawal of the mob psychology charge. There are two main reasons for this lack of clarity. The first is a misunderstanding on Kuhn's part about the objectivist claim that scientific change is a rational matter; and the second is great confusion on the part of the objectivists over *which* exact scientific judgments about the relative merits of rival scientific theories are dictated by "reason."

First, Kuhn (and some Kuhnians) often writes as if it were a surprise that holdouts like Priestley or Brewster produced *arguments* for their position and as if this on its own refuted the view that reason has dictated certain theoretical changes in science. But no defender of the objectivist approach, I take it, claims (or has ever claimed) that those who hold out for the older paradigm in what turns out to be a revolution will resort to simple dog-in-the-manger "Yah, boo, hiss!" tactics. *Of course* these holdouts will *argue* for their position—that is, give, in the straightforward sense, *reasons* for it. No one even denies, I take it, that some of these could be considered prima facie good reasons—in the sense that they cannot simply be dismissed without investigation as appeals to emotion rather than reason. Creationists, after all, have produced long books arguing their case; Jehovah's Witnesses will happily engage you for hours in arguments that purport to show that Nature has a design that bespeaks God's hand. Arguing *against* creationists, say, is, indeed, less easy than some other people think (as is shown by the hash that is sometimes made of it).[11] The objectivist-rationalist, however, claims that it *may* nonetheless turn out that the creationist's case disintegrates (or is shown to be extraordinarily weak) *on careful analysis*. This would mean that such a creationist *could* be judged "unscientific" or "irrational." But if this latter epithet is applied, then it should be understood as meaning that the person is not persuaded by *scientifically cogent* reasons, *not* that she has no reasons at all for holding the views she does. Of course, she *will* have such reasons—at least her prior belief in God and the 'literal truth' of the *Bible*; but also perhaps 'scientific evidence' that she holds to be strong and that may require a good deal of analysis before being revealed as bogus, *or* simply as of little weight when compared with the evidence in favor of rival views. (As I shall discuss in more detail later, Kuhn also writes as if his "rationalist" opponents were committed to the view that *all* the evidence standardly points in favor of the new theory in a scientific revolution; but it is surely clear that his opponents need only claim that *on balance* the evidence objectively favored the new theory.)

I said there are two obstacles in the way of clarifying the exact disagreement between Kuhn and those in the objectivist philosophical tradition. The second is that the upholders of this tradition have either not been clear, or have been in clear disagreement, about *exactly* which judgments they see as sometimes dictated by objective rational considerations.

In Brewster's case, as we saw, his "choice" of theory was not a simple matter: he expressed several quite different views about different aspects of the wave-particle rivalry, different views that might merit different responses concerning their rationality. In general, there are at least three judgments that a scientist might make about a particular theory and that ought to be distinguished. *First*, the judgment that that theory is *presently best favored* by the known evidence; *second*, the judgment that the theory is *true*, or, perhaps, "approximately" or "essentially" true; and *third*, the judgment that it is the best theory to *work on*, in that the general ideas underlying it provide the best opportunities for further scientific advance.

There are, of course, important connections between the three judgments. Indeed, if the development of science were "essentially" cumulative, at *all* levels—theoretical as well as empirical—there would hardly be any urgency in separating the three judgments. Suppose that—as "older" philosophers of science are usually accused of holding—new theories in science always included older theories (where "inclusion" is allowed to mean "inclusion with minor modifications"). Even in that case, there would, of course, be no question of our theories being *demonstrated* by the empirical evidence. But at least there would be nothing in the history of science that told *against* the view that in accepting a theory, a scientist accepts it as *true* (or essentially true). Similarly since we should then presumably make the inductive assumption that essential accumulation would continue to hold in the case of *future* changes in science, it would seem to follow that the only rational choice of theory to try to develop further would be the *presently accepted* theory. This is because any theory that was accepted in the future would be (inductively) guaranteed to be an extension (or "essential extension") of that theory presently accepted.

No knockdown refutation can be expected of this essential accumulation view of scientific development: because of the vagueness introduced by the modifier "essential." Nonetheless comparison with anything like an accurate history of science makes the view extremely implausible. Consider for example the history of optics. Even if this history is run only from the late seventeenth-century onwards, it contains a succession of quite different theoretical ideas. The idea that light consists of material particles was widely held in the eighteenth-century, until it was superseded, following Fresnel's work, by the idea that light consists of periodic disturbances transmitted through an all-pervading elastic medium. Particles in a void and waves in a medium certainly look as close to "chalk and cheese" as do chalk and cheese themselves. The wave theory was in turn superseded by Maxwell's theory of light as a disturbance transmitted through a disembodied electromagnetic field. Maxwell himself, of course, was convinced that a mechanical ether underlay the field, but a whole series of attempts to produce a mechanical model failed and left the field as an irreducible, or at any rate unreduced, primitive—meaning that here too there was radical discontinuity at the theoretical level: it is again hard to think of two things more different than an elastic disturbance and an electric (displacement) current. Finally, as part of the quantum revolution, the theory of the constitution of light was again fundamentally altered—according to this theory, light consists of photons obeying a new, entirely nonclassical mechanics.

Of course (and despite what Kuhn and Feyerabend at one time seemed to be claiming) there *has* been "essential accumulation" at the *empirical* level. Successive theories, despite being separated by "revolutions," dealt with an ever wider range of phenomena. The material particle theory could *at best* deal adequately only with simple reflection and refraction; Fresnel's wave theory added interference, diffraction, and polarization; Maxwell added various phenomena concerned with the interaction between light and electricity and magnetism; the photon theory added the photoelectric effect and many others. In this process no *empirical* explanatory power was lost, except perhaps momentarily, even though the explanations were radically altered.[12]

In this optical case (as I believe in *most* cases), more of the older theory enters the new than simply its empirical success: the mathematical equations, and hence, if you like, the *structure*, of the older theory are preserved as well (perhaps as limiting cases). An especially clear instance is provided by the transition from Fresnel to Maxwell. Fresnel's equations, which yield the relative intensities of reflected and refracted light beams in various circumstances, are preserved *entirely* intact within Maxwell's more general theory. However, this continuity is purely *structural* or *syntactic*. The equations remain the same, but the *interpretation* of the fundamental theoretical term involved in them changes completely. Of course, a theory-neutral term, like "optical disturbance," can easily be introduced to do service for both theories; but this should not be allowed to obscure the fact that the optical disturbance in Fresnel's theory represents the distance a particle of the ether has been moved from its equilibrium point, while in Maxwell it is a disturbance in a disembodied, nonmechanical electromagnetic field.

The picture, then, seems clearly to be one of theoretical discontinuity coupled with "essential" empirical (and indeed structural) continuity.[13] The need to distinguish the three judgments about scientific theories mentioned above becomes apparent. Acceptance of a theory as presently most favored by the evidence need *not* involve accepting that theory as *true*, or even "approximately true." It may empirically be the case that many scientists do believe in the truth (or approximate truth) of the latest scientific theories. And, of course, they *may* be correct: it is logically *possible* that, after a series of "failures" (glorious failures), science has now hit on the truth. But those scientists who are more historically aware are surely more likely to be persuaded by the so-called pessimistic induction that even our best current basic theories will one day be replaced by quite different ones. Similarly in the case of heuristic advice, the claim that the only rational course is to try to develop that theory that is presently best favored by the evidence depends crucially on the assumption that science is (and will continue to be) cumulative.

Kuhn never explicitly defines his term "theory choice." On his construal it does however *seem* to involve taking the theory fully to one's breast, believing it and working on it to the exclusion of all others. He takes it that at any rate the *original* aim of the philosophy of science was to construct an algorithm for theory choice. And this seems to imply that the original view in philosophy was that the rational scientist must always choose, that is, believe and seek to develop, that theory that is already most favored by the evidence.

It is true that many philosophers of science—both subjectivist and objectivist—have talked in terms of rational degrees of belief in a theory, and that it is difficult to see what this can mean except for belief that the theory is *true*. However, there certainly is also a long-established anti-realist (or better: structural realist) tradition in philosophy of science (represented by Duhem and Poincaré, as well as more recent writers) and a fallibilist, or *conjectural* realist tradition (represented by Popper, Lakatos, and others). Duhem, Poincaré, Popper, Lakatos, and many others have all explicitly insisted that one can "rationally accept" a theory *without* believing it to be true. Moreover, so far as the *heuristic* question goes, those who think they can identify an orthodoxy or "received view" in twentieth-century philosophy of science all agree that an integral part of that orthodoxy was the distinction between justification and discovery, and the insistence that philosophy or logic of science was concerned only with the former. This would imply, of course, that theory appraisals have no heuristic consequences whatever. There are certainly some more recent philosophers (such as

Lakatos) who have held that an assessment of the heuristic power of a theory (or paradigm or research program) is an important part of the appraisal of its *present merits*. But Lakatos was careful to insist that there still can be no direct inference from "theory T is currently most favored by the evidence" to "the only rational course of action is to try to develop T" (or, "it would be irrational to try to develop any rival theory T*"). No sophisticated analysis is needed to see the complete untenability of any position that *was* committed to any such straightforward inference. Such a position would entail that the great geniuses of science acted irrationally: the wave theory of light, for example was certainly not unambiguously the best available theory when Fresnel started to work on it in the early nineteenth-century; it was Fresnel's work that *turned it into* overwhelmingly the best available theory.

As I see it, then, the objectivist philosophical tradition was never committed (or, at any rate, ought never to have been committed) to the view that the only rational course of action for a scientist was to "choose" (in Kuhn's sense) that theory that is presently objectively most favored by the evidence. The tradition *is*, I shall take it, committed to the view that there is always an objective ordering of the available theories. There is no reason why this should *always* be a strict ordering, but the objectivist is, I think, also committed to the view that what generally happens in scientific revolutions is that the previously entrenched theory is deposed by one that is strictly superior to it.

Over and above these two core views there has been little agreement between different proponents of the objectivist tradition. They standardly agree on the preference ordering of a given set of rival theories at a given stage of their developments, but often disagree about the general principles that underlie such orderings. More importantly for present purposes, they often disagree about what exactly these orderings of *theories* require from the rational *theorist* (beyond, of course, acceptance of the ordering itself). Against this background, Kuhn's arguments purporting to show that there is no "objective algorithm for theory choice" need carefully to be separated into two different groups. Those in the *first* concern the ranking of theories in terms of their objective merits and, in particular, claim to show that the new upstart theory in some scientific revolution was *not* objectively superior to its previously entrenched rival. The arguments in the *second* group point to difficulties in connecting objective rankings of theories with rational action and rational belief on the part of theorists. The two sets of arguments have very different statuses. Those in the first set would, if successful, knock out the core objectivist thesis. But those in the second set can only, I think, serve to clarify the *open* question of what exactly it is rational or irrational to do, given that one accepts that the scientific evidence currently favors a particular theory over all known rivals. (Since Kuhn himself does not make this distinction, I shall need to take the liberty in what follows of recasting his arguments slightly.)

3(b) Kuhn, Theory Appraisal, and the Objective Superiority of the Wave Theory Circa 1830

The objectivist holds that there is always an objective ranking of rival theories, basically in terms of the evidence in their favor. He also holds that, at any rate generally, a "scientific revolution" consists of the replacement of one theory by one objectively strictly superior to it. Two main arguments can be found in Kuhn's work that, if successful, would tell directly against these theses.

One, remember, is this. Kuhn gives a whole list of criteria, which he is ready to concede are "objective" or, rather, shared by all scientists. The list includes empirical accuracy (that is, detailed fit with the data), empirical scope, consistency (both internal and with other accepted theories), simplicity, and fruitfulness. One reason why these criteria do not supply a choice algorithm is that in live cases of theory choice, and, in particular, during scientific revolutions, these different criteria seldom, if ever, tell in the same direction. Much *later*, once the revolutionary theory has been developed and improved, it may outscore its older rival on all counts—but this happens *as a result* of the revolution and therefore can't form its rationale. For example, as I already indicated, Kuhn points out that, if the Ptolemaic and Copernican theories are compared, not as they stood *after* the work of Kepler, Galileo, and Newton, but at, say, the time when Kepler and Galileo were actually choosing to work on the Copernican theory, then the two factors of consistency and simplicity (or harmony) told in opposite directions. The Copernican theory, in its basic form, undoubtedly gave simpler explanations of, for example, the planetary stations and retrogressions and the limited elongation of Mercury and of Venus. But, the Copernican theory clashed wildly with the prevailing, Aristotelian physics and cosmology, while the Ptolemaic theory was, of course, an integral part of the Aristotelian worldview.

Although Kuhn clearly does not establish it in *every* case of fundamental theory change, his historical claim seems to me likely to be correct. Turning back to my own example, if the wave and emission theories of light are compared as they stood in 1830, then a case can certainly be made out that, whatever the other merits of the wave theory, the emission theory still outscored it in terms of mathematical manipulability. (Classical particle mechanics had long been fully articulated mathematically, while continuum mechanics remained partially undeveloped—despite having recently made major advances, often in tandem with wave optics.)

Assume, for the sake of argument, that Kuhn's historical claim is indeed correct in general. His argument against the objectivist nonetheless goes through only if we accept the initial assumption that the objectivist can do no better than supply a "laundry list" of objective factors, and is therefore left entirely without recourse when two factors from the list pull in opposite directions.[14] But I know of no objectivist who would accept Kuhn's list as it stands and none who would be happy to leave *any* such list unstructured. For example, for Duhem, Poincaré, Lakatos, and many others, there is a *basic* criterion: that of predictive empirical success. When this criterion is properly understood, it informs most of those on Kuhn's list. The basic idea behind this proper understanding is that a theory achieves predictive success by yielding an empirical fact *without* any prior tinkering specifically aimed at making the theory yield that fact.[15] So, stations and retrogressions, for example, "fall out" of the basic Copernican heliocentric idea, but have to be deliberately built into the Ptolemaic geocentric theory by suitable choice of auxiliary assumptions. Thus prediction properly understood need not involve a hitherto unknown fact—Copernican theory *predicted* the already well-known phenomena of stations and retrogressions.[16]

"Simplicity" and "unity"—in the scientifically most important senses of these terms—are closely related to predictive success. There are surely no clear-cut intuitions about when one *basic* theory in science is simpler than a rival. Is, for example, the idea that light consists of material particles more or less simple than the idea that it consists of waves in a medium? I don't see how even to begin answering the question. Where we *do* have clear intuitions is in cases where a basic theory has been so hedged around with qualifications and split into so many unrelated subcases that it clearly becomes too complex, not sufficiently simple, to be scientifically acceptable. But in all such cases the complexity and disunity have been introduced under the pressure of initially independent or recalcitrant experimental results. The basic theory has enjoyed no predictive success: it has either turned out to be silent about some phenomenon clearly in its field, or to yield an incorrect prediction. Special cases and exceptions have therefore had to be introduced to accommodate the facts—at the cost of increased complexity and decreased unity. This is clearly what had happened in the case of Ptolemaic astronomy; it also happened, as we shall see, in the case of the corpuscular theory of light.

"Fruitfulness" too is intimately connected to predictive success. A general theoretical approach (a paradigm or research program) shows its fruitfulness by supplying ideas for developing specific theories *independently of empirical results*. Such an approach will be judged barren (as Lakatos put it, the research program's "heuristic" will have "run out of steam") only when all these ideas have been tried *without predictive success*; and hence the approach has been reduced to tagging along *behind* the empirical data, always accommodating it *post hoc* rather than predicting it in advance.

By the early to mid-1830s, for example, the emission approach to optics had very definitely proved barren. The ideas supplied by the general claim that light is a Newtonian particle had all been tried in the attempt to produce specific theories that dealt with optical phenomena. Particles were, of course, subject to forces; forces could be attractive or repulsive: all the apparent deviations from rectilinear propagation—reflection, refraction, interference, and diffraction—

might be explained by having ordinary "gross" material objects exert forces on the light particles. The idea that these are strictly point particles always had to be an idealization — the finite dimensions of the real particles might come in useful: it might for example be assumed that the particles have sides or poles and revolve with respect to these poles as they move along. Various isolated results could be explained (at any rate in outline) on the basis of these assumptions — but, when it came to anything like details, the "natural" assumptions about the forces and the polar revolutions unambiguously failed and instead the required theoretical assumptions had always to be "read off" the *already given* facts. There was never any correct prediction of a different phenomenon. Instead each new phenomenon required further elaboration of the theoretical assumptions (perhaps another complication in the field of force set up by the diffracting or refracting body or yet another axis of revolution in the particles). As Humphrey Lloyd put it in a famous report on the "Progress and Present State of Physical Optics":

> An unfruitful theory may . . . be fertilized by the addition of new hypotheses. By such subsidiary principles it may be brought up to the level of experimental science, and appear to meet the accumulating weight of evidence furnished by new phenomena. But a theory thus overloaded does not merit the name. It is a union of unconnected principles. . . . Its very complexity furnishes a presumption against its truth. . . . The theory of emission, in its present state, exhibits all these symptoms of unsoundness, . . . (1833, p. 296)

Similarly, by the early years of this century, ether-based classical physics was no longer fruitful. Instead of the general idea of an ether that fills space suggesting new specific theories, the ether had become an embarrassment — ad hoc explanations having to be provided one after the other for why otherwise expected manifestations of the ether failed to show up empirically.

As for the other "objective factors" on Kuhn's list, the philosophers I have mentioned would all, I think, either deny them *any* role or relegate them to subsidiary roles.

This is particularly true of consistency (that is, consistency with other, already accepted theories). It is surely a *virtue* in a theory, rather than a vice, if it clashes with some well-entrenched claim — *provided* that there is strong evidence for the theory in the form of predictive empirical success. The inconsistency of Copernican theory with accepted Aristotelian physics supplied interesting and demanding problems for further research. Scientists will, no doubt correctly, downgrade (or more usually ignore) new theories that clash with well-established ones — but *only* when there is no independent evidence for the new theory. The fact that various current hypotheses concerning the "paranormal" clash with accepted theories is currently regarded as an important argument against them, but again only because those hypotheses have showed no empirical predictive success. It is predictive success that flips inconsistency with other well-accepted theories over from a vice to a virtue.

Kuhn's demonstration that there are important historical cases in which different objective factors pulled in opposite directions need not, then, trouble this sort of objectivist. He will happily pronounce Copernicus's theory scientifically preferable to Ptolemy's in 1543, while admitting that Copernicus's theory was inconsistent with other previously accepted theories (and even, as we shall see, while admitting that the Ptolemaic still had, to some degree, superior established empirical accuracy). And, in my own example, such an objectivist will happily pronounce the wave theory of light well ahead in 1830, while acknowledging the emission theory's superior mathematical power. This is because the criterion of predictive success is dominant for him. And on that score, the wave theory of light (as Brewster himself more or less clearly acknowledged, as we saw) was simply miles ahead of its rival by 1830. In over 150 years, the emission theory had failed to produce anything remotely capable of standing alongside Fresnel's success in predicting in minute detail the sizes and separations of diffraction fringes, let alone his success with the "white spot" at the center of the shadow of a small circular disk, the emergence of circularly polarized light from a Fresnel rhomb, internal and external conical refraction, and so the list goes on.

A second argument is to be found in Kuhn against the idea that the winning theory in a scientific revolution is generally objectively superior to the older theory. The first argument was based, as we just saw, on the assumption that each objective factor tells unambiguously in favor of one of the rival theories, but then went on to claim that different objective factors may tell unambiguously in *different* directions. But Kuhn also argues that this initial assumption itself is often false: scientists may reasonably disagree over the way that *single* objective factors point. For example, simplicity told in favor of Copernicus over Ptolemy only when understood in a very special sense. In other senses, Copernicus's theory was by no means clearly the simpler. Similarly, although it is often assumed that the Copernican theory was better than the Ptolemaic in terms of empirical accuracy, in fact, as the theories stood in 1543, this criterion delivers no clear preference.

One major problem here is again Kuhn's nonanalytical, acritical approach to admission onto his list of objective virtues. The result—from the point of view of the analytic philosopher—is often a confused amalgam of various quite different ideas about criteria of scientific merit. It is then no news that such confused "criteria" supply no clear-cut judgments. Everyone would, for example, surely concede to Kuhn that a whole variety of notions of theoretical simplicity are to be found in science and philosophy. The moral seems to be that careful analysis is needed to sort out the really important notion or notions. (As I already indicated, on my view the important sense of simplicity is intimately related to predictive success.)

The case of Kuhn's criterion of empirical accuracy is similar. He describes this as "the most nearly decisive" of all the objective factors. But his notion of empirical accuracy is an unfortunate amalgam of two criteria that should be kept separate: predictive success and overall, detailed fit with all the known, relevant empirical data.

Kuhn points out that, contrary to widespread belief, the Copernican theory did *not*, as it stood in 1543, exhibit unambiguously better empirical accuracy than the Ptolemaic: the former did *not* account for every detailed empirical datum accounted for by the latter (plus some more) — instead *each* theory enjoyed empirical successes not shared by the other. Copernican theory did *eventually* come to dominate the older theory empirically, but only as a *result* of Kepler's and Galileo's decisions to "choose" Copernicanism. Similarly in the case of the chemical revolution, and again contrary to widespread present-day belief, there were empirical phenomena that the phlogiston theory could account for, but for which Lavoisier's theory could give no account. This is, in other words, the historical phenomenon, or alleged historical phenomenon, of "Kuhn loss." It acts as a further important source of reasonable subjectivism for Kuhn: if a scientist happens to give special weight to a phenomenon whose theoretical codification is "lost" in the switch to the new theory, then that scientist may reasonably resist the switch.

Is Kuhn loss a genuine historical phenomenon? Kuhn's own examples of lost content tend to be unconvincing. The example he tends to cite in the case of the chemical revolution for instance concerns the (alleged) fact that metals are "more similar" to one another than are metallic ores. But this is a curious empirical phenomenon — certainly it cannot stand as an observation report on a par with such things as "the needle in apparatus A pointed to near '5' on the scale" or "the measured angle of elevation of telescope T at time t was θ'." Moreover the phlogiston theory's alleged explanation of this curious fact is more curious still. The "explanation" is that metals are more similar to one another "because," unlike the ores, they all contain a common ingredient: phlogiston. This "explanation" relies of course on the implicit assumption that any two things that share a common constituent are "more similar" to one another than any two other things which do not. This *either* makes no real sense (apart from the notorious multiple ambiguity of "similarity," there is also the question of how deep we go in the search for common ingredients — after all, we now think that *everything* is "made out of" elementary particles) *or* it is arguably false (whatever exactly Kuhn had in mind, it seems difficult to argue that, say, a piece of coal is more similar to the Koh-i-noor diamond than, say, oxygen gas is to hydrogen gas).[17]

The loss allegedly involved in the Copernican revolution is also unconvincing — though for a different reason of more general significance. Kuhn's brilliant analysis of this revolution showed that, while the new Copernican theory gave genuine explanations of various important qualitative phenomena (such as planetary stations and retrogressions and the limited elongation of Mercury and of Venus), which had only been forced into the Ptolemaic framework post hoc, the Ptolemaic theory could give detailed quantitative accounts of phenomena that the Copernican theory, *as it stood in 1543*, could not match. This, however, only illustrates the importance of keeping quite distinct the criteria of empirical *predictive* success and overall empirical content, rather than conflating them into one notion of empirical accuracy.

There is a crucial difference between the success enjoyed by the Copernican theory and the, admitted, extradetailed empirical content of the Ptolemaic. Namely that the latter, but not the former, can be achieved simply by hard work. Ptolemy's theory had, of course, been developed and applied for centuries when Copernicus challenged it. In general it is not at all surprising if the entrenched theory has detailed acounts of phenomena that the new upstart theory cannot yet match. No matter how successfully predictive a new theory might have been, there would always be some areas where it needed detailed emendation and elaboration. But this is largely a question of hard work ("normal science"). The Ptolemaic system had been developed by letting the already known data guide the construction of the required auxiliaries within the general geocentric framework: no predictive success having been achieved in the process beyond that secured by simple inductive extrapolation.[18] It was surely clear already in 1543 that, just as the detailed phenomena had been worked into the Ptolemaic framework by suitable elaboration of auxiliary assumptions and mathematical devices, so they could, with sufficient effort, be accommodated within a heliocentric (or, rather, heliostatic) framework. Given a general theoretical framework, specific theories with ever greater empirical content can generally be developed simply through hard work. What *cannot* by definition be achieved in this way is the sort of qualitative predictive success that made Kepler and Galileo think that that sort of hard work on the Copernican framework was worthwhile. These predictive successes occur precisely when the empirical result "falls out" of the general theory *without* any tinkering. The Ptolemaic theory had had no such success. So nothing was "lost" in this case that could not clearly be regained.

Our own optical example might seem to supply a rather more convincing example of Kuhn loss. Brewster, as we saw, made a good deal of the fact that the wave theory could explain neither dispersion nor selective absorption. But this case is not clear-cut either.

As Airy and Powell insisted in their replies to Brewster, neither of these phenomena was properly explained on *either* theory. Certainly neither was (in my sense) *predicted* by the emission theory—neither fell directly out of that theory in the way that diffraction patterns fall out of Fresnel's theory, or that the bending of light rays, say, falls directly out of the general theory of relativity. Indeed not only was neither phenomenon predicted by the emission theory, no *full* emissionist account of either dispersion or selective absorption was *ever* given—even post hoc. The most that the emissionist could argue is what Brewster did in fact argue: that it was easier to see how, *in general conceptual terms*, an explanation of the phenomena *might* be produced within the corpuscular theory, than it was to see how such an explanation might be produced within the wave theory.

Concentrating for simplicity just on the case of selective absorption: the wave theory, as Brewster forcefully argued, was bound to have great difficulty in conjuring this discrete, "black-and-white" phenomenon out of its underlying assumptions that were unambiguously assumptions of continuity. An infinitesimal change in a continuous parameter—the length of the wave—would somehow have to make all the difference between free passage through the ether within a selective medium and no passage at all. The emission theory, on the other hand, made light consist of *different* particles: if the emission theory made some effect depend on the value of some parameter associated with these particles, there was no need for it to assume that all possible values of this parameter were instantiated. It could always indeed explain any apparent continuity, for example of the "degrees of refrangibility" associated with the solar spectrum, as an illusion, deriving from the inability of our coarse senses to detect slight but nonetheless existent differences. No precise emissionist account of selective absorption was in sight in the 1830s; but such an outline account could readily be seen to be a conceptual possibility. It could, for example, readily be conceived that two different light particles, while having almost identical degrees of refrangibility, might nonetheless differ in some other important respect, which accounted for one of them being absorbed by the medium while the other passed through. Brewster's own suggestion was that the phenomenon might be *chemical* in nature—that the different light particles have different chemical constitutions, which might then explain why one is absorbed and the other not. Brewster had no more than this to say—hence his suggestion was certainly, as it stood, vague and untestable. But even if the wave theory had *almost* nothing to match in this regard, it must be admitted that in 1830 it could *not* match it.

Does this mean that, at any rate for the wave optics revolution, I have conceded Kuhn's case and accepted that empirical accuracy did *not* tell unambiguously in favor of the wave theory? Such an inference, clearly encouraged by Kuhn, would be an obvious *non sequitur*. Kuhn's argument again seems to presuppose that those philosophers who hold that theory change in science is generally a rational affair are committed to the claim that *nothing* ever tells in favor of the superseded theory. In fact, of course, such philosophers have long recognized the need to *weigh* evidence. An "objectivist-rationalist" clearly need not hold that the theory superseded in a revolution had *no* virtues, nor even the view that it had no virtues not shared by the superseding theory. It is enough if *on balance* the superseding theory is clearly better. This applies in particular to *empirical* or evidential virtues.

This simple point allows us to bring together much of the foregoing discussion. In the Copernican case, even if the objectivist acknowledged that the accounts it provided of detailed empirical phenomena favored the Ptolemaic theory, he would certainly hold that the qualitative *predictive* successes scored by the Copernican theory favored that theory *much more highly*. No attempt to explain theory change as rational can hope to succeed if it fails to give extra theory-confirming weight to predictive success over post hoc accommodation. Indeed since such post hoc accommodation can *always* be achieved at least in principle (this follows from Duhem's point that the central framework-supplying theories in science have *in isolation* no empirical consequences), some philosophers give *zero* confirming weight to empirical data that have simply been worked into a theoretical framework. For such philosophers, a genuine case of Kuhn loss would need to involve the loss of some *genuinely predicted* content: that is, a case in which some phenomenon "fell out" of the older theory, but not out of the newer theory. To my knowledge no such case has been presented. But even *without* adopting this extreme view, and even if there *are* genuine cases of Kuhn loss, the objectivist need not be in trouble. Let's accept (as I believe we should) that dispersion and selective absorption (weakly) favored the emission theory in the 1830s. Still, everyone accepts (including Brewster, as we saw) that by then a long list of phenomena that (strongly) favored the wave theory could readily be produced. This list includes several phenomena that had been genuinely predicted by the wave theory – such as various diffraction patterns, various results about circularly and elliptically polarized light, and the phenomena of internal and external conical refraction. Airy and Powell were right that the *only* theory to enjoy any genuinely predictive success was the wave theory.

On any account, then, and being as generous to the emission theory as one likes, the evidence, *on balance*, strongly favored the wave theory. Thus the objectivist could (rather generously) concede that there was a Kuhn loss involved in this revolution concerning dispersion and selective absorption, without threat to her position. This is because the Kuhn loss involved in *not* making the switch to waves would have been enormous. Philosophers of science have not achieved any great measure of agreement over the general principles involved in weighing evidence, but everyone surely agrees on the need to weigh. And it is clear that no adequate account of weight of evidence could fail to have the balance coming down with a mighty bang in favor of the wave theory in the 1830s.

I claim to have shown so far that nothing in either Brewster or Kuhn tells against the view that, by the 1830s, the wave theory was objectively superior to its emissionist rival. There remains, then, what I have insisted must be treated as a separate question: that of what acceptance of this appraisal requires from the "rational scientist." As I suggested, I shall use the example of Brewster in an investigative way to attempt to illuminate the murky issue of just how strong an implication for rational belief and conduct our theory appraisals ought to have. So: did any of Brewster's theoretical views place him beyond the "rational" pale?

(3c) Was Brewster Irrational to Hold Out Against the Scientifically Superior Wave Theory?

Let me first clear away a possible misunderstanding of a purely linguistic kind over the terms "rational" and "irrational." Brewster was clearly a clever man, who dealt in arguments, who accepted all well-tested experimental data, made all the usual inductive generalizations of such data, and who did nothing to transgress the rules of deductive logic. Moreover, he clearly accepted that, in terms of predictive success, the wave theory had greatly outscored the Newtonian theory as things stood in 1830. If, despite all this, we end up saying that some of his views were "irrational," then this should clearly be understood in a rather special sense: one that carries no suggestion that Brewster is to be put on a par with Russell's famous "lunatic" (who believed that he was a poached egg), nor with anyone who, aiming to get down safely to the ground floor, proposes to take the window rather than the elevator, claiming that all evidence that this is foolhardy is evidence purely about the past. Without being irrational in any such blatant sense, Brewster might still have contravened best scientific practice in his attitudes towards the rival theories available to him.

It is (presumably) trivial that the objectivist-rationalist *would* pronounce Brewster irrational ("mistaken" would be better) if he denied that the empirical evidence currently favored the wave theory over its emissionist rival. But his acknowledgment of the "unequalled" predictive and explanatory success of the wave theory is surely tantamount to accepting this appraisal. The worry is that if—as Lakatos, for example, explicitly stated—this acknowledgment is *all* that is required from the rational scientist, then our rules of rationality say perilously little. Indeed Feyerabend has claimed that if all that Lakatos's methodology requires is an admission of the present score between the rival theories, then it is really "anarchism in disguise." It is wrong, I think, to underestimate the importance of simply keeping the objective score. (Try, for example, to get a creationist to say that he accepts that the evidence currently strongly favors the Darwinian theory, but that he nonetheless is working on the creationist approach, hoping eventually to reverse the evidential tables.) Nonetheless it is difficult not to yearn for a somewhat stronger theory of rationality.

Let's then look at Brewster's views about what follows (or fails to follow) from the acceptance that, in terms of empirical success, the wave theory was well ahead of its rival by the 1830s. Can any of these views plausibly be categorized as irrational and, if so, on what grounds?

As I already indicated, three of Brewster's views raise interesting questions in this connection. They are:

(1) Despite all its success in explaining and predicting optical phenomena, the wave theory is not *true* as a fully realistically interpreted "physical theory";

(2) Despite all its problems, the Newtonian theory might yet "stage a come-back" and ultimately turn the tables of scientific superiority on its rival.
(3) Therefore, Airy and Powell's view—that the only reasonable response to the difficulties facing the wave theory was to try to solve them *within* the general wave approach—was not correct.

I consider each of these in turn.

(3ci) Brewster's Disbelief in the Wave Theory as a Physical Truth

Brewster argued that the impressive explanatory and predictive success of the wave theory does not logically entail its truth as a fully fledged, realistically inter-preted physical theory. But Brewster went beyond this obviously correct logical claim and clearly held that, as a physical theory, the wave theory was actually false. Indeed he predicted that "after it has hung around for another hundred years," that theory will give way to a completely different physical theory.

Given that the wave theory was undoubtedly the best-supported theory avail-able to him, was it irrational of Brewster to believe it to be false? This would seem a harsh judgement to make in view of the fact that Brewster's prediction was correct—indeed he was overgenerous to the wave theory, which lasted at best an-other 70 years, rather than another hundred.

It is true—and importantly true—that many of the mathematical equations sup-plied by the wave theory still live on in science; and it is true—and importantly (if rather obviously) true—that repeatable (and repeated) experiments do not change their results, so that all the correct empirical consequences of the wave theory are still, of course, correct. Nonetheless, at the theoretical level there has been radical, ineliminable change. The ether—at any rate in anything like the form understood by Fresnel—has been entirely rejected by present-day science; photons traveling through empty space, despite their so-called wave*like* charac-teristics, could hardly be more different than they are from waves in a mechani-cal, space-filling medium.

If current scientific theories are correct, then so was Brewster correct in be-lieving the wave theory to be false. We surely should *not* then require the rational scientist to believe in the truth of the currently best-available theory. What if the rationality requirement is watered down so that it requires from the rational scien-tist only belief in the *approximate* truth of the current best theory? The notion of approximate truth has proved extremely resistant to precise analysis, but it does seem reasonably clear intuitively that, however approximate truth is eventually analyzed, Brewster's complete rejection of a real ether is inconsistent with an ascription to him of belief even in the approximate truth of the wave theory. It would, again, however, surely be difficult to find him guilty of irrationality on this score. After all he was surely *right*; no scientific realist will ever, I fear, pro-duce an acceptable account of approximate truth that would yield the judgment that if the photon theory of light is true, then the classical wave theory is approximately true. "To a large degree empirically adequate"—yes; "to some degree *structurally* accurate"—no doubt; but "approximately true"—no.

Many scientists do seem to believe the presently best-available theory to be not just highly empirically adequate but actually *true* (or "very nearly" true); and the number of believers not surprisingly increases with the continuing empirical success of that theory. Certainly there were many scientists in the early nineteenth century (Fresnel himself among them) who believed in a real, mechanical ether and the full truth of the basic wave theory. On the other hand there have always been other scientists—perhaps more aware of the *history* of science—who find belief in the *truth* of the latest theory impossible. This may, as perhaps in Brewster's case, be motivated by prior metaphysical beliefs, but it may also be motivated, as in, say, Poincaré's case, by general methodological and historical considerations about scientific theory. Both because the agnostics often turn out eventually to be right and because to do otherwise would be to prejudge a live *philosophical* debate, it would surely be wrong to brand them irrational. Whatever consequence the judgment that T is the best available theory has about rational belief, it is not the consequence that the only rational course is to believe T to be true (or even approximately true).

As I indicated earlier, many of the problems that Kuhn raises about "theory choice" and the lack of an "objective algorithm" for it arise from his implicit assumption that in choosing a theory a scientist must take it fully to his bosom and believe it to be true. *Of course* there are always holdouts in this sense. For example, anyone who holds a thoroughgoing instrumentalist view of scientific theory will *always* be such a holdout—no matter how she ranks the specific scientific theories available to her. Although instrumentalism certainly tends to be more popular at times when even the best scientific theory is in clear difficulties, it is not *invariably* adopted in this defensive way. Interestingly enough, the two main British advocates of the wave theory, Airy and Baden Powell, themselves might have to be classified as holdouts in *this* sense: both adopted explicitly uncommitted views of the ether. I quoted Airy *above* (p. 321) claiming that Fresnel's wave theory had the same status as Newton's gravitation theory, both being "certainly true." However, later in this passage he asserted:

> This character of certainty I conceive to belong only to what may be called the *geometrical* part of the theory: the hypothesis, namely, that light consists of undulations depending on transversal vibrations, and that these travel with certain velocities in different media. . . . The *mechanical* part of the theory, as the suppositions relative to the constitution of the ether . . . though generally probable, I conceive to be far from certain.

Similar sentiments were expressed by Baden Powell. Yet both were fully committed to the wave theory as superior to all known rivals.

(3cii) Brewster's Continued Belief That the Emission Theory Might Eventually Prove Triumphant

Brewster seems to have held that the emission theory — if diligently developed by scientists of talent — was likely *eventually* to prove superior to the wave theory (or at any rate *might* eventually do so). In general, on Kuhn's view, the main source of the resistance to new paradigms by elderly holdouts is the assurance that they feel that "the older paradigm will ultimately solve all its problems, that nature can be shoved into the box the [older] paradigm provides" (1962 151–2). Kuhn insists that this assurance — while it may irritate the revolutionaries — cannot be faulted as "illogical" or "unscientific" or "irrational." Is Kuhn right?

Suppose first that the claim that nature *can* be "shoved" into the older "box" is merely one about logical possibility. Then the claim is of course correct: put in more orthodox terms, and following Duhem (1906), the core theory underlying the older theoretical system will not be testable in isolation, and so it follows from deductive logic that there must be *some* assumptions that are consistent with that core theory and that, together with that core theory, entail *any* given experimental results. More strikingly, in our historical case (as well as others) the outlines of how actually to shove *most* known optical phenomena into the older box had been constructed by the 1820s and 1830s. Among such shovable phenomena were ones like interference and diffraction, which more superficial, later treatments presented as crucial phenomena — predicted by the wave theory, but quite beyond the scope of the emission theory.[19]

On this weak interpretation, Kuhn's assurance that the older paradigm *can* accommodate all the phenomena certainly exists. But, of course, it does *not*, as Kuhn seems to think, automatically make resistance to the new theory rational. *That* would require (at least) a quite different assurance — the assurance that the phenomena can be accommodated within the older paradigm *in a scientifically acceptable way*.[20] But this much stronger assurance does not follow from Duhem's point about untestability in isolation. We know that some sort of account could be given of *diffraction*, for example, within the emission theory — not only because of logical considerations, but because the outlines of such accounts were constructed. But those accounts *were* uniformly awful — the assumptions that had to be made about different masses of the different light particles, their rotations about various axes, and about the dependence of the "diffracting force" on the phase of rotation and on the distance from the diffracting object were simply "read off" the facts. Those assumptions had to be made more and more complex as more facts were taken into account. Even the most assiduous of emissionists, like Biot, quietly gave up long before *all* the facts had been accommodated. Deductive logic alone certainly does not, of course, *guarantee* that any nonawful account even *could be* constructed within the emission theory.

The only assurance really available to the holdout fails to explain his resistance as rational. But it is of course a further question whether or not it can actually become *irrational* to hold that an older theory will eventually provide scientifically acceptable explanations of presently recalcitrant phenomena. Was it irrational, for example, to believe in 1830 that the emission theory could eventually *adequately* explain diffraction?

As I just indicated, such a belief certainly does not run counter to deductive logic. If one is satisfied with deductive rationality, then the answer must be that it was *not* irrational to believe that the emission theory would eventually adequately explain diffraction. Nonetheless there were strong reasons for regarding this belief as false. Every explanatory avenue open to the emissionist had been tried and had failed to produce a scientifically adequate account of diffraction. It had to be *logically possible* to make sufficiently complicated assumptions about the diffracting forces exercised by gross matter on the light particles, about differences between the particles themselves, and about "fits" undergone by the particles so as to accommodate all the known facts. But actually putting this into effect had proved *practically* impossible. Moreover, any such theoretical accommodation was clearly going to involve extremely implausible general assumptions — for example, that the diffracting forces are quite independent of the chemical constitution of the gross diffracting object (two straightedges made respectively of, say, cardboard and copper, produce the *same* diffraction pattern). It is simply, as I see it, a *fact* about the emissionist approach in 1830 that it would require the incorporation of some radically new idea before diffraction could ever be *adequately* accounted for. To close one's eyes to this fact would surely be no less irrational than to close one's eyes when confronted with Galileo's telescope.

Brewster's optimistic remarks about the corpuscular theory's prospects are not specific enough to make it clear whether he did close his eyes to this fact, or whether he simply held the much weaker view that working on the general Newtonian approach *might* produce just the right radically new idea to revitalize the approach. Assume that Brewster's view was only this weaker one. Was *it* irrational?

Well, whatever its disadvantages, it again certainly does not contravene deductive logic: *of course* working on the Newtonian approach *might* have produced the required idea, if we are talking merely about *logical* possibilities. However, and quite unlike most heuristic judgments made in science, this view expresses no more than a pious hope. A mid-eighteenth-century Newtonian could quite plausibly remain unconcerned by diffraction fringes (which had, after all, been around since Grimaldi in 1665 and had been extensively investigated by Newton himself). The phenomena of reflection and refraction already established for him that gross matter was capable of exercising forces on the light corpuscles, forces that diverted those particles from their naturally rectilinear paths. It was not at all surprising if similar deviations occurred when those corpuscles passed close by the edge of a gross object. It was just a matter of investigating the diffracting forces in detail and thus building up a full theory of the phenomenon. An eighteenth-century corpuscularian who expressed the view that working on his research program was likely to produce an account of diffraction was not, then, simply expressing a pious hope—he could indicate in an abstract but reasonably precise way the approach to be adopted. But absolutely no success was, as a matter of fact, achieved in this way, despite a good deal of effort through the eighteenth century. Moreover, a great many fundamental problems accumulated as this effort was made: Quite unlike the refracting force which differed from substance to substance (with refractive index), the diffracting force (assumed by most Newtonians to be another manifestation of the same force) seemed to be quite independent of the constitution of the diffracting object. The diffracting forces had been assumed to switch from attractive to repulsive and back again with bewildering rapidity as the distance from the diffracting object increased; the effect of the diffracting force had been made to depend in various complicated ways on the phase of the light particle's periodic "fits of easy transmission/reflection" (conjecturally associated with periodic revolutions of the particles). All this and still no satisfactory theory was remotely in sight. No reasonably well-articulated idea existed that had not been tried and found wanting. Thus, as I suggested, by the 1830s Brewster's view was nothing more than an expression of a hope that some new idea could be conjured out of the blue, a new idea that fitted in with the general Newtonian approach, and that solved the problems with diffraction.

Well, one can *always* hope; but, there are, so far as I can tell, no historical cases in physics in which a theory subsequently recovered scientific credibility having earlier been in straits as dire as those the emission theory was in by 1830. No doubt there are cases of very general metaphysical ideas that have had a checkered history; once incorporated into a program that steadily degenerated, they have then much later been revived by incorporation into a different program that progressed. Atomism is often cited in this connection. But if we look for cases, not at this very general level but at the level of specific Kakatosian research programs (or Kuhnian paradigms), then I, at any rate, don't see any in the history of mathematical physics.[21]

Assuming for the moment that this is indeed the lesson of history, the question arises of whether it should be written into our account of scientific rationality, so that that account pronounces it irrational to try to resuscitate a theory that has degenerated beyond a certain point. If it is, there will be problems specifying exactly what that certain point is. Moreover, such a rationality principle would have nothing even resembling an a priori justification, but would simply rest on an inductive extrapolation from future to past—this time an inductive extrapolation of a methodological kind.

Let me postpone further discussion of this point until after considering

Brewster's third controversial view, which is closely related to the one we have just been discussing and raises similar methodological issues.

(3ciii) Brewster and the Wave Theory's "Monopoly"

As I noted earlier, Brewster held that work on the emission theory was still likely to prove fruitful, and accordingly he regarded as ill advised the "monopoly" that he believed the wave theory had come to exert. He saw the experimental difficulties with dispersion and selective absorption as enough to deny the wave theory "our implicit assent" and enough to justify a less committed approach. Airy and Powell held that, on the contrary, given the established success of the wave theory, "none but the most ignorant or perverse" would do other than commit himself still more wholeheartedly to the wave approach in an effort to solve its difficulties. Were they rational and Brewster irrational?

The first point in favor of Airy and Powell was that, so far as dispersion went, there already existed within the general wave theoretical approach some hopeful lines of attack on the problem. Fresnel's theory, in its initial version, certainly entails no dispersion. But this initial version was based on a very simple theory of the ether—one that involved the assumption that its parts *strictly* obey Hooke's law of the proportionality of restoring force to displacement. Several general ideas were already around about how a somewhat less simple theory involving a slightly more complicated expression for the restoring force could be constructed that might yield dispersion. Though none of these had yet borne unambiguous fruit, equally they had not all unambiguously run into sand. This is again, as I see it, just a *fact* about the wave approach: it *already* possessed potential explanatory resources with respect to dispersion that had *not* been exhausted.

Airy and Powell went on to back this up with an argument that is explicitly inductive in nature. Airy in particular pointed out that, not only was the wave theory already far ahead in terms of explanatory and predictive success, but the way it had dealt with earlier experimental difficulties (especially those posed by polarized light) had already shown that it had the capacity to be modified and extended in a scientifically fruitful way. Moreover, there was only one alternative theoretical approach—the emissionist one—and it faced many more difficulties and had shown no capacity successfully to overcome them. Airy is claiming in my preferred terminology that, when confronted by the choice between a research program that has long been degenerating and a program that has been highly progressive, then the only reasonable course of action is to work on the second.

Should this principle be incorporated into our theory of scientific rationality? First let's be a little clearer about what exactly the principle is. Certainly *within* research programs there will often be alternative strategies available, and it may even not be too farfetched to regard the program as specifying rough probabilities for each alternative strategy's paying off. In such a case "let a thousand flowers bloom" and "don't put all your research eggs in one basket" will be the watchwords: even if there is a "most probable" strategy for advance, it will equally clearly not be irrational to pursue a different one. Indeed, as has often been pointed out, we would *like* some members of the scientific community to pursue high-risk strategies. In such cases — again as has often been pointed out — rationality applies (at any rate most directly) to decisions taken by a community collectively rather than to those of individuals.[22] Even *between* research programs, there are cases that are far from clear-cut — at around 1700, say, I would not want to say that working on *either* the wave *or* the Newtonian approach was in any sense irrational. But in just the sort of case we have been discussing — the sort of case exemplified by optics around 1830 and in general the sort of case covered by Kuhn's notion of a scientific revolution — the situation is quite different, I believe. So far as I can tell, once a truly progressive program has been developed in science, the way forward has *always* been to follow that program, ignoring its degenerating rivals. Kuhn suggests (and has been echoed by others) that it may be good for the community for some diehards to remain, since they may produce problems for the revolutionaries to solve. This suggestion, as plausible as it might sound prima facie doesn't, so far as I can see, wash historically. Certainly in the optics case, the post-Fresnel diehards were no more than a distraction. (Brewster did important experimental work, but *none of it* was informed by his emissionist views — even though most of his results were forced into the language of a sort of instrumentalized version of the emissionist theory.)

The general principle, then, that I tentatively propose as a candidate for incorporation into the theory of scientific rationality is something like this: When the choice is between a highly progressive program and a highly degenerate one, the only rational course is to pursue the progressive one until such a time as degeneration sets in there too.

But, assuming that such a principle were to be incorporated into the theory of scientific rationality, what would the grounds for such incorporation be?

(3d) How Strong Should the Theory of Scientific Rationality Be?

Brewster in particular and — presumably — Kuhn's holdouts in general cannot be faulted on the grounds of consistency with experimental results and deductive logic alone. Long before the appearance of *The Structure of Scientific Revolutions*, Duhem had fully recognized that those grounds alone leave the theoretical scientist with enormous freedom; that is, they produce only a very weak theory of scientific rationality. Duhem went on, however, to applaud the "good sense" that was enjoyed by the best theoretical scientists and that in effect greatly curtailed this freedom. He was reluctant to incorporate the general principles underlying this good sense into what he called the "logic of science." Although he was not entirely clear about his grounds for this reluctance, they seem to have been essentially that the logic of science should somehow be self-justifying, while the principles of scientific good sense clearly involve substantive, and therefore challengeable, assumptions. Duhem's good sense, indeed, consists basically in following the types of procedures that have paid off for science in the past. Hence, as I have just tried to explain, if good sense is incorporated into our logic of science or general theory of scientific rationality, while it certainly strengthens that general theory, it also brings with it certain inductive assumptions. If these are challenged—"Why mightn't it happen, *just next time*, that pursuing a highly degenerate program suddenly pays off handsomely?"—it is difficult to see how they could be further defended.

But surely it cannot be a good idea to incorporate pure assumptions into our theory of rationality? I think we have to face up to the fact (as perhaps Duhem did not) that they are already there. Such assumptions are involved even at the level of theory appraisal (Duhem's "*logic* of science"). No acceptable system of appraisal can, for example, do without some principle that downgrades ad hoc explanations compared to non-ad hoc ones. For a nineteenth-century Newtonian astronomer, for instance, to respond to the difficulties with Mercury's orbit by saying "All bodies in the universe are Newtonian except for Mercury and its motion is described by the following empirical law . . . " cannot be scientifically acceptable. But who says that God's blueprint of the universe did not specify that there be one exception to every general rule? So far as I can see, we simply *assume* that this is not true. Similarly, most of us would presumably regard the person who proposes to get down safely to the ground by taking the window as irrational (though thankfully he won't be irrational for long). But long discussions of Hume's problem seem to me, at any rate, only to have revealed that even this judgment relies on a pure assumption about the uniformity of nature (even if it is an assumption that is genetically "hard wired").

To sum up, then: As should always have been clear, deductive logic alone, even when coupled with acceptance of *low-level* ("crude") observation results, supplies only a very weak theory of rationality. Making that theory stronger requires committing oneself to substantial assumptions: both general metaphysical assumptions and some frankly inductive assumptions based on past scientific success. It does seem to be a fact about the history of physics that no one who has stuck to a *highly* degenerating program when a progressive alternative was available has ever managed to reverse the situation. Hence, those willing to make the necessary inductive assumption provide themselves with a theory of rationality that says that the only rational course of action in such a situation is to follow and develop the progressive program, at any rate for the time being.

This would mean that when Kuhn claimed that there is no point at which "resistance becomes illogical, or unscientific" he conflated two claims: one right, one wrong. Applying the claim to the particular case of Brewster, there was certainly nothing "illogical" about his theoretical views. But this, as I just pointed out, is no surprise in view of the (of course crucial) but extremely weak requirements imposed by deductive logic alone. Brewster *was*, however, unscientific or irrational, *not* in any sense that suggests he was a danger to himself or others, but simply in the sense that he did not follow a procedure that seems invariably to have paid off in science. His belief in the continued viability of the emission approach and in the ill-advisedness of the wave theory's monopoly *were* contrary, not to any eternal rules of deductive logic, but to what appears to be best scientific practice. On the other hand, his resistance to believing in the full *truth* of the wave theory was not irrational. The history of science provides no inductive grounds for believing in the truth of the fully fledged, realistically interpreted versions of accepted theories. At any rate in some scientific fields, the truth is rather that, on the contrary, history supports the recently much-discussed "pessimistic induction," which concludes that every fundamental scientific theory, no matter how firmly entrenched it might appear for a time, is eventually rejected and replaced by another theory inconsistent with it.

So how, finally, might a "rationally reconstructed" Brewster have viewed the wave-particle rivalry? (Using this stronger theory of rationality as the basis for reconstruction.) Well, roughly as follows.

> Like my unreconstructed counterpart, I just can't bring myself to believe that there is an ether that fills space from my eye "to the remotest verge of the starry heavens." Nonetheless, the wave theory of light seems somehow or other to have latched onto part of the structure of the universe. As my real counterpart admitted, that theory's striking predictive success surely means that "it must contain among its assumptions . . . some principle which is inherent in . . . the real producing cause of the phenomena of light. . . . " Moreover, there is a good deal of "heuristic steam" left in the wave approach. It cannot possibly do any harm to pursue that approach wholeheartedly. I have no doubt that the approach will *eventually* be superseded by a theory based on a more believable metaphysics. But the lesson of history seems clearly to be that, at any rate *structurally*, the wave theory, like all predictively successful theories, will "live on" within that future theory, perhaps as a limiting case. The metaphysics underlying that future theory may well turn out to be closer to that underlying the present emission approach. But this approach *as it stands* is played out. And again the lesson of history seems clear: that no amount of flogging will revive a horse as dead as this one.

This rationally reconstructed Brewster would surely have found favor with "committed" wave theorists like Airy and Powell—who would have recognized his position as, to all scientific intents and purposes, indistinguishable from their own.

Notes

1. Kuhn (1962, 151). Kuhn elaborates on his conception of the role of the "objective factors" in his (1977, especially chap. 13).

2. The issue is somewhat complicated by the fact that the orthodox eighteenth-century Newtonian was much more likely to talk about the "parts" or "elements" of light than about material light corpuscles; and indeed, if pressed, would tend to deny any commitment to the material nature of light. Brewster, very much the orthodox Newtonian, did adopt this position and seems to have believed that it implied that his view of light was, if not entirely theory free, then certainly theory neutral. Hence he sometimes presented himself, *not* as defending one theory of the nature of light against a rival theory, but rather as defending one scientific method—an empiricist one, which allegedly stuck closely to the facts—against a rival scientific method, which not only sanctioned highly theoretical entities, in particular the ether, but gave them full scientific honors. This view of his own position (which certainly derives from Newton himself) is confused: the "parts" of light, whether or not one abstains from a theory of their "ultimate nature," are assumed to retain their own identity as they travel, if undisturbed, along their rectilinear paths; if they fail to travel rectilinearly, this is assumed to be due to some action that was, in all but name, a force acting on a particle. The whole idea of "parts" of light, far from being theory neutral, is directly inconsistent in a number of respects, with the classical wave theory. (Although it is true that this was not generally clearly realized even by some of the early nineteenth-century defenders of the wave theory.) I have therefore indulged in a (surely permissible) "rational reconstruction" and treated Brewster as defending one theory against another (as, logically speaking, he surely was). The methodological aspects of the controversy will nonetheless be apparent. (For a detailed account of the "parts" of light and the inconsistency of this instrumentalized version of the emission theory with the classical wave theory, see my (1989).)

3. Brewster is clearly referring (a) to Airy's modification of the Newton's rings experiment in which a metal plate was used as second reflecting surface—Airy showed that, as predicted by the wave theory, the central spot was, for certain angles of incidence, light instead of dark; and (b) to Hamilton's prediction of the totally new phenomena of conical refraction on the basis of Fresnel's equation for the wave surface within biaxial crystals, a prediction experimentally confirmed by Humphrey Lloyd in 1833.

4. Airy and Powell both insisted that any advantage the emission theory had in regard to dispersion and selective absorption was marginal, since nothing like a fully adequate emission account could be given of either phenomenon. See also *below*, pp. 338–39.

5. This was by no means the only major theoretical problem associated with the switch to the elastic solid ether. Another was that a periodic disturbance in an ordinary elastic solid produces a wave with *both* transverse *and longitudinal* components. The longitudinal component seemed to play no role whatsoever in any optical effect. What happened to it? (Much of the history of attempts to solve the problems of the elastic solid ether theory is charted in Schaffner (1972).)

6. Shapere (1966), p. 67.

7. Lakatos (1970), p. 178.

8. Kuhn (1977), chap. 13. Unadorned page numbers *below* in this section refer to this work of Kuhn's.

9. Not all of those who think of themselves as Bayesians accept the principle of conditionalization as a constraint on rational belief. However, unless such Bayesians are ready to specify other constraints governing changes in belief, then the contrast that I want to emphasize between their position and that of the "objectivists" becomes still more marked.

10. They include some "objective Bayesians" who argue that there have to be rules about which prior probability values (or at any rate which *ranges* of probability values) are "rationally permissible."

11. Someone who does *not* make a hash of arguing against the creationists is Philip Kitcher in his (1982).

12. The empirical continuity which, contrary to much presently received opinion, I see in the development of science occurs at the level of what Poincaré called "crude facts" and Duhem "practical facts." Science is undoubtedly *not* cumulative at the level of "scientific" or "theoretical facts"—but this is (a) not surprising and (b) not the major difficulty for a general empiricist view that it is often taken to be. (For more details see my [1978] and especially my [1982].) The history of the simple law of reflection affords an instructive example. If it is regarded as an empirical fact that light is reflected from plane mirrors at an angle equal to the angle of incidence—which, despite its universal character, would not, I admit, strain ordinary usage—then the development of Fresnel's wave theory undeniably did lead to the *rejection* of previously accepted "empirical facts." Not only are the laws of geometrical optics *always strictly wrong* according to Fresnel's theory (on that theory there is just no such thing, strictly speaking, as a ray in the sense of geometrical optics), but also Fresnel's theory (correctly) predicts an *observable* divergence from the simple reflection law in the case of *very narrow* mirrors. But the actual *results* here were not the (very useful) *idealizations* of geometrical optics, but rough and ready ray tracings and the like—*all of which are yielded equally well by Fresnel's theory*. No one before Fresnel had experimented with mirrors narrow enough to produce the observable deviation from the geometrical law predicted by him. (Indeed prior to this prediction there would have been no interest in doing so.)

13. This picture was already completely clear to Duhem and Poincaré. It formed the basis of their insistence that the "explanatory" or "metaphysical" part of science (which had proved subject to change) is valueless, and that it is only the "representative" part of science that really counts (this representative part invariably being carried over into successor theories). For a clarification and defense of Duhem's and (especially) Poincaré's "structural realist" (*not* anti-realist) view of scientific theories, see my (1989).

14. The structure of Kuhn's argument here does seem implicitly to commit him to this "laundry list" view of the objectivists' position. It should be acknowledged, however, that he *does* on occasion nod in the direction of weighting the different criteria; and he is also, as we shall see, sometimes inclined to make empirical accuracy the single most important factor.

15. For further elaboration and references see my (1985).

16. In his (1957), Kuhn took this as an instance of Copernican theory's greater "harmony"; Lakatos and Zahar (1976) show that it is more perspicuously viewed as a genuine prediction (even though of an effect that had long been known to occur when Copernicus formulated his theory).

17. The examples that Paul Feyerabend cites to exemplify "Kuhn loss" are even more curious: the "fact" that the Brownian particle is a perpetual motion machine of the second kind (a fact "lost" in the statistical-kinetic revolution) and even sometimes the "fact" that phlogiston is given off or absorbed in certain circumstances! The level at which mature science is cumulative is that of Poincaré's "crude facts" ("the needle pointed to somewhere close to 5 on the scale"). Of course, scientists often talk of much higher-level statements as factual or empirical *depending on which theories they take for granted as parts of "background knowledge."* Thus, given certain auxiliary theories the above crude fact may be rendered "the current in the wire was 5 amps." This is an example of a "scientific fact" for Poincaré. Of course if certain very high-level theories are taken for granted, then we can get *very* high-level empirical facts—a description of the individual situation as seen in the light of presently accepted theories ("certain free electrons move through the wire so as to create a current of a certain strength," or whatever). It hardly needs to be said that, in view of historical changes in accepted high-level theories, and even in "background knowledge" auxiliary theories, *such* facts described in these highly theoretical terms may well be "lost" as a result of theory change in science. (See my (1978) for further details.)

18. The predictive success enjoyed by Fresnel's and Copernicus's theories and not, so far as I can tell, by the emission theory or by Ptolemy's, concerns the prediction of general *types* of phenomena:

planetary stations and retrogressions, the diffraction pattern for small circular opaque disks, or whatever. Almost every theory, of no matter how "cobbled up," "degenerate" or ad hoc a kind, will of course enjoy "predictive success" of a straightforward inductive, extrapolative kind. Having fixed its various epicyclic parameters on the basis of the observation of an orbit (or series of orbits) of the planets, the Ptolemaic theory will, of course, go on to predict the future orbits of those planets.

19. For an account of some of the details of the corpuscular-theoretic accounts of interference and diffraction see my (1976).

20. See my (1985) for details. Kuhn's talk of "shoving" phenomena into the older paradigm's "box" seems to imply that he is implicitly aware of this point. But, so far as I can tell, he never explicitly considers how distinguishing between shoving a phenomenon into a theoretical framework and having a phenomenon "fall out" of that framework (and giving more epistemic weight to a theory in the latter case) might alter his view of the empirical justification for theory change.

21. Certainly the photon theory does not constitute such a revival of the Newtonian corpuscular theory in any significant sense. Photons are of course "something like" material particles – but then any two things, no matter how dissimilar, are something like each other in some respects; here the *dis*similarities are overwhelming.

22. See, for example, Musgrave (1976) and some very interesting, forthcoming work by Philip Kitcher.

References

Airy, G. B. 1831. *A Mathematical Tract on the Undulatory Theory of Light*. Quoted from the reprint in his *The Undulatory Theory of Optics*, 2d ed., London, 1866.

——. 1833. Remarks on Sir David Brewster's Paper "On the Absorption of Specific Rays, &c." *The Philosophical Magazine*, 3d ser., 2: 419–24.

Brewster, D. 1833a. A Report on the Recent Progress of Optics. In *British Association for the Advancement of Science, Report of the First and Second Meetings 1831 and 1832*. London. (Brewster's report was delivered at the first meeting of 1831.)

——. 1833b. Observations of the Absorption of Specific Rays, in Reference to the Undulatory Theory of Light. *The Philosophical Magazine*, 3d ser., 2: 360–63.

——. 1838. Review of *Cours de Philosophie Positive*, by Comte. *Edinburgh Review*, 67: 279–308.

Feyerabend, P. 1974. *Against Method*. London: New Left Books.

Kitcher, P. 1982. *Abusing Science*. Cambridge: MIT Press.

Kuhn, T. S. 1957. *The Copernican Revolution*. Cambridge: Harvard University Press.

——. 1962. *The Structure of Scientific Revolutions*. 2d, enlarged ed., 1970. Chicago: University of Chicago Press.

——. 1977. *The Essential Tension*. Chicago: University of Chicago Press.

Lakatos, I. 1970. "Falsification and the Methodology of Scientific Research Programmes." In *Criticism and the Growth of Knowledge*, eds. Lakatos and Musgrave. Cambridge: Cambridge University Press.

——. 1978. *The Methodology of Scientific Research Programmes*, vol. 1 of *Philosophical Papers*. Cambridge: Cambridge University Press.

Lakatos, I., and Zahar, E. 1976. Why Did Copernicus's Programme Supersede Ptolemy's? Reprinted in Lakatos (1978).

Lloyd, H. 1834. Report on the Progress and Present State of Physical Optics. *British Association for the Advancement of Science Reports* 4.

Musgrave, A. E. 1978. "Can the Methodology of Scientific Research Programmes Be Rescued from Epistemological Anarchism?" In *Essays in Memory of Imre Lakatos*, eds. Cohen, Feyerabend, and Wartofsky. Dordrecht: D. Reidel.

Powell, B. 1841. *A General and Elementary View of the Undulatory Theory as Applied to the Dispersion of Light and Some Other Subjects*. London.

Schaffner, K (ed). 1972. *Nineteenth Century Aether Theories*. Oxford: Pergamon.

Shapere, D. 1966. "Meaning and Scientific Change." In *Mind and Cosmos*, ed. Colodny. Pittsburgh: Pittsburgh University Press.

Worrall, J. 1976. "Thomas Young and the 'Refutation' of Newtonian Optics." in *Method and Appraisal in the Physical Sciences*, ed. C. Howson. Cambridge: Cambridge University Press.

———. 1978. "Is the Empirical Content of a Theory Dependent on Its Rivals?" In *The Logic and Epistemology of Scientific Change*, eds. Niiniluoto and Tuomela. Amsterdam: North-Holland.

———. 1982. The Pressure of Light: the Strange Case of the Vacillating Crucial Experiment. *Studies in the History and Philosophy of Science* 13: 133–71.

———. 1985. "Scientific Discovery and Theory-Confirmation." In *Change and Progress in Modern Science*, ed. J. Pitt. Dordrecht: Reidel.

———. 1989. Structural Realism: the Best of Both Worlds? *Dialectica* 43:99–124.

Richard Boyd

Realism, Approximate Truth, and Philosophical Method

1. Introduction

1.1. Realism and Approximate Truth

Scientific realists hold that the characteristic product of successful scientific research is knowledge of largely theory-independent phenomena and that such knowledge is possible (indeed actual) even in those cases in which the relevant phenomena are not, in any non-question-begging sense, observable (Boyd 1982). The characteristic philosophical arguments for scientific realism embody the claim that certain central principles of scientific methodology require a realist explication. In its most completely developed form, this sort of abductive argument embodies the claim that a realist conception of scientific inquiry is required in order to justify, or to explain the reliability with respect to instrumental knowledge of, all of the basic methodological principles of mature scientific inquiry (Boyd 1973, 1979, 1982, 1983, 1985a, 1985b, 1985c; Byerly and Lazara 1973; Putnam 1972, 1975a, 1975b).

The realist who offers such arguments is not committed to the view that rationally applied scientific method will always lead to progress towards the truth, still less to the view that such progress would have the exact truth as an asymptotic limit (Boyd 1982, 1988). Nevertheless it would be difficult to defend scientific realism without portraying the central developments of twentieth-century physical science, for example, as involving a dialectical and progressive interaction of theoretical and methodological commitments (Boyd 1982, 1983).

A defense of realism along these lines requires that two things. In the first place, the realist must be able to defend a historical thesis regarding the recent history of relevant sciences according to which their intellectual achievements involve *approximate* theoretical knowledge and according to which theoretical progress within them has been (to a large extent) a process of (not necessarily converging) *approximation*. No realist conception that does not treat theoretical knowledge and theoretical progress as involving approximations to the truth is even prima facie compatible with the actual history of science. The realist must, therefore, employ a conception of approximate theoretical knowledge and of theoretical progress through approximation that makes historical sense of the recent development of scientific theories.

Secondly, the realist must be able to establish that her historical appeal to approximate theoretical knowledge and to theoretical progress by successive approximation is appropriate by philosophical as well as by historical standards. Neither the realist's historical account nor her appeal to it in the defense of scientific realism as a philosophical thesis should be undermined by any of the distinctly philosophical considerations characteristic of anti-realist positions in the philosophy of science. Important challenges to scientific realism arise from doubts that a realist conception of approximate truth and of the growth of approximate knowledge is available that satisfies both of these constraints. The appropriate realist responses to these challenges and the philosophical implications of those responses are the subject of the present essay.

1.2. Challenges to a Realist Treatment of Approximation

A number of philosophers (realists included) have had serious concerns about the realist's ability to provide an adequate account of the development of scientific theories as involving the growth of approximate theoretical knowledge. The *locus classicus* of objections to realism reflecting these concerns is surely Laudan 1981 (see also Fine 1984). That there should be such concerns is, in significant measure, a reflection of the striking difference between the depth of our understanding of the notion of (exact) truth and that of our understanding of approximate truth.

Since the work of Tarski in the 1930s we have had a systematic, general, and topic-and-context-independent mathematical and philosophical theory of (exact) truth. By contrast there is no generally accepted general and systematic theory of approximate truth. We have available from the various special sciences a very large number of well-worked-out examples of particular instances of approximation but the details in theses cases depend not only on contingent and often esoteric facts about the relevant natural phenomena, but also upon the particular context of application within which the approximate theories and models are to be applied. In part because of the complexities created by such topic and context dependence, we do not have as clear a general understanding of what the epistemological relevance of appeals to approximate truth should be. Moreover, as we shall see, the dependence of the relevant details upon a posteriori theoretical claims raises special problems of philosophical method when an appeal to conception of approximate truth is to be made in the course of a defense of scientific realism.

I have argued elsewhere (Boyd 1982, 1983, 1985a, 1985b, 1985c, 1988) that the scientific realist must adopt distinctly naturalistic conceptions of philosophical methodology and of central issues in epistemology and metaphysics. My aim in the present paper will be to show how the distinctly naturalistic arguments for realism that I have developed in the papers cited can be extended to provide an adequate realist treatment of approximate truth.

Instead of replying to particular anti-realist arguments in the literature, I shall respond to four objections that capture, I believe, the deep philosophical concerns that the realist's conception of approximate theoretical knowledge properly occasion. My expectation is that the responses to those objections will provide an adequate basis for a realist's response to other objections regarding her conception of approximate truth and approximate knowledge. The objections I shall consider are these:

1. (*The historical objection*) Realists are simply mistaken as a matter of historical fact: many important scientific advances seem to have been grounded in what (by realist standards) were deep errors in background theories. Approximately true background theoretical knowledge is thus not required to explain reliability of scientific practices.

2. (*The triviality objection*) The the realist might reply (following Hardin and Rosenberg 1982, for example) about many of the advances in question that the relevant background theories were *to some extent* or *in some respects* approximately true.

Here the realist's philosophical project is in danger of being reduced to *triviality*. The problem is that we lack altogether a general theory of approximation: we have no general characterization of what it is for a sentence to be approximately true, to be approximately true to a specified degree or in a specified respect, or to be more nearly true (in specific respects or in general) than some other sentence. If we had such a general theory then the realist could appeal to it in refining the thesis that the relevant historical episodes reflect some respects of approximation to the truth. As it is we are faced with the fact that *any* consistent theory is approximately true in some respects or other, and *any* sequence of such theories will reflect progress towards the truth in some respects or other.

3. (*The contrivance objection*) The realist might next reply by distinguishing between relevant and irrelevant respects of approximation to the truth regarding matters theoretical, and by claiming that the growth of scientific knowledge characteristically involves the former. Here the realist avoids triviality at the expense of a contrived or ad hoc conception of approximate truth, indeed at the expense of both contrivance (objection 3) and circularity (see objection 4).

The contrivance in question arises from the important difference just mentioned between extant theories of truth and of approximate truth respectively. In the case of truth *simpliciter* Tarski's strategy for defining truth (Tarski 1951) provides a uniform treatment that is largely independent of the particular subject matter or of the particular historical episodes or context of application under consideration. By contrast, our conception of relevant approximation reflects considerations specific to the particular theory or theories, historical settings, and contexts of application under consideration.

Thus, for example, if the realist sees relativistic mechanics as growing out of previously acquired approximate theoretical knowledge her conception of the relevant respects of approximation reflected in Newtonian mechanics will emphasize numerical accuracy for systems of particles with relative velocities low with respect to that of light, the identification of, and the development of reliable measurement procedures for, various physical magnitudes, and the central role assigned to certain fundamental laws. It will de-emphasize, for example, numerical accuracy for high relative velocities, or of soundness of the Newtonian theoretical conception of space and time.

Here the distinctions between relevant and irrelevant respects of approximation reflect judgements, based on current theoretical conceptions, about the respects in which Newtonian mechanics happened to be approximately true, and similarly theory-dependent judgments about the role that such approximations played in the successful development of relativistic mechanics. Since we lack a general theory of approximation, the realist's appeal to relevant respects of approximation in response to the triviality objection will always have to be grounded in just this sort of topic-and-episode-sensitive conception.

We can now see why the realist's treatment of respects of approximation will involve an ad hoc or contrived element. For each of the episodes of scientific inquiry typically considered by philosophers of science there is a standard realist picture (or, at any rate, a narrow range of such pictures) of how the relevant approximations to the truth have gone and what contributions, if any, they have made to the subsequent growth of scientific knowledge. The realist, in defining the relevant sense(s) of approximation, will rely on such a picture. But such a picture merely reflects the realist research tradition in the history and philosophy of science. Since there is no topic-and-episode-neutral conception of relevant approximation with respect to which her proposed definitions may be assessed, the realist will simply be presupposing the soundness of the "findings" of her own tradition when she defines the difference(s) between relevant and irrelevant respects of approximation. It is no surprise—and certainly no basis for an abductive argument for realism—that the realist can construct a realist account of approximate truth when she is permitted to beg questions in so thoroughgoing a way.

4. (*The circularity objection*). There is some precedent in scientific inquiry, especially historical inquiry, for explanatory concepts that lack topic-and-episode-neutral general specifications of the sort alluded to above: sometimes theoretical considerations that resist incorporation into a fully general definition can justify the (topic-and-episode-nonneutral) ways in which such concepts are applied in particular cases. Let us suppose for the sake of argument that this is the case with respect to the employment of the concept of approximate truth in the various historical explanations of scientific progress (or its absence) that are

offered in the realist tradition. Even if the realist's accounts of the relevant episodes are thus methodologically acceptable *as explanations in the history of science,* they will involve an unacceptable circularity if they are understood to address the *philosophical* issue between scientific realists and anti-realists.

Here's why: Any realist explanation of the growth of knowledge and of reliable methodology in a particular field must involve an account of the kinds of epistemically relevant causal interactions that exist(ed) between members of the relevant scientific community and the features of the world that were (or are) the alleged objects of their study. Thus for example, a realist account of such developments in atomic theory will incorporate a causal account of how scientists gain(ed) epistemic access to various subatomic particles and the realist's claim that atomic theory is about such unobservable theory-independent particles will depend on that account (see sections 2.1.3 and 2.1.4). The realist's account of epistemic access to subatomic particles will be grounded in the best available theory of such particles together with related contemporary physical theories.

Suppose now that the realist's explanation of the development of some field, including the relevant account of epistemic access, is advanced in defense of realism as a philosophical thesis. Plainly the resulting defense of realism is cogent only if the realist's explanation, and her account of epistemic access in particular, are understood *realistically.* For example, only if the account of epistemic access to subatomic particles is understood realistically is the realist's case that atomic theory has an unobservable and theory-independent subject matter advanced. But, on the realist's own account, her explanation and the account of epistemic access it incorporates are ordinary scientific theories themselves grounded in the very research tradition regarding which a defense of realism is sought. To insist on a realistic interpretation of the realist's explanation would thus *presuppose* realism regarding the tradition in question. Thus the realist's appeal to her explanation of the development of instrumentally reliable methodology in an abductive argument for realism as a philosophical thesis is question-beggingly circular.

1.3. An Argumentative Strategy

The challenges we are considering seem to fall into two classes: The first three represent an essentially prephilosophical critique of the realist's historical explanations: they deny that the realist's conception of the role of approximate truth regarding theoretical matters in the growth of scientific knowledge represents the best explanation for the relevant episodes in the history of science. The fourth offers a distinctly philosophical challenge: it argues that even if the *realist's* account of the growth of scientific knowledge does provide the best explanation, inductive inference to *realism* begs the philosophical question at issue.

After some philosophical preliminaries, I propose to respond to the challenges in two distinct stages corresponding to these two classes . In the first stage of my response, I treat the characteristic realist explanatory appeal to approximate truth as an ordinary piece of historical explanation. I identify a general methodological problem of *parametric specification* in explanatory contexts of which the deeper problems raised by the first three challenges are special cases, and I identify the generally appropriate solution to that problem. I then indicate why it is plausible that the realist's explanatory appeal to approximate truth satisfies the methodological demands dictated by the solution in question.

With respect to the fourth challenge, I assume for the sake of argument that the realist's historical explanations have been confirmed and I inquire whether they are to be understood realistically or whether instead such an understanding — which is essential to the realist's case — begs the question against the anti-realist. Here too I argue that the methodological question regarding the realist's appeal to approximate truth — in this case a question about *philosophical* method — is a special case of a more general methodological question about the appropriate interaction between philosophical considerations and empirical findings in the philosophy of science. I define the notion of a large-scale *philosophical package* and I indicate why the incorporation of realistically understood scientific theories into the realist philosophical package is compatible with (and indeed required by) an adequate *and noncircular* defense of the realist package against rival philosophical conceptions.

On now to the philosophical preliminaries.

2. Philosophical Preliminaries

2.1. The Abductive Argument for Scientific Realism

The challenges we are considering arise in the context of a class of abductive arguments for realism according to which we must recognize approximate knowledge of unobservable (and appropriately mind-independent) "theoretical entities" in order to adequately explain the growth of even instrumental knowledge in recent science. To assess the realist's arguments and the appeals to the notion of approximate truth embodied in them, we need an understanding of just what those arguments are. In what follows of this section I'll indicate, in broad outline, how the abductive arguments for realism go.

2.1.1. Objective Knowledge from Theory Dependent Method

By the "instrumental reliability" of a scientific theory I mean the extent of its capacity to make approximately true observational predictions about observable phenomena — the extent of its approximate empirical adequacy. By the "instrumental reliability" of some body of methods I mean the extent to which their practice is conducive to the acceptance of instrumentally reliable theories. The abductive arguments for scientific realism take place in a dialectical situation in which scientific realists and their philosophical opponents largely agree that the methods of actual recent scientific practice are significantly instrumentally reliable.

The abductive arguments for realism are in the first instance directed against the empiricist who denies the possibility of "theoretical" knowledge — knowledge of "unobservables." Against the empiricist the realist argues that only by accepting the reality of approximate theoretical knowledge can we adequately explain the (uncontested) instrumental reliability of apparently theory-dependent scientific methods. In the present paper I shall focus my attention primarily on the dispute between realists and empiricists, reserving attention to the corresponding dispute between realists and constructivists largely to a later paper. I discuss the realism-constructivism dispute briefly in section 2.4 and briefly discuss the distinctly constructivist version of the circularity objection in section 4.3.

The case for realism lies largely in the recognition of the extraordinary role that theoretical considerations play in actual (and patently successful) scientific practice. To take the most striking example, scientists routinely modify or extend operational "measurement" or "detection" procedures for "theoretical" magnitudes or entities on the basis of new theoretical developments. The reliability and justifiability of this sort of methodology is perfectly explicable on the realist's conception of measurement and of theoretical progress. Accounts of the revisability of operational procedures that are compatible with an empiricist position appear inadequate to explain the way in which theory-dependent revisions of "measurement" and "detection" procedures make a positive methodological contribution to the progress of science.

There are two important consequences of the realist explanation for the reliability of the methodology in question. First, scientific research, when it is successful, is *cumulative by successive (but not necessarily convergent) approximations to the truth.* Second, this cumulative development is possible because *there is a dialectical relationship between current theory and the methodology for its improvement.* The approximate truth of current theories explains why our existing measurement procedures are (approximately) reliable. That reliability, in turn, helps to explain why our experimental or observational investigations are successful in uncovering new theoretical knowledge, which, in turn, may produce improvements in measurement techniques, etc.

Theory dependence of methods and the consequent dialectical interaction of theory and method are entirely general features of all aspects of scientific methodology — principles of experimental design, choices of research problems, standards for the assessment of experimental evidence and for assessing the quality and methodological import of explanations, principles governing theory choice, and rules for the use of theoretical language. In all cases there is a pattern of dialectical interaction between accepted theories and associated methods of just the sort exemplified in the case of the theory dependence of measurement and detection procedures. Moreover, this pattern of theory dependence contributes to the reliability of scientific methodology rather than detracting from it (Boyd 1972, 1973, 1979, 1980, 1982, 1983, 1985a, 1985b, 1985c; Kuhn 1970, Putnam 1972, 1975a, 1975b; Van Fraassen 1980).

According to the realist, the only scientifically plausible explanation for the reliability of a scientific methodology that is so theory dependent is a thoroughgoingly realistic explanation: Scientific methodology, dictated by currently accepted theories, is reliable at producing further knowledge *precisely because, and to the extent that, currently accepted theories are relevantly approximately true.* Scientific method provides a paradigm-dependent paradigm-modification strategy: a strategy for modifying or amending our existing theories and methods in the light of further research that is such that its methodological principles at any given time will themselves depend upon the theoretical picture provided by the currently accepted theories. If the body of accepted theories is itself relevantly sufficiently approximately true, then this methodology operates to produce a subsequent dialectical improvement both in our knowledge of the world and in our methodology itself. It is not possible, according to the realist, to explain even the instrumental reliability of actual recent scientific practice without invoking this explanation and without adopting the realistic conception of scientific knowledge that it entails (Boyd 1972, 1973, 1979, 1982, 1983, 1985a, 1985b, 1985c).

2.1.2. Projectability, Evidence, Theoretical Plausibility and the Evidential Indistinguishability Thesis

If the realist's abductive argument is correct, a dramatic rethinking of our notion of scientific evidence is required. Consider the question of the "degree of confirmation" of a theory given a body of observational evidence. To a very good first approximation, a theory receives significant evidential support from a body of successful predictions (or other evidentially favorable observations) just in case (a) the theory is itself "projectable" (see Goodman 1973), (b) the observations in question pit the theory's predictions (or, in other contexts, its explanations) against those of its projectable rivals; and (c) in the relevant experiments or observational settings, there have been suitable controls for those possible artifactual influences that are themselves suggested by projectable theories of those settings (Boyd 1982, 1983, and especially 1985a).

Central to the realist's argument is the observation that projectability judgments are, in fact, judgments of theoretical plausibility: we treat as projectable those proposals that relevantly resemble our existing theories (where the determination of the relevant respects of resemblance is itself a theoretical issue). The reliability of this conservative preference is explained by the approximate truth of existing theories, and one consequence of this explanation is that *judgments of theoretical plausibility are evidential.* The fact that a proposed theory is plausible in the light of previously confirmed theories is some evidence for its (approximate) truth. Judgments of theoretical plausibility are matters of inductive inference from (partly) theoretical premises to theoretical conclusions; precisely these inferences justify, and explain the reliability of, "inductive inference to the best explanation" (Boyd 1972, 1973, 1979, 1982, 1983, 1985a, 1985b, 1985c) .

The claim that judgments of theoretical plausibility are evidential affords the realist a reply to the deepest empiricist argument against realism. The empiricist appeals (tacitly or explicitly) to a principle that I have called the *evidential indistinguishability thesis*. In its most plausible form it holds that for any two empirically equivalent total sciences, the empirical support or disconfirmation that one receives, given a given body of observational data, will be just the same as that received by the other. The empiricist's conclusion that knowledge of unobservables is impossible is a straightforward application of this thesis, which can be thought of as an empiricist analysis of the claim that all scientific knowledge is empirical knowledge. The realist accepts the latter claim but rejects the empiricist analysis. Instead, the realist holds, evidential considerations regarding theoretical plausibility are indirectly experimental and can serve to distinguish total sciences that embody or naturally extend the current total science (that are favored by those considerations) from empirically equivalent total sciences which significantly depart from the prevailing total science (which such considerations reject as unprojectable). [See Boyd 1982, 1983, and section 2.2.]

2.1.3 . Natural Definitions

Locke speculates at several places in Book IV of the *Essay* (see, e.g., IV, iii, 25) that when kinds of substances are defined, as empiricism requires, by purely conventional "nominal essences," it will be impossible to have a general science of, say, chemistry. There is no reason to believe that kinds defined by nominal essences will reflect actual causal structure and thus be apt for the formulation or confirmation of general knowledge of substances. Only if we are able to sort substances according to their hidden real essences will systematic general knowledge of substances be possible.

Locke was right (at any rate so the realist thinks). Only when kinds (properties, relations, magnitudes, etc.) are defined by natural rather than conventional definitions is it possible to obtain the theory-dependent solutions to the problem of projectability just described (Putnam 1975a; Quine 1969a; Boyd 1979, 1982, 1983). It is thus central to the realist's abductive argument that most scientific terms be seen as possessing natural rather than conventional definitions. Such terms are defined in terms of properties, relations, etc., that render the kinds (etc.) to which they refer appropriate to particular sorts of scientific or practical reasoning. In the case of such terms, proposed definitions are always in principle revisable in the light of new evidence or new theoretical developments, and it is possible for people to refer to the same kind (property, magnitude, etc.) by a term while disagreeing about what its correct a posteriori natural definition is. This last consequence of the naturalistic conception of definitions is essential to the realist's dialectical conception of the development of scientific knowledge and methods. The realist will (at least typically) need to portray developments in which mature scientific communities change their conception of the definitions of kinds, relations, magnitudes, etc., as dialectical advances (or, if things go badly, setbacks) rather than as changes of subject matter (Putnam 1972, 1975a, 1975b; Boyd 1979, 1980, 1988). (For more on naturalistic definitions see section 2.5.)

2.1.4. Reference and Epistemic Access

If the traditional empiricist account of definition is to be abandoned for scientific terms in favor of a naturalistic account, then a naturalistic conception of reference is required for such terms. An account of the appropriate sort is provided by recent causal theories of reference (see, e.g., Feigl 1956, Kripke 1972, Putnam 1975a). The reference of a term is established by causal connections of the right sort between the use of the term and (instances of) its referent.

The connection between naturalistic theories of reference and of knowledge (see section 2.2) is quite intimate: reference is itself an epistemic notion and the sorts of causal connections that are relevant to reference are just those that are involved in the reliable regulation of belief (Boyd 1979, 1982). *Roughly,* and for nondegenerate cases, a term t refers to a kind (property, relation, etc.) k, just in case there exist causal mechanisms whose tendency is to bring it about, over time, that what is predicated of the term t will be approximately true of k. In such a case, we may think of the properties of k as regulating the use of t, and we may think of what is said using t as providing us with socially coordinated *epistemic access* to k. t refers to k (in nondegenerate cases), just in case the socially coordinated use of t provides significant epistemic access to k, and not to other kinds (properties, etc.) (Boyd 1979, 1982). The mechanisms of reference *just are* the mechanisms of reliable belief regulation .

Thus, just as the realist conception requires, two different terms, or the same term in two historically different settings, may afford epistemic access to, and thus may refer to, the same kind (property, etc.) even though the definitions associated with them by the relevant linguistic communities are quite different or even inconsistent.

One further feature of the naturalistic conception of reference is important to an understanding to the realist's conception of the growth of approximate knowledge. In many scientifically important cases the use of a term may afford epistemic access to more than one kind (property, relation, . . .), but our knowledge may be insufficient for us to recognize that this is so, and we may consequently have a conception of, as it seems to us, one kind (etc.) that conflates information regarding several distinct kinds.

Field (1973, 1974) calls the relation thus established between a term and several kinds (etc.) *partial denotation,* and he calls the revision of language usage to eliminate such cases of ambiguity *denotational refinement*. On the realist's conception of the growth of approximate knowledge one sort of approximate knowledge is that represented by a body of sentences involving a partially denoting term when what is predicated of that term in these sentences represents methodologically important approximations to the truth regarding one or more of the relevant *partial denotata* considered individually. In such cases, one characteristic form of subsequent improvement in approximation is the discovery of the ambiguity and the consequent denotational refinement (see Boyd 1979).

2.2. Naturalism and Radical Contingency in Epistemology

Modern epistemology has been largely dominated by "foundationalist" conceptions: all knowledge is seen as grounded in certain foundational beliefs that have an epistemically privileged position. Other true beliefs are instances of knowledge only if they can be justified by appeals to foundational knowledge. It is an a priori question which beliefs fall in the privileged class. Similarly, the basic inferential principles that are legitimate for justifying nonfoundational knowledge claims can be justified a priori; it is moreover an a priori question about a given inference whether it meets the standards set by those principles or not. We may fruitfully think of foundationalism as consisting of two parts, *premise foundationalism* which holds that all knowledge is justifiable from an a priori specifiable core of foundational beliefs, and *inference foundationalism*, which holds that principles of justifiable inference are reducible to inferential principles that are *a priori justifiable* and whose application is *a priori checkable*.

Recent work in "naturalistic epistemology" (see, e.g., Armstrong 1973; Goldman 1967, 1976; Quine 1969b) strongly suggests that foundationalism is fundamentally mistaken. For the typical case of perceptual knowledge, there seem to be neither premises nor inferences; instead perceptual knowledge obtains when perceptual beliefs are produced by epistemically reliable mechanisms. Even where premises and inferences occur, it seems to be the reliable production of belief that distinguishes cases of knowledge from other cases of true belief. A variety of naturalistic considerations suggest that there are no beliefs that are epistemically privileged in the way foundationalism seems to require.

I have argued (see Boyd 1982, 1983, 1985a, 1985b, 1985c) that the abductive defense of scientific realism requires an even more thoroughgoing naturalism in epistemology and, consequently, an even more thoroughgoing rejection of foundationalism. In particular *all* of the significant methodological principles of inductive inference in science are profoundly theory dependent. They are a reliable guide to the truth only because, and to the extent that, the relevant background theories are relevantly approximately true. They are not reducible to some more basic rules whose reliability as a guide to the truth is independent of the truth of background theories. Since it is a contingent empirical matter which background theories are approximately true, the justifiability of scientific principles of inference rests ultimately on a contingent matter of empirical fact, just as the epistemic role of the senses rests upon the contingent empirical fact that the senses are reliable detectors of external phenomena. Thus inference foundationalism is radically false; there are no a priori justifiable rules of nondeductive inference, and it is an a posteriori question about any such inference whether or not it is justifiable. The epistemology of empirical science is an empirical science (Boyd 1982, 1983, 1985a, 1985b).

One consequence of this radical contingency of scientific methods is important to the realist's conception of the growth of approximate knowledge. The emergence of successful modern scientific methodology as we know it depended upon the logically, epistemically, and historically contingent emergence of a relevantly approximately true theoretical tradition. It is not possible to understand the initial emergence of such a tradition as the consequence of some more abstractly conceived scientific or rational methodology that itself is theory independent. There is no such methodology. The theoretical innovations that established the first successful paradigm within a particular scientific discipline must be thought of as the beginnings of successful methodology within the field, not as consequences of it (for a further discussion see Boyd 1982).

Note that radical contingency in epistemology is central to the realist's case against empiricism. Against the evidential indistinguishability thesis the realist argues that plausibility judgments grounded in the current total science afford evidential distinctions between empirically equivalent total sciences. But, according to the realist's account, it is not the *currency* of the current total science that makes plausibility judgments with respect to it epistemically reliable but its approximate truth. That a time should have arisen in which total sciences embodied relevant approximations to the truth is of course radically contingent. Thus central to the realist's rebuttal to empiricism are the epistemological principles that reflect that contingency.

2.3. Metaphysics and 'Metaphysics'

Logical positivists employed the term 'metaphysics' for the sort of inquiry about "unobservables" that verificationism led them to reject. Most of what has traditionally fallen under that term was 'metaphysics' in the positivists' sense, but so was inquiry about, e.g., the atomic structure of matter. If scientific realism is right, then it follows that scientists routinely do successful 'metaphysics'. With respect to metaphysics (as philosophers and others ordinarily use the term) the situation is more complex.

If scientific realism is true for any of the standard reasons then scientists have discovered the real essences of chemical kinds (Kripke 1971, 1972) and have thus done some real metaphysics. Moreover, the fact that scientific knowledge of unobservables is possible makes it a serious question whether or not scientific findings have (or will have) resolved some traditional metaphysical questions. Certainly the recent near consensus in favor of a materialist conception of mind reflects a realist understanding of the possibility of experimental metaphysics. Nevertheless it does not follow from scientific realism that scientists routinely tend to get the right answers to the distinctly metaphysical questions that are the special concern of philosophers even when their methods lead them to adopt theories that reflect answers to such question.

In particular, when a scientific realist proposes to explain the reliability of the scientific methods employed at a particular historical moment by appealing to the approximate truth of the background theories accepted at that time, she need not hold that the metaphysical conceptions embodied in those theories represent a good approximation by philosophical standards. Two examples will illustrate the point.

Consider the way in which the reliability of the methods by which Darwin's account in the *Origin* was assessed is to be explained by reference to the approximate truth of much of the prevailing background biological theory. A great deal was known, for example, about species—not just facts about particular species, but about anatomical, behavioral, genetic and biogeographical generalizations that can only be formulated in terms of the notion of a species. The realist will hold that the approximations to the truth embodied in this lore of species is part of what explains the reliability of the research methods in biology employed by Darwin and his contemporaries.

Prior to Darwin's work the prevailing conception made species membership in the first instance a property of individuals; after Darwin we have correctly seen a species as in the first instance a family of populations. The background biological theories of Darwin's era got it profoundly wrong about the metaphysics of species. Nevertheless, the classificatory practices of pre-Darwinian biologists were reliable enough to serve to establish the rich and significantly accurate lore about species upon which the reliability of methodology in early evolutionary theory crucially depended—or, at any rate, so the realist may reasonably maintain.

Similarly, the realist will want to explain the reliability of the methods by which physicists assessed early developments in quantum theory by appealing to respects in which the prequantum theory of, say, atoms and subatomic particles was approximately true. She will appeal to the correct identification of various subatomic particles and of (many of) the fundamental physical magnitudes, to the availability of reliable procedures for the detection of those particles and for the measurement of various of their physical properties, and to the classical insights reflected both in the formulation of the equation for the time-evolution of quantum mechanical systems and in the techniques employed in practice in picking the appropriate Hamiltonian for quantum mechanical systems.

Indeed she will want to portray much of the early development of quantum theory as the gradual extension of the range of phenomena for which an adequate quantum mechanical treatment had been provided. On such an account, at any given stage in the early development of quantum theory, the proposed models for physical systems were always a mixture of distinctly quantum mechanical components together with essentially classical (or relativistic) components awaiting later quantum mechanical reformulation. The realist will want to explain the reliability and justifiability of this sort of development by appealing to the respects of approximation to the truth of classical mechanics itself and of the successive stages in the development of the quantum theory.

Consider now the classical conception of atomic phenomena understood as a contribution to philosophical metaphysics. Arguably the metaphysical component of that conception is some sort of mechanistic atomism: a picture of discrete and unproblematically individuated particles and their associated fields interacting in a deterministic fashion without action at a distance. Our current quantum mechanical conception of matter rejects each component of this picture: for the atomist's discrete particles we substitute entities with wavelike features for which particlelike individuation is sometimes impossible; we reject determinism; and we acknowledge that there are nonlocal effects that would surely be precluded by the classical philosophical rejection of action at a distance. Classical conceptions of the atomic world were, let us agree, poor approximations to the truth in metaphysics. Does this preclude their having been good enough approximations in other respects to sustain the realist's account of the development of quantum theory?

Plainly not. Whatever other objections there may be to the realist's account, it is not a cogent objection that the classical conception that her account treats as relevantly approximately true is not good metaphysics. All she need do is to explain how the metaphysical errors in the classical conception failed to vitiate the methodological contribution of its genuine insights. To this end she might, e.g., appeal to the respects in which subatomic particles are (classical) particlelike, to the determinism of the time-evolution of quantum mechanical systems prior to measurement, and to the wide variety of phenomena that do not significantly exhibit the effects of nonlocal "action at a distance." Perhaps in the case of the development of evolutionary theory and certainly in the case of quantum mechanics, the realist's account will have scientists doing 'metaphysics' with some significant success; in neither case must she portray them as doing good metaphysics.

The cases just discussed illustrate an additional point. In each case, if the metaphysical criticism of the earlier theoretical tradition is sound, then it embodied, in addition to metaphysical errors, errors about the logical form of certain key propositions. Conspecificity is a relation between populations, not individuals; so pre-Darwinian biology embodied a mistake about the logical type of propositions regarding species membership. Similarly, quantum mechanics requires that we think of the classically acknowledged physical magnitudes as corresponding to Hermitian operators rather than to vector- or scalar-valued functions; in consequence classical mechanics is mistaken about the logical form of, e.g., attributions of position or momentum to particles. Neither error undermines the contribution that the approximate truth of the earlier theory is said to have made to the methodology by which the latter theory was developed and confirmed. The realist need attribute to successful background theories neither metaphysical success nor logical exactitude. Approximation need not be philosophically clean. (Note that the distinctly realist naturalistic semantic conceptions are operative in this discussion. What evolutionary theory and quantum mechanics have taught us is that, as we might say, "there are no classical species" and "there are no classical particles." Only naturalistic alternatives to the empiricist conceptions of definitions and reference permit the realist to say—as the account just given requires—that nevertheless Darwinian species and the particle-like phenomena acknowledged by quantum mechanics were the subjects of the relevant classical investigations.)

2.4. Realism Causation and Mind Independence

The realist conception of science contrasts with various neo-Kantian constructivist conceptions according to which when scientific theories address fundamental questions there is a deep element of social construction of reality reflected in what they say. It is sometimes said that realists and constructivists differ over the extent to which the reality studied by scientists is "mind independent" or is "theory independent." In order to understand the demands placed on the realist by what we have called the "circularity objection," we require some understanding of what is distinctly realist about the realist's explanatory appeal to approximate truth of theoretical presuppositions, given that the constructivist shares with the realist the conviction that scientific progress involves theoretical as well as instrumental knowledge and that scientific methods are deeply theory dependent. In the present essay I'll touch on this issue only briefly.

2.4.1. Defining Mind Independence

The realist and the constructivist each reject the Humean and verificationist claim that reference to hidden mechanisms, essences, and causal powers is, on "rational reconstruction," eliminable from the findings of science. They agree scientists' methods and conceptions are determined by ineliminably metaphysical conceptions about the basic sorts of mechanisms, processes, and forces that operate to produce the phenomena under study and that this dependence is not merely a psychological quirk of the "context of invention" to be rationally reconstructed away in the "context of confirmation." They agree too in rejecting the eliminative Humean or regularity account of the causal powers and relations discovered by scientific inquiry. So where does the difference lie, what is the import of the question of the mind or theory independence of reality given that both parties reject empiricism?

The answer, subject to an important qualification, is that the realist denies, while the constructivist affirms, that the adoption of theories, paradigms, conceptual frameworks, perspectives – or the having of associated interests, intentions, purposes, etc. – in some way constitutes, or contributes to the constitution of, the causal powers of, and the causal relations between, the objects scientists study in the context of those theories, interests, etc. Of course (here is the qualification) the realist does not deny that the adoption of theories, etc., and the having of projects or interests, are themselves causal phenomena and thus contribute *causally* to the establishment of, for example, those causal factors that are explanatory in, for example, the history, philosophy and sociology of science and that in consequence the adoption of a theory in such a discipline could contribute causally to the causal powers and relations that are the subject matter of the theory itself. What the realist denies is that there is some further sort of contribution (logical, conceptual, socially constructive, or the like) which the adoption of theories or the having of interests makes to the establishment of causal powers and relations.

Thus the realist denies the *noncausal* contribution of minds and (the adoption of) theories to the establishment of causal powers and relations, whereas the constructivist insists that such a contribution is fundamental. While the present paper focuses primarily on the realist's abductive argument against empiricism, it is important to note two constraints that a suitably developed realist explanation of the reliability of scientific methods must meet if there is to be any prospect of its serving as the basis for a rebuttal to constructivism. In the first place, the definitions of natural kinds, categories, etc., to which the realist's explanation makes essential reference are, in a certain sense, interest dependent. The properties and causal powers that are relevant to explanation or prediction depend on the practical or theoretical projects being undertaken. Thus appropriateness of definitions and conceptual frameworks depends upon the interests with respect to which they are to be employed. The realist must acknowledge this fact in a way which is compatible with denying that the interest dependence in question involves any noncausal contribution of the adoption of interests or projects to the causal powers of the objects of scientific study.

Similarly, as Quine and others have reminded us, even when an agenda of interests and projects is fixed, there may be several ways of defining kinds and categories – of "cutting the world at its joints" – that are equally adequate to the task of reflecting explanatorily significant causal relations (even as the realist understands those relations). It may sometimes happen that the theoretical commitments of two such frameworks will appear to involve conflicting metaphysical conceptions. The choice between such frameworks will be, for the realist, arbitrary. Thus the realist's account of approximation must not treat one such framework as more nearly approximately true than the others, despite apparent metaphysical conflicts; certainly it must not treat the adoption of one rather than another as contributing noncausally to the establishment of causal relations or to similar settling of matters metaphysical.

It is by no means uncontroversial that arbitrariness and the interest dependence of kinds can be treated in the way the realist requires. For the purposes of the present essay I'll assume that an appropriate realist treatment is possible, while acknowledging that, in an essay in which constructivism rather then empiricism was the primary target, the question would require more extensive treatment. Two other issues regarding mind independence deserve our brief attention.

2.4.2. Mind Independence and the Causal Role of Minds

We have seen that the realist acknowledges the causal role of mental phenomena (since, e.g., she explains the reliability of scientific method by reference to the causal powers of approximately true beliefs) and differs from constructivists only in that she denies such phenomena a noncausal role in constituting causal structure. Nevertheless there are cases in which the attribution of a plainly causal role to mental phenomena has been seen as supporting constructivism. Two such cases deserve attention. First, scholars who are impressed by the social role of ideology often claim that "human nature" and the "natures" of various socially defined groups are "social constructions," and often they appear to mean by this, at least in the first instance, that the actual psychological capacities and tendencies exhibited by people generally or by members of socially defined groups are significantly determined by the ideologically established beliefs about psychological tendencies and capacities that are accepted in their own culture – determined in such a way as to tend to make their psychologies conform to the ideology.

Interestingly, many who make such claims seem to take this mode of social construction to be appropriate to a constructivist conception of reality and of knowledge. Plainly this is not so. Whatever the independent evidence for constructivism, the fact that culturally transmitted stereotypes causally influence the actual psychological makeup of those stereotyped provides no evidence of the sort of *non*causal determination of causal structure by minds or theories that the constructivist requires.

The second case concerns solutions to the problem of defining the notion of measurement in quantum mechanics. According to one important conception, part of what characterizes measurements is that they are epistemically relevant interactions so that measurement is defined in terms of knowledge – that is in terms of something (one component of which is) mental – and it is a special sort of interaction with a knowing system that produces discontinuous changes in physical state and results in sharp values for measured quantities. It is sometimes added that the explanation for the fact that measurements are not governed by

Schrödinger's equation is that they involve interactions between a physical system (whose isolated time evolution is governed by that equation) and a nonphysical mind. Whether or not the second suggestion is adopted, it is sometimes suggested that the special role of knowing systems thus identified refutes realism because it shows that the phenomena studied by scientists – in particular the results of their experimental measurements – are mind dependent. Reflection shows that this interpretation (even in its dualist version) simply assigns a distinctive causal role to certain mental phenomena. No noncausal social construction of causal structure is suggested. Indeed, the development of quantum mechanics might well be cited as the most dramatic recent demonstration of our *inability* to define causal reality in accordance with our conceptual schemes (for an excellent discussion see McMullin 1984).

2.5. Homeostatic Property-Cluster Definitions, Realism and Bivalence

There is an established practice of identifying realism regarding a body of inquiry with the view that all of the sentences in the vocabulary employed within it have determinate mind-independent truth values and such a conception of realism places a significant constraint on any realist account of the growth of approximate knowledge. We have just seen that the requirement of mind independence must be carefully qualified. Moreover, the role in approximation that the realist assigns to partial denotation and to denotational refinement (see 2.1.4) precludes any understanding according to which scientific statements must have determinate truth value: statements involving partially denoting expressions might be true on one denotational refinement and false on another.

There is a quite different way in which a realist conception of scientific language predicts failures of bivalence, and it is important to our understanding of the realist's explanatory project both because it reflects another dimension of dialectical complexity in the realist's account of approximation and because it provides the philosophical machinery for a deeper analysis of the underlying notion of scientific rationality.

The sorts of essential definition of substances anticipated by Locke and reflected in the currently accepted natural definitions of chemical kinds by molecular formulas (e.g., "water = H_2O") appear to specify necessary and sufficient conditions for membership in the kind in question. Recent *non*naturalistic property-cluster or criterial attribute theories in the "ordinary language" tradition suggest the possibility of definitions that do not provide necessary and sufficient conditions. Instead, some terms are said to be defined by a collection of properties such that the possession of an adequate number of those properties is sufficient for falling within the extension of the term. It is supposed to be a conceptual (and thus an a priori) matter what properties belong in the cluster and which combinations of them are sufficient for falling under the term. However, it is usually insisted that the kinds corresponding to such terms are "open textured," so that there is some indeterminacy in extension legitimately associated with property-cluster or criterial attribute definitions. The "imprecision" or "vagueness" of such definitions is seen as a perfectly appropriate feature of ordinary linguistic usage, in contrast to the artificial precision suggested by rigidly formalistic positivist conceptions of proper language use.

I doubt that there are any terms whose definitions actually fit the ordinary-language model, because I doubt that there are any significant "conceptual truths" at all. I believe however that terms with somewhat similar definitions are commonplace in the special sciences that study complex phenomena. Here's what I think often happens (I formulate the account for monadic property terms; the account is intended to apply in the obvious way to the cases of terms for polyadic relations, magnitudes, etc):

(i) There is a family F of properties that are contingently clustered in nature in the sense that they co-occur in an important number of cases.

(ii) Their co-occurrence is, at least typically, the result of what may be metaphorically (sometimes literally) described as a sort of *homeostasis*. Either the presence of some of the properties in F tends (under appropriate conditions) to favor the presence of the others, or there are underlying mechanisms or processes that tend to maintain the presence of the properties in F, or both.

(iii) The homeostatic clustering of the properties in F is causally important: there are (theoretically or practically) important effects that are produced by a conjoint occurrence of (many of) the properties in F together with (some or all of) the underlying mechanisms in question.

(iv) There is a kind term *t* that is applied to things in which the homeostatic clustering of most of the properties in F occurs.

(v) *t* has no analytic definition; rather all or part of the homeostatic cluster F, together with some or all of the mechanisms that underlie it, provide the natural definition of *t*. The question of just which properties and mechanisms belong in the definition of *t* is an a posteriori question—often a difficult theoretical one.

(vi) Imperfect homeostasis is nomologically possible or actual: some thing may display some but not all of the properties in F; some but not all of the relevant underlying homeostatic mechanisms may be present.

(vii) In such cases, the relative importance of the various properties in F and of the various mechanisms in determining whether the thing falls under *t*—if it can be determined at all—is a theoretical rather than a conceptual issue.

(viii) Moreover, there will be many cases of extensional vagueness that are such that they are not resolvable even given all the relevant facts and all the true theories. There will be things that display some but not all of the properties in F (and/or in which some but not all of the relevant homeostatic mechanisms operate) such that no rational considerations dictate whether or not they are to be classed under *t*, assuming that a dichotomous choice is to be made.

(ix) The causal importance of the homeostatic property cluster F together with the relevant underlying homeostatic mechanisms is such that the kind or property denoted by t is a natural kind (see section 2.1.3).

(x) No refinement of usage that replaces t by a significantly less extensionally vague term will preserve the naturalness of the kind referred to. Any such refinement would either require that we treat as important distinctions that are irrelevant to causal explanation or to induction, or that we ignore similarities that are important in just these ways.

(xi) The homeostatic property cluster that serves to define *t* is not individuated extensionally. Instead, the property cluster is individuated like a (type or token) historical object or process: certain changes over time (or in space) in the property cluster or in the underlying homeostatic mechanisms preserve the identity of the defining cluster. In consequence, the properties that determine the conditions for falling under *t* may vary over time (or space), *while* t *continues to have the same definition.* The historicity of the individuation criterion for the definitional property cluster reflects the explanatory or inductive significance (for the relevant branches of theoretical or practical inquiry) of the historical development of the property cluster and of the causal factors that produce it, and considerations of explanatory and inductive significance determine the appropriate standards of individuation for the property cluster itself. The historicity of the individuation conditions for the property cluster is thus essential for the naturalness of the kind to which t refers.

The paradigm cases of natural kinds – biological species – are examples of homeostatic-cluster kinds. The appropriateness of any particular biological species for induction and explanation in biology depends upon the imperfectly shared and homeostatically related morphological, physiological and behavioral features that characterize its members. The definitional role of mechanisms of homeostasis is reflected in the role of interbreeding in the modern species concept; for sexually reproducing species, the exchange of genetic material between populations is thought by some evolutionary biologists to be essential to the homeostatic unity of the other properties characteristic of the species, and it is thus reflected in the species definition that they propose (see Mayr 1970). The *necessary* indeterminacy in extension of species terms is a consequence of evolutionary theory, as Darwin observed: speciation depends on the existence of populations that are intermediate between the parent species and the emerging one. Any "refinement" of classification that artificially eliminated the resulting indeterminacy in classification would obscure the central fact about speciation upon which the cogency of evolutionary theory depends.

Similarly, the property-cluster and homeostatic mechanisms that define a species must be individuated nonextensionally as a processlike historical entity. It is universally recognized that selection for characters that enhance reproductive isolation from related species is a significant factor in phyletic evolution, and it is one which necessarily alters over time the species's defining property cluster and homeostatic mechanisms (Mayr 1970).

It follows that a consistently developed scientific realism *predicts* indeterminacy for those natural kind or property terms that refer to complex homeostatic phenomena; such indeterminacy is a necessary consequence of "cutting the world at its (largely mind-independent) joints" (contrast, e.g., Putnam 1983 on "metaphysical realism" and vagueness). Realists' accounts of approximation need not honor bivalence even when partial denotation is not at issue. Similarly, scientific realism predicts the existence of nonextensionally individuated definitional clusters for at least some natural kinds, and thus it treats as legitimate vehicles for the growth of approximate knowledge linguistic practices that would, from a more traditional empiricist perspective, look like diachronic inconsistencies in the standards for the application of such natural kind terms.

Moreover, the homeostatic-cluster conception of definitions may permit a more perspicuous formulation of the central explanatory thesis of scientific realism. I have argued elsewhere (Boyd 1979, 1982, 1983) for an understanding of knowledge and of reference according to which (although I did not use this terminology) the relations 'x knows that y' and 'x refers to y' possess homeostatic property-cluster definitions. I will suggest in section 3.7 that scientific rationality has a homeostatic property-cluster definition and that the realist's explanation for the reliability of scientific methods is best understood as the crucial component in an explanation of the homeostatic unity of scientific rationality.

Not all challenges to realism that arise from considerations about bivalence require in rebuttal an appeal to the possibility of actual bivalence failure. For example, the measurement problem in quantum mechanics is sometimes put by saying that quantum systems lack determinate values of classical magnitudes prior to measurement, and the problem is to characterize the interactions that relieve the indeterminacy with respect to a particular magnitude. Sometimes the alleged indeterminacy prior to measurement is seen as an indication of the failure of realism. Realism is seen as predicting determinacy for (premeasurement) values of classical magnitudes.

In response the realist need not appeal to the possibility of a realist explanation for failures of bivalence. There are two ways of understanding the claim about a physical system that it possesses a determinate value of a classical magnitude, a determinate component of orbital angular momentum, for example. On the first understanding, that claim is understood to incorporate the classical *mis*conception of the logical status of statements about angular momentum, in which case the statement is always false, in however many respects special cases of such statements may also have been usefully approximately true. Alternatively, the statement may be interpreted as attributing to the system an eigenstate of the relevant operator, in which case it need not be false, but it has, depending on the system in question, some determinate truth value. On careful analysis there is no bivalence failure here to explain.

3. Approximate Truth and Parametric Specification: The Realist's Explanation as Ordinary Science

3.1. The Status of the Realist's Explanation

Recall that the argumentative strategy proposed in section 1.3 calls for us to first assess the evidence for the realist's explanation for the instrumental reliability of scientific methods considered as an ordinary scientific hypothesis. If the realist's explanation appears well confirmed, then there will remain the further and more distinctly philosophical task of determining whether or not, with respect to the realist's explanation itself, it is legitimate to adopt the realist interpretation without which no defense of a realist position in the philosophy of science is forthcoming.

This approach presupposes that the realist's explanation has the form of an ordinary causal explanation in science subject to confirmation or disconfirmation by ordinary scientific standards. Two considerations might suggest that it does not. First, some philosophical explanations of epistemic matters seem noncausal; this is true, for example, of some transcendental explanations and of some "ordinary language" analyses of notions like "evidence," "reliable," "justification," and the like. Secondly, there are ways of thinking of the notions of truth and approximate truth (disquotational analyses, for example) that make them noncausal.

The realist's conception of the epistemology and semantics of scientific theories does not raise any of these problems. Truth is definable from "primitive denotation" (Field 1974), and denotation, on the realist's account, is an epistemic and thus a causal matter; truth is correspondence truth and correspondence is a matter of complex causal interactions. Similarly, to talk of respects of approximation to the truth is to talk of respects of similarity and difference between actual causal situations and certain possible ones. It is philosophically challenging to give a general account of the nature of such comparisons with counterfactual possibilities, but such comparisons are so routine a feature of ordinary causal reasoning in science (including reasoning about the reliability of particular methods) that there is no reason to suppose that they raise difficulties in the present context.

Likewise the explanatory claims of the realist are perfectly ordinary causal claims. Under certain sorts of historical and social circumstances individually and socially held beliefs are said to exhibit a particular causal power—a tendency to generate methods that are (causally) conducive to the establishment of approximate knowledge—when they are in causally relevant ways approximately true. However controversial, this is an ordinary causal thesis about the interactions of scientific communities and the rest of the world. We may reasonably inquire about how it fares by ordinary scientific standards of evidence. It is to this issue that we now turn our attention.

3.2. Does the History of Science Immediately Refute the Realist's Explanation?

According to the historical objection, the realist's explanation for the reliability of scientific method is refuted by the fact that there have been episodes in the history of science during which methodological practices were successful, but during which the relevant background theories were not, by contemporary standards, approximately true as the realist's explanation requires. The realist's response comes in two parts.

First, the realist's explanation does not require that scientists, even during periods of mature inquiry, be especially good at doing metaphysics. The realist need not necessarily show about any episode in the history of science that the relevant background theories are close to the truth on metaphysical matters. The realist's position is not compromised by any respects of error in earlier background theories that do not undermine her appeal to the specific respects of approximation regarding unobservable phenomena that are crucial to her explanation of the reliability of methods during that episode.

Second, the realist's account of the methods of science predicts that there will be early stages in the history of *any* currently mature science in which the relevant background theories will have been too far from the truth to ensure the sort of reliability of methods that is characteristic of mature sciences. This conclusion is a consequence of the radical contingency in epistemology dictated by the realist explanation for the reliability of scientific methods and, in particular, of the claim that it is, in an epistemically important sense, accidental that the earliest relevantly approximately true theories arise within any scientific discipline. Of course I do not mean that no historical explanations are possible for particular early successes, but only that, according to the realist, the explanation cannot involve appeal to the operation of rational methods with anything like the reliability of the methods of (what from the contemporary point of view are) theoretically more mature stages in the same sciences.

In sum, the realist's explanation is vulnerable to straightforward refutation by the phenomenon of successful science guided by deeply false background theories only if (a) the relevant historical episodes involve the operation of methods that exhibit the profound and routine reliability of judgments of projectability and related matters characteristic of the most mature sciences in the twentieth century, and (b) the respects of falsity in the relevant background theories are not merely deep but such as to preclude an explanation of that reliability by appeal to the respects in which those theories are approximately true. The tendency in recent empiricist philosophy of science towards realism reflects precisely the opposite conception: philosophers were tempted by realism precisely because they thought they could see how to offer a realist explanation of the reliability of methodological practices in highly successful science, and they lost their confidence in alternative empiricist "rational reconstructions" of those methods. In any event what I envision as the realist's reply to the historical objection is simply that there aren't actual cases satisfying (a) and (b). Realism is, after all, supposed to be an empirical thesis, and here is one of the empirical claims upon which it rests.

3.3. Triviality, Contrivance, and the Methodology of Parametric Specification

Against the charge of immediate historical falsification, the realist replies by insisting, as the logic of her explanations dictates anyway, that her thesis is that background theories in mature sciences must be seen as approximately true in relevant respects. As we saw in section 1.2, the realist now faces the challenge that her explanations are trivial: that any consistent theory is true in some respects, and that she has offered no general theory of the relative importance of respects of approximate truth. Here the reply is the obvious one that the respects of approximation that are important are those that are required to sustain the realist's distinctive explanation of the reliability of scientific methods and that it is with respect to these that approximations to the truth are claimed. The reply is successful just in case the charge of contrivance can be met: just in case, that is, the realist can argue that, even in the absence of a general context- and episode-neutral account of degrees of approximation, her appeal to respects of approximation appropriate to her own theoretical project does not constitute an ad hoc and thus methodologically inappropriate contrivance.

In order to assess the prospects for the realist's explanations we need to know what distinguishes such contrivances from methodologically appropriate appeals to context-specific specifications of causal variables. Fortunately the question is not esoteric; frequently, especially in the context of historical explanations, we confirm theories that appeal to context-dependent specifications of causal parameters and the methodology for avoiding ad hoc theorizing is well understood. Consider for example explanations in evolutionary theory. There are a variety of possible evolutionary mechanisms—individual selection, kin selection, genetic drift, selection for pleiotropically linked traits, etc.—for no one of which does evolutionary theory provide a context-independent prediction of its influence in any particular evolutionary episode. Moreover, in particular evolutionary episodes several of these factors may operate, and there is no context-independent way of predicting their relative influence. Still, the modern evolutionary explanation for the diversity of life is well confirmed. What methodological principles permit us to treat the explanations provided by evolutionary theory as appropriate rather than ad hoc, and as appropriate for "inductive inference to the best explanation"?

The answer is pretty clear. What we require of the various individual evolutionary explanations for particular features of living organisms is that they cohere not only with each other but with the independent results of inquiry in the related scientific disciplines: geology, genetics, developmental biology, animal behavior, atmospheric sciences, oceanography, anthropology, etc. This requirement of integration of the various particular explanations into the broader framework of scientific knowledge constitutes our methodological safeguard against the possibility that the apparent explanatory successes of evolutionary theory are reflections of mere contrivance. This pattern is quite general: particular explanations provide evidence for a broader theory whose explanatory resources they exploit just in case theory-dependent evidential standards, including requirements of theoretical integration, dictate the acceptance of the particular explanations, and just in case the success of those individual explanations lends inductive support for the causal claims of the broader theory (Boyd 1985b).

Exactly the same standards apply, of course, to the realist's broad explanation for the reliability of scientific methods. The charge of contrivance is met just in case the realist's explanations for the reliability of methods in particular episodes, including the context-dependent specifications of respects of approximation they contain, are independently supported by scientific evidence, and in particular that they pass the test of coherence with the rest of established scientific theory, and (this is the easier part) just in case these particular realist explanations lend inductive support to the broader realist explanatory picture of scientific epistemology.

3.4. The Local Coherence of Realism

Are the individual realist explanations for the reliability of specific scientific methods well confirmed and do they in particular cohere appropriately with the rest of science? Do they inductively support the general realist conception of the growth of approximate knowledge? At an important level of analysis the answer to both questions must be "obviously yes."

The particular realist explanations of the reliability of methods fall roughly into two categories. In the first category are the theoretical explanations for the reliability of particular measurement and computational procedures and for the reliability of various sorts of controls and other features of the design of experimental and observational studies. In the second category are the theoretical explanations for the reliability of the judgments of projectability which determine the broader outlines of rational experimental and observational inquiry. Explanations in either category may be either static or dialectical. By a static explanation I understand an explanation that explains the reliability of some piece of methodology by appeal to the approximate truth of some theories that have been long established at the time of the relevant methodological judgments; dialectical explanations explain the reliability of some novel feature of methodology or of some revision of a previously established methodological practice by appealing to changes in theoretical outlook that bring about a closer approximation to the truth along relevant lines.

At any given time in the history of recent science, individual realist explanations in the first category both static and dialectical look just like well-established pieces of boringly normal science: they are the sorts of claims that are routinely made explicit in the methods sections of papers in the empirical sciences, in which scientists explain the appropriateness of research design. Most explanations of the static sort and almost all of the dialectical ones will embody reference to context-specific degrees and respects of approximation in the current theoretical conception or its immediate predecessors. Those explicit pieces of scientific theorizing are not produced in service of any philosophical or historical project, realist or otherwise. In the better established sciences they are apparently as well confirmed as anything gets; certainly there is no evidence that they fail to cohere with the rest of established science. That, after all, is what made such pieces of ordinary science so disturbing to empiricists. The prospect that they are vulnerable to the contrivance objection is vanishingly remote.

Scientists seem rarely to investigate explicitly the causal question of the reliability of particular projectability judgments *under that description.* They do however offer justifications for their own methodological judgments, critiques of such judgments by others, and proposals for changes in such judgments. Such justifications are made explicit in published papers, in referees' reports, in grant proposals, in the introductory parts of experimental papers, and in theoretical papers and books and the judgments they justify are *in fact* judgments of projectability of the sort to which the realist explanation refers. It is all but the consensus position among students of the logic of scientific inference (e.g., Hanson 1958, Kuhn 1970, Quine 1969a; Van Fraassen 1980) that ordinary scientific standards of reasoning treat these projectability judgments as inductive inferences from background theories, just as realism requires. Here again the justifications in question routinely appeal to context-specific respects of approximation, especially in cases in which they mirror realist explanations of the dialectical sort. There is again no prospect that scientists' reasoning in such cases is contrived to serve a philosophical purpose nor is there any reason to hold that the requirement of coherence with the rest of science is not honored in such reasoning—indeed it is in reasoning of this sort that the requirement of coherence finds its expression in ordinary science!

Here then is the phenomenon of *local coherence*: the explicit and near-explicit findings of ordinary science examined synchronically seem to strongly confirm, if only tacitly, the particular explanations for the reliability of projectability judgments on which the realist's explanatory enterprise rests and they appear to do so in a way that subjects the context-dependent judgments of relevant respects of approximation which they contain to the appropriate requirement of coherence.

Do the particular realist explanations we are considering, taken together, inductively support the realist conception of scientific epistemology developed in part 2? Here we cannot defer to any particular science except philosophy, but we can observe that the whole tendency to take realism seriously as an alternative to logical empiricism from the mid-1950s on reflects the extremely widespread judgment among philosophers of science that the actual practices of science *appear* to require a realist explanation. I conclude that, if we examine the question *pre*philosophically, there appears to be very good reason to hold that the realist's explanation for the reliability of scientific methods is well confirmed as a scientific hypothesis and in particular, that there is no reason to think that the realist's approach to the problem of parametric specification is any more in doubt than, say, that of the evolutionary biologists who must also rely on specifications not given antecedently by a context-independent formula.

We turn now to the question of what the distinctly philosophical dimension is to the confirmation of the realist's explanatory hypothesis. The elaborate machinery rehearsed in part 2 indicates that a lot is going on philosophically. Some of it is relevant only to the question of circularity, but much is relevant also to a defense of realism as a scientific thesis in the methodological climate created by the philosophical disputes over realism.

3.5. What's Distinctly Philosophical? I: Diachronic Patterns of Inference and of Language Use

Central to the realist's abductive argument for realism is the claim that no alternative exists that adequately explains the reliability of scientific methods or justifies their use. It is possible to *imagine* that a case along these lines for realism — or at any rate against the verificationist insistence that knowledge of unobservables is impossible — could be made by the synchronic examination of only a few episodes in the history of science for which only realist explanations and justifications seem available. Nevertheless the deep plausibility of empiricist epistemological principles, especially the evidential indistinguishability thesis, is so great that it is doubtful that realism about a few isolated cases would, even as a scientific hypothesis, be rationally acceptable. Instead the plausibility of any individual realist explanation seems to rest upon diachronic considerations that provide additional and crucial support for the general realist explanation of the reliability of scientific methods. In effect, the role of these diachronic considerations is to establish that the individual synchronic-realist explanations can be coherently integrated into a scientifically acceptable historical conception of the reliability of scientific methodology.

In particular there are two patterns in the history of science whose recognition is a distinctive contribution of philosophers and historians in making the case for the realist's explanations. In the first place, there is the utterly commonplace phenomenon of *mutual ratification* between consecutive stages in the development of scientific disciplines. It is routine in the case of theoretical innovations that (a) the new and innovative theoretical proposal is such that the only justification scientists have for accepting it, given the relevant evidence, is that it resolves some scientific problem or question *while preserving certain key features of the earlier theoretical conceptions;* and (b) the new proposal ratifies the earlier conception as approximately true in just those respects that justify their role in its own acceptance. Moreover the patterns of mutual ratification are characteristically seen to be *retrospectively sustained*: although later theoretical innovations typically require a revision in our estimates of the degrees and respects of approximation of both the earlier innovative proposals and their predecessors, the initially discernable relation of mutual ratification is typically sustained as a very good first approximation to the evidentially and methodologically important relations between the innovation and its predecessor theories. It is the ubiquity of this sort of *retrospectively sustained mutual ratification* and the difficulty in "rationally reconstructing" it away with respect to the justification of theoretical innovations that has made the case for realism so plausible.

A second pattern concerns the use of scientific language. The realist conception of projectability requires that the categories that scientists employ in formulating general laws and causal claims typically reflect underlying causal structures rather than conventionally specified nominal essences, and many of the changes in classificatory practice for which individual realist explanations are forthcoming seem to indicate an attempt to obtain a fit between categories and causal structure. It is essential to the case for realism that this pattern in scientific language use be sustained: that the diachronic linguistic behavior of scientists involves an apparent disposition to take the definitions of scientific kinds, relations, magnitudes, etc., to be revisable in the light of new data and new theoretical developments. Thus the identification of just such a pattern of *apparent essentialism* in the actual linguistic practices in scientific communities is an important distinctly philosophical contribution to the case for the realist's explanation of the reliability of scientific method.

3.6. What's Distinctly Philosophical? II: Epistemological, Metaphysical, and Semantic Underpinnings

The ubiquitous patterns of retrospectively sustained mutual ratification and apparent essentialism constitute philosophical reasons to accept the realist's explanation, and the recognition of those patterns was a central factor in the emergence of contemporary scientific realism. Still, their effect would not have been so significant were it not for more theoretical attempts to understand their philosophical import. The obvious examples here are causal theories of reference and associated naturalistic conceptions of definition. Had it not proven possible to articulate these distinctly philosophical theories, then it might have been rational to hold that the apparently rational theory-and-evidence-driven revision of definitions in science was only apparent, or only apparently rational. The initial case for the realist explanation would have been crucially undermined.

Analogous considerations hold for the epistemological dimension. Both the realist explanations for the reliability of scientific methods in particular cases and the view that the ubiquity of the pattern of mutual ratification supports the broader realist explanation entail that evidential considerations in science are deeply theory dependent. Were it not possible to provide a realist epistemological framework that incorporates this conclusion—and in particular were it not possible to articulate that framework so as to refute the evidential indistinguishability thesis and make palatable the consequent abandonment of foundationalism—then it would have not been rational to take either the particular explanations or the pattern of mutual ratification as significant support for the realist explanation. Thus the development of a nonfoundationalist realist treatment of projectability judgments and the incorporation of that treatment into an independently developing tradition of nonfoundationalist naturalism in epistemology proves to have been essential for the rational acceptance of the realist explanation.

On to metaphysics. The causal theory of reference and the naturalistic conceptions within epistemology with which realist anti-foundationalism can be profitably assimilated all appear to reflect a distinctly non-Humean conception of causal relations. The cogency of these fundamental elements in the defense of the realist's explanation depend therefore (at least prima facie) on the successful articulation of a non-Humean conception of causation (e.g., Boyd 1985b; Mackie 1974; Shoemaker 1980).

Acceptance of the realist's explanation as a scientific theory does not entail the acceptance of scientific realism, since the realist's explanation might itself be interpreted nonrealistically. What I have been suggesting is that nevertheless the realist's explanation is sufficiently novel in its apparent epistemological, semantic, and metaphysical implications that the articulation of just the sort of broader realistic and naturalistic conceptions of (scientific and other) knowledge, of language, and of metaphysics indicated in part 2 is essential for the defense of that explanation.

I think that the picture just presented captures the current case that the realist's explanation for the reliability of scientific methods is a well-confirmed scientific theory, context-dependent specifications of respects of approximation notwithstanding. An even broader philosophical setting for that case is available if we exploit the distinctly naturalistic conception of homeostatic property-cluster definitions outlined in section 2.5

3.7. Realism and the Homeostatic Character of Scientific Rationality

I argued in section 2.5 that lots of natural kinds, properties, etc. possess homeostatic property-cluster definitions, and I suggested that knowledge and reference are among them. I want now to suggest a similar homeostatic cluster treatment of scientific rationality itself. Ordinarily we think of scientific rationality as being exhibited in two different features of the practice of science: the high level of deliberative rationality in the reasoning of researchers, and the spectacular successes of scientific research in understanding and predicting natural phenomena. If foundationalism is mistaken, as it surely seems to be, then the first of these features does not logically entail the second, and the realist explanation may be thought of as explaining why (and when) they reliably co-occur. Here is a kind of homeostasis of the two distinct components of scientific rationality.

Once it is recognized that this co-occurrence is a causal matter, then it is easy to see that at a finer level of analysis there is a family of similar sorts of co-occurrences requiring explanation. The methodological norms in a particular subdiscipline are set not only by the background theoretical findings in that subdiscipline but as well by findings from other subdisciplines and from quite different disciplines altogether. That the methodological norms determined by such a wide range of theories should be unified enough to be a practical guide to successful scientific research requires explanation. Why aren't the resulting methodological norms characteristically irreconcilably conflicting, for instance?

Similarly, scientists working largely independently within different disciplines frequently converge on the same solution to a problem they may not have recognized that they have in common. Why should this happen? Likewise, it often happens that largely independently developing disciplines become ripe for interdisciplinary work, and their largely independently developed theories and methodologies prove (with some difficult but not impossible negotiation) to be integrable. Why is this so frequently possible?

What I propose is that we think of scientific rationality as being defined by the homeostasis of all of these various components of scientific practice and that we should think of the realist explanation of the coincidence of deliberative rationality and theoretical and empirical success as the first step toward a more general realist explanation of the relevant homeostasis. It is even possible that this project could be extended fruitfully to incorporate a naturalistic conception of moral rationality (Boyd 1988; Brink 1984, 1989; Miller 1984b; Railton 1986; Sturgeon 1984a, 1984b).

If the proposal of the present section were to prove successful it would prima facie provide further support for the realist explanation and for the philosophical naturalism that underwrites it. However, we still need to know whether the realist's explanation should itself be understood realistically or whether instead, as the circularity objection suggests, that would simply beg the question against the anti-realist.

4. Meeting the Circularity Objection

4.1. Circularity and Philosophical Packages

According to the circularity objection, the realist's explanation for the success of scientific methods, even if well confirmed, cannot without begging the question be interpreted realistically and thus cannot without circularity be treated as confirming scientific realism. The problem posed by this objection is faced not only by the particular defense of realism under consideration but by almost any plausible defense of scientific realism.

The reason is simple: in all but the most trivial cases the defense of realism regarding one or more theories or traditions will require the defense of a theory of epistemic contact that spells out the sort of epistemically relevant causal relations that are supposed to obtain between the subject matter of the theories or traditions and the behavior of the relevant inquirers. Because the realist thesis and the theory of epistemic contact that supports it are causal theses, their confirmation will always depend upon the confirmation of theories (or, for very simple cases, commonplaces) about the causal powers of the entities that are the putative subject matter of the theory or tradition in question. The confirmation of specific theories of epistemic contact will, in turn, depend in part upon theoretical considerations grounded in the best available theories of the relevant subject matter. Such theories will be a vital background assumption against which the evidence for the realist thesis is judged. As we have seen, the theory of epistemic contact, and (thus) the theories upon which its confirmation in turn depends, will themselves have to be understood realistically if they are to help to validate the realist thesis itself. But of course these theories will, in any plausible case, be subject to the same anti-realist assessments as the theory or tradition about which realism is initially in question. Indeed if that theory is a well-established contemporary theory it may *itself* provide the foundations for the relevant theory of epistemic contact! Is this not a point at which the defense of realism begs the question against anti-realists?

Here the answer is "no." If theories of epistemic contact by themselves constituted the sole argument of the realist against anti-realism, if for example the *sole* argument in favor of realism in atomic theory consisted of the articulation of an apparently well-confirmed theory of epistemic contact between scientists and atoms, their properties and their constituent parts, then the question would indeed be begged by the assumption that that theory itself should be understood realistically. The actual role of theories of epistemic contact is quite different.

The issue of realism arises in the form we have been discussing only in the case of a theory or tradition of inquiry about which there is a prima facie case that it possess a theory-independent (even if unobservable) subject matter. The prima facie case for realism will thus rest upon the apparent confirmation of a (realistically understood) theory of epistemic contact. In the special case of realism defended along the lines proposed here, that theory of contact is the one embodied in the realist's explanation for the reliability of scientific methods. The defense of realism, however, depends not upon the theory of epistemic contact *alone* but upon the ability of realists to incorporate suitably elaborated versions of it into an epistemological, semantic and metaphysical conception of the theory or tradition in question (a *philosophical package*) that is superior to that those available to defender of the various anti-realist conceptions.

Thus, for example, the defense of realism regarding the tradition of atomic theory depends upon the best-confirmed atomic theories providing the basis for an apparently realistic theory of epistemic contact, but it depends as well upon additional, more explicitly philosophical considerations, which legitimize the realist treatment of such a theory. On the version of scientific realism presented here, these additional considerations are of two sorts. First, it is argued that only on a realist construal of atomic theory generally, and of the relevant theory of epistemic contact in particular, is it possible to avoid skepticism about the possibility of purely instrumental knowledge in physics and chemistry: knowledge of a sort acknowledged by empiricists and constructivists as well as by realists. Secondly, it is argued that the picture that emerges from a realist treatment of atomic theory is consonant in its departures from foundationalism and in its treatment of scientific language with other quite independently defensible developments in epistemology and semantic theory.

In such a dialectical setting, the dependence of the realist theories of epistemic contact upon a realist understanding of the theory or tradition in question (or of some closely related theory or tradition) need not constitute begging the question against the anti-realist. Fairness to the case for realism requires that realism be understood in a context provided by a realist interpretation of the apparently best-confirmed realist theories of epistemic contact and of the apparently best-confirmed substantive theories of the alleged (theory-independent) subject matter in question.

Importantly, just the same understanding of the issue is required by fairness to the case *against* realism. Both the empiricists' and the constructivists' anti-realist arguments depend upon the assumption that the realist accepts the prevailing theoretical conception and its associated methodology. The realist is understood to take the properties of the putative socially unconstructed referents of the terms of a theory or tradition to be, at least approximately, those required by (a realist understanding of) the apparently best-confirmed theories of the presumed subject matter and to accept the methodology dictated by them as approximately reliable. On those assumptions (but not without them) the empiricist can reason that the realist's position commits her to the possibility of investigating the properties of unobservable phenomena, and thus to an epistemological position against which the empiricist has very powerful arguments.

The constructivist anti-realist similarly assumes that the realist accepts a realist interpretation of the prevailing theoretical and methodological conceptions. Only on such an understanding is it clear that the realist is committed to the possibility of investigating a theory-independent reality using theory-dependent methods – just the possibility that the constructivist critique of realism rejects. Thus an adequate treatment of the controversy between realists and either of their standard opponents requires that we accept that the philosophical package offered in defense of realism contains the apparently best-confirmed theories of the alleged subject matter, realistically understood, and in particular that it be understood as incorporating an associated realistically understood conception of epistemic contact.

Once it is seen that no question is begged against the anti-realist by adopting a realist interpretation of the realist's explanation for the reliability of scientific methodology, we are left with the question: Suppose that the realist's explanation is well confirmed, then why would a realist philosophical package incorporating a realist version of that explanation be superior to an empiricist package incorporating the explanation instrumentally interpreted or to a constructivist package incorporating the realist's explanation understood as a piece of social construction? My main aim in the present essay is to show that the realist's appeal to a distinctively realist explanation for the growth of approximate knowledge, incorporating an appropriate context-and-episode-dependent account of relevant respects of approximation, does not involve any triviality, contrivance, or begging of the question – not to finish once and for all the task of defending realism. I will therefore indicate only briefly the outlines of the considerations that seem to me to justify a preference for the realist package over the two alternatives in question.

4.2. Against the Empiricist Package

The key argument for scientific realism according to the program presented here is that realism as a scientific hypothesis presents the only scientifically acceptable explanation for the realiability of scientific methods. The empiricist might be unimpressed by the demand for explanation in this case (Fine 1984, Van Fraassen 1980). Still the realist can also argue that accepting the realist explanation provides as well the only justification we have for accepting the instrumental findings of science (Boyd 1983, 1985a). One possible empiricist response is that we can justify accepting the inductive deliverances of an apparently realistic scientific method as a result of the second-order induction about induction whose conclusion is that reasoning like a realist in science is instrumentally reliable.

Since this conclusion is only about observables, the empiricist can accept it and employ it to justify accepting currently accepted theories as empirically adequate.

Against this rebuttal I have argued (Boyd 1983, 1985a) that the induction in question is demonstrably just as theory dependent as any other in science and is thus unavailable to the empiricist who is adopting the proposed strategy. Here is a possible reply: We justify the second-order induction by a third-order induction about inductions about induction, the third-order induction by appeal to a fourth-order induction, etc. For the nth case the justification for the relevant projectability judgments is provided not by apparently realistic theoretical considerations but by the n + first-order induction.

If I am right this last response is what the incorporation of the realist's explanation into an empiricist philosophical package would require if that package were to provide any even remotely plausible account of the justification of (instrumental) scientific knowledge. I claim that the resulting philosophical package would prove to be only remotely plausible in consequence. Here we have not just infinite regress but infinite ascent: each level of inductive inference is justified by appeal to a more abstract and problematical level of inductive inference. Given that the realist's package already incorporates an alternative, less speculative, and independently justified naturalistic epistemology I predict that it will prove superior.

4.3. Against the Constructivist Package

Response to the sort of constructivist philosophical package that might be constructed so as to include the realist's explanation for the reliability of scientific methods is substantially more difficult. Constructivism is a richer philosophical program than empiricism, and at the same time it incorporates features (often just the ones that add to its richness) whose consistency is disputable. Rather than even beginning to sort out all of the issues that a thoroughgoing realist response to constructivism would have to address, I will just indicate briefly how two quite standard objections to constructivism might be brought to bear on the proposed package.

In the first place, any adequate philosophical package will have to incorporate versions of most of the apparently best-established scientific and methodological findings. The suggestion outlined in section 2.4, that the establishment of social institutions and linguistic conventions does not contribute noncausally to the causal powers of the objects studied by participants in those institutions and conventions, has very deep roots in quite diverse features of our understanding both of causation and of social phenomena. Thus any constructivist philosophical package will be prima facie vulnerable at any point at which it incorporates a distinctly constructivist conception of the social construction of causal relations. The proposed constructivist package would incorporate this doubtful feature into its version of the naturalistic account of the reliability of scientific methods and thus in to the very center of its basic epistemology. It is doubtful therefore that the proposed package will afford as satisfactory a treatment of absolutely central epistemological issues as its realist rivals.

A second standard objection to constructivism is that the historical fact of anomalies indicates that the world scientists study does not have a structure logically, socially, or conceptually determined by the paradigms or theories they accept. It is beyond the scope of this essay to examine the variants on this objection and the range of possible replies. It cannot be doubted however that it does pose a serious challenge to the acceptability of any constructivist package. Since there are anomalies in methodological matters that exactly parallel those in theoretical matters, the incorporation of a doctrine of social construction of the reliability of scientific method seems hardly to strengthen the constructivist philosophical package.

I conclude that the resources exist for a spirited defense of a realist philosophical package against empiricist and constructivist alternatives, and in particular that the incorporation of a realist interpretation of the realist's explanation of the reliability of scientific methodology strengthens rather than (as the circularity challenge suggests) weakens the realist package.

References

Armstrong, D.M. 1973. *Belief, Truth and Knowledge*. Cambridge: Cambridge University Press.

Boyd, R. 1972. Determinism, Laws and Predictability in Principle. *Philosophy of Science* 39: 431–50.

———. 1973. Realism, Underdetermination and a Causal Theory of Evidence. *Noûs* 7: 1–12.

———. 1979. "'Metaphor and Theory Change." In *Metaphor and Thought*, ed. A. Ortony. Cambridge: Cambridge University Press.

———. 1980. "Materialism Without Reductionism: What Physicalism Does Not Entail." In *Readings In Philosophy of Psychology*, vol.1, ed. N. Block. Cambridge: Harvard University Press.

———. 1982. Scientific Realism and Naturalistic Epistemology. In *PSA 1980. Volume Two*, eds. P.D. Asquith and R.N. Giere. East Lansing: Philosophy of Science Association.

———. 1983. On the Current Status of the Issue of Scientific Realism. *Erkenntnis* 19: 45–90.

———. 1985a. "Lex Orendi est Lex Credendi." In *Images of Science: Scientific Realism Versus Constructive Empiricism*, eds. Churchland and Hooker. Chicago: University of Chicago Press.

———. 1985b. "Observations, Explanatory Power, and Simplicity." In *Observation, Experiment, and Hypothesis In Modern Physical Science*, eds. P Achinstein and O. Hannaway. Cambridge: MIT Press.

———. 1985c. The Logician's Dilemma. *Erkenntnis* 22: 197–252.

———. 1988. "How to be a Moral Realist." In *Moral Realism*, ed. G. Sayre McCord. Ithaca: Cornell University Press.

Brink, D. 1984. Moral Realism and the Skeptical Arguments from Disagreement and Queerness. *Australasian Journal of Philosophy* 62.2: 111–25.

———. 1989. *Moral Realism and the Foundations of Ethics*. Cambridge: Cambridge University Press.

Byerly and Lazara. 1973. Realist Foundations of Measurement. *Philosophy of Science* 40: 10–28.

Carnap, R. 1934. *The Unity of Science* Trans. M. Black. London: Kegan Paul.

Feigl, H. 1956. "Some Major Issues and Developments in the Philosophy of Science of Logical Empiricism." in *Minnesota Studies in the Philosophy of Science,* vol. 1, *The Foundations of Science and the Concepts of Psychology and Psychoanalysis,* eds. H. Feigl and M. Scriven. Minneapolis: University of Minnesota Press.

Field, H. 1973. Theory Change and the Indeterminacy of Reference. *Journal of Philosophy* 70: 462–81.

———. 1974. Tarski's Theory of Truth. *Journal of Philosophy* 69: 347–75.

Fine, A. 1984. "The Natural Ontological Attitude." In *Scientific Realism,* ed. J. Leplin. Berkeley: University of California Press.

Goldman, A. 1967. A Causal Theory of Knowing. *Journal of Philosophy* 64: 357–72.

———. 1976. Discrimination and Perceptual Knowledge. *Journal of Philosophy* 73: 771–91.

Goodman, N. 1973. *Fact Fiction and Forecast,* 3d ed. Indianapolis and New York: Bobbs-Merrill.

Hanson, N. R. 1958. *Patterns of Discovery.* Cambridge: Cambridge University Press

Hardin, C., and Rosenberg, A. 1982. In Defense of Convergent Realism. *Philosophy of Science* 49: 604–15.

Kripke, S.A. 1971. "Identity and Necessity." In Identity and Individuation, ed. M.K. Munitz. New York: New York University Press.

———. 1972. Naming and Necessity. In *The Semantics of Natural Language,* eds. D. Davidson and G. Harman. Dordrecht: D. Reidel.

Kuhn, T. 1970. *The Structure of Scientific Revolutions,* 2d ed. Chicago: University of Chicago Press.

Laudan, L. 1981. A Confutation of Convergent Realism. *Philosophy of Science* 48:218–49.

Mackie, J. L. 1974. *The Cement of the Universe.* Oxford: Oxford University Press.

Mayr, E. 1970. *Populations, Species and Evolution.* Cambridge: Harvard University Press.

Mc Mullin, E. 1984. "A Case for Scientific Realism." In *Scientific Realism,* ed. J. Leplin. Berkeley: University of California Press.

Miller, R. 1984a. *Analyzing Marx.* Princeton: Princeton University Press.

———. 1984b. Ways of Moral Learning. Philosophical Review 94:507–56.

Nagel, E. 1961. *The Structure of Science.* New York: Harcourt Brace.

Putnam, H. 1962. "The Analytic and the Synthetic." In *Minnesota Studies in the Philosophy of Science, vol. 3, Realism and Reason,* eds. H. Feigl and G. Maxwell. Minneapolis: University of Minnesota Press.

———. 1972. "Explanation and Reference." In *Conceptual Change,* eds. G. Pearce and P. Maynard. Dordrecht: Reidel.

———. 1975a. "The Meaning of Meaning." In *Mind, Language and Reality,* ed. H. Putnam. Cambridge: Cambridge University Press.

———. 1975b. "Language and Reality." In *Mind, Language and Reality.* Cambridge: Cambridge University Press.

———. 1983. "Vagueness and Alternative Logic." In *Realism and Reason.* Cambridge: Cambridge University Press.

Quine, W.V.0. 1969a. "Natural Kinds." *Ontological Relativity and Other Essays.* New York: Columbia University Press.

———. 1969b. "Epistemology Naturalized." In *Ontological Relativity and Other Essays.* New York: Columbia University Press.

Railton, P. 1986. Moral Realism. *Philosophical Review* 95:163–207.

Rawls, J. 1971. *A Theory of Justice.* Cambridge: Harvard University Press.

Shoemaker, S. 1980. "Causality and Properties." *Time and Cause,* In ed. P. van Inwagen. Dordrecht: D. Reidel.

Sturgeon, N. 1984a. "Moral Explanations." In *Morality, Reason and Truth,* eds. D. Copp and D. Zimmerman. Totowa, N.J.: Rowman and Allanheld.

——. 1984b. Review of *Moral Relativism* and *Virtues and Vices*, by P. Foot. Journal of Philosophy 81: 326–33.

Tarski, A. 1951. "The Concept of Truth in Formalized Languages." In *Logic, Semantics and Metamathematics.* New York: Oxford University Press.

Van Fraassen, B. 1980. *The Scientific Image.* Oxford: Oxford University Press.

Elliott Sober

Contrastive Empiricism

I

Despite what Hegel may have said, syntheses have not been very successful in philosophical theorizing. Typically, what happens when you combine a thesis and an antithesis is that you get a mishmash, or maybe just a contradiction. For example, in the philosophy of mathematics, formalism says that mathematical truths are true in virtue of the way we manipulate symbols. Mathematical Platonism, on the other hand, holds that mathematical statements are made true by abstract objects that exist outside of space and time. What would a synthesis of these positions look like? Marks on paper are one thing, Platonic forms another. Compromise may be a good idea in politics, but it looks like a bad one in philosophy.

With some trepidation, I propose in this paper to go against this sound advice. Realism and empiricism have always been contradictory tendencies in the philosophy of science. The view I will sketch is a synthesis, which I call Contrastive Empiricism. Realism and empiricism are incompatible, so a synthesis that merely conjoined them would be a contradiction. Rather, I propose to isolate important elements in each and show that they combine harmoniously. I will leave behind what I regard as confusions and excesses. The result, I hope, will be neither contradiction nor mishmash.

II

Empiricism is fundamentally a thesis about *experience*. It has two parts. First, there is the idea that experience is necessary. Second, there is the thesis that experience suffices. Necessary and sufficient for what? Usually this blank is filled in with something like: knowledge of the world outside the mind. I will set the

This paper is dedicated to the memories of Geoffrey Joseph and Joan Kung – two colleagues and friends from whom I learned a lot. Each influenced the way my views on scientific realism have evolved. I will miss them both.

issue of knowledge to one side and instead will focus on the idea that experience plays a certain role in providing us with justified beliefs about the external world. Never mind what the connection is between justified belief and knowledge.

These two parts of empiricism have fared quite differently in the past 200 years or so. The idea that experience is necessary has largely lapsed into a truism. No one thinks that a priori reflection all by itself could lead to reasonable science. Later, I'll identify a version of this necessity thesis that is more controversial.

The other thesis — that experience somehow suffices — has been slammed pretty hard, at least since Kant. Percepts without concepts are blind. Or as we like to say now, there is no such thing as an observation language that is entirely theory-neutral. Although positivists like the Carnap of the *Aufbau* tried to show that this empiricist thesis could be made plausible, it is now generally regarded as mistaken or confused.

One vague though suggestive metaphor for what empiricism has always aimed at is this: our knowledge cannot *go beyond* experience. Pending further clarification, it is unclear exactly how this idea should be understood. But the basic thrust of this idea has also come in for criticism. The standard point is that pretty much everything we believe about the external world goes beyond experience. Even a simple everyday claim about the commonsense characteristics of the physical objects in my environment goes beyond the experiences I have had or can ever hope to have. A consistent empiricism, so this familiar line of criticism maintains, ultimately leads to a solipsism of the present moment.

How should realism be understood? There are many realisms. Realism is often described as a thesis about what truth is or as a thesis about what is true. Neither of these is the realism I will address.

Realism as a view about the nature of truth is a semantical thesis; a realist interpretation of a set of sentences will claim that those sentences are true or false independently of human thought and language.[1] The sentences are said to describe a mind-independent reality and to depend for their truth values on it. The standard opponent of this semantical thesis has been verificationism, which either rejects the notion of truth or reinterprets it so that truth and falsity are said to depend on us in some way. This semantical issue will not concern me further. The issue between realism and empiricism that I want to examine concedes that truth is to be understood realistically.

Realism is sometimes described as a thesis about how we should interpret the best scientific theories we now have. We should regard them as true and not simply as useful predictive devices that tell us nothing about an unobservable reality. There really are genes and quarks, so this sort of realist says.

Putnam (1978) has challenged this realist position by claiming that our present theories will probably go the way of all previous theories — future science will find them to be false. He uses this inductive argument to say that we are naive if we regard current science as true. Realism of this sort, he claims, is predicated on the unscientific expectation that the future will not resemble the past.

Putnam's argument strikes me as overstated. I don't think that *all* previous scientific theories have been found to be false in every detail. Rather, historical change has preserved some elements and abandoned others. Nevertheless, I think his skepticism about labeling all our best current theories as true is well taken. A realist in my sense may decline to say that this or that present theory is true.

Realism, in the sense at issue here, is not a thesis about what truth is; nor is it a thesis about what is true. Rather, it is a thesis about the goals of science. This is the realism that Van Fraassen (1980) singled out for criticism. Science properly aims to identify true theories about the world. Realists may refuse to assert that this or that current theory is true, though they perhaps will want to say that some theories are our current best guesses as to what is true.

What would it be to reject this thesis about the proper goal of science? Empiricism, in Van Fraassen's sense, holds that the goal of science is to say which theories are empirically adequate. Roughly, empirical adequacy consists in making predictions that are borne out in experience.

How could the search for truth and the search for empirical adequacy constitute distinct goals? Consider two theories that are *empirically equivalent*, but which are not merely notational variants of each other. Though they disagree about unobservables, they have precisely the same consequences for what our experience will be like. Not only are the theories both consistent with all the observations actually obtained to date; in addition they do not disagree about any possible observation. A realist will think that it is an appropriate scientific question to ask which theory is true; an empiricist will deny that science can or should decide this question.

The idea of empirical equivalence has had a long history. Descartes's evil-demon hypothesis was constructed to be empirically equivalent with what I'll call "normal" hypotheses describing the physical constitution of the world outside the mind. In the last hundred years, the idea of empirical equivalence has played a central role in the philosophy of physics. Mach, Poincaré, Reichenbach, and their intellectual heirs have used this idea to press foundational questions about the geometry of space and about the existence of absolute space.

In problems of this sort, realists appeal to criteria that discriminate between the two competing hypotheses and claim that those criteria are scientifically legitimate. Perhaps we should reject the evil-demon hypothesis and accept the normal hypothesis because the latter is more parsimonious, or because the former postulates the existence of an unverifiable entity. The same has been said about the existence of absolute space and the existence of universal forces, which Reichenbach (1958) introduced to play the role of an evil demon in the problem of geometric conventionalism. Realists argue that criteria of this sort provide legitimate grounds for claiming that some theories are true and others are false. Empiricists disagree, arguing that these criteria are merely "aesthetic" or "pragmatic" and should not be taken as a ground for attributing truth values.[2]

I mention parsimony and verifiability simply as examples. Realists may choose to describe the criteria they wish to invoke in an entirely different way. The point is that realists claim that scientific reasoning is *powerful* in a way that empiricists deny.[3]

III

The development of empiricism has been guided by the following conditional: If our knowledge cannot go beyond experience, then it should be possible to delimit (i) a set of *propositions* that can be known and (ii) a set of *methods* that are legitimate for inferring what is true. For this reason empiricists have felt compelled (i) to draw a distinction between observation statements and theoretical ones and (ii) to develop a picture of the scientific method whereby the truth of theoretical statements can never be inferred from a set of observational premises.

The result is generally thought to have been a two-part disaster. The theory/observation distinction has been drawn in different ways. But each of them, I think, has either been too vague to be useful, or, if clear, has been epistemologically arbitrary. Maxwell (1962) and Hempel (1965), among many others, asked why the size of an object should determine whether it is possible to obtain reasonable knowledge about it. Apple seeds are observable by the naked eye, but genes are not. Hempel asked "So what?"—a question that empiricism's critics have continued to press.

Empiricist theories of inference have fared no better. If empiricists are to block theoretical conclusions from being drawn from observational premises, they must narrowly limit the rules of inference that science is permitted to use. Deduction receives the empiricist seal of approval, and maybe restricted forms of "simple induction" will do so as well. But there are scientific arguments from observational premises to observational conclusions that do not conform to such narrow strictures. Rather, they seem to require something philosophers have liked to call "abduction"—inference to the best explanation.[4] However, once these are admitted to the empiricist's organon of methods, the empiricist's strictures dissolve. The point is that inference to the best explanation also seems to allow theoretical conclusions to be drawn from observational premises. This now-standard argument against empiricism recurs so often that it deserves a name; I call it *the garden-path argument*.

These familiar problems affect Van Fraassen's (1980) constructive empiricism just as much as they plague earlier empiricisms. Van Fraassen says that it is appropriate for science to reach a verdict on the truth value of statements that are strictly about observable entities. But when a statement says something about unobservable entities, no conclusion about its truth value should be drawn. In this case, the scientist should consider only whether the statement is empirically adequate.

Van Fraassen takes various facts about our biology to delimit what sorts of entities are observable. Observable means observable *by us*. But the question then arises as to why no legitimate forms of scientific inference can take us from premises about observables to conclusions about unobservables. As with previous empiricisms, constructive empiricism seems to impose an arbitrary limit on the kinds of inferences it deems legitimate.

There is an additional problem with Van Fraassen's approach. It concerns the concept of aboutness. The appropriate scientific attitude we should take to a statement is said to depend on what that statement is about. But what is aboutness? I see no reason to deny that the statement "All apples are green" is about everything in the universe; it says that every object is green if it is an apple. If the universe contains unobservable entities, then the generalization is about them as well. Pending some alternative interpretation of "aboutness," constructive empiricism seems to say that science should not form opinions about the truth value of any generalization (Musgrave 1985, 208; Sober 1985).

Modern empiricism has frequently been plagued by semantical problems. Carnap tried to divide theoretical from observational statements by a verificationist theory of meaning. Van Fraassen abandons this empiricist semantics, but his theory is undermined by a semantical difficulty all the same. It is aboutness, not verificationism, that causes the problem.

Realism appears strongest when it deploys criticisms of empiricism of the kinds just mentioned. The best defense is a good offense. But when one looks at the positive arguments that realists have advanced, their position looks more vulnerable. Indeed, the problems become most glaring when the issue of empirically equivalent theories is brought to the fore.

Before Putnam lapsed from the realist straight and narrow, he sketched an argument for realism that struck many philosophers as very powerful. It is encapsulated in Putnam's (1975) remark that "realism is the only philosophy that doesn't make the success of science a miracle." I now want to consider this miracle argument for realism.

The idea is this. Suppose a theory *T* is quite accurate in the predictions it generates. This is something on which the realist and the empiricist can agree. The question is then: *Why* is the theory successful? What explains the theory's predictive accuracy?

The miracle argument seeks to show that the hypothesis that the theory is true (or approximately true) is the best explanation of why the theory is predictively accurate. If the theory postulates unobservables, then the theory's predictive success is best explained by the hypothesis that the entities postulated by the theory really exist and their characteristics are roughly as the theory says they are. Here is an example of an abductive argument that leads from observational premises to a nonobservational conclusion. The empiricist must block this argument.

Fine (1984, 84–85) has argued that the realist cannot employ abductive arguments of this sort, since they are question begging. The issue, he says, is precisely whether abduction is legitimate. Boyd (1984, 67) rejects this criticism. He claims that scientists use abductive arguments, and so it is quite permissible for a philosopher to use abduction to defend a philosophical thesis about science.

My assessment of the miracle argument differs from both Fine's and Boyd's. I have no quarrel with philosophical abductive arguments, as long as they conform to the standards used in science. The problem with the miracle argument is not that it is abductive, but that it is a very weak abductive argument.

When scientists wish to assess the credentials of an explanatory hypothesis, a fundamental question will be: What are the alternative hypotheses that compete with the one in which you are really interested? This is the idea that theory testing is a contrastive activity. To test a theory T is to test it against at least one competing theory T'.

The miracle argument fails to specify what the set of competing hypotheses is supposed to be. The hypothesis of interest is that T is true or approximately true in its description of unobservables. If the problem is to choose between T and T', where T' is a theory that is not predictively equivalent with T, then the miracle argument might make sense. That is, if the choice is between the following two conjectures, there clearly can be good scientific evidence favoring the first:

(*One*) T is true or approximately true.

(*Two*) T' is true or approximately true.

But if I vary the contrasting alternative, matters change. What scientific evidence can be offered for favoring hypothesis (*One*) over the following competitor:

(*Three*) T is empirically adequate, though false.

If (*Three*) were true, it would not be surprising that T is predictively successful.

So what becomes of the thesis that realism is the only hypothesis that doesn't make the success of science a miracle? Strictly speaking, it is false. A realist interpretation of the theory T is given by (*One*); if true, it would explain what we observe—that T is predictively successful. But the same holds of (*Three*); if it were true, that also would explain the predictive success of T.

Notice that my critique of the miracle argument does not proceed by artificially limiting science to a discussion of observables. Nor does it reject the legitimacy of abduction. Both T and T' may talk about unobservables; and the choice between hypotheses (*One*) and (*Two*) may be an unproblematic case of inference to the best explanation. The criticism just sketched differs from the empiricist's standard position. Rather, it is characteristic of the view that I call Contrastive Empiricism.

It may be objected that hypothesis (*Three*) is really no explanation at all of the predictive success that *T* has enjoyed. If Holmes finds a corpse outside of 221B Baker Street, the hypothesis that Moriarty is the murderer is one possible explanation. But is it an explanation to assert that Moriarty is innocent, though the crime looks just as it would have if Moriarty had done the dirty deed? Does this remark explain why the murder took place?

This question is a subtle one for the theory of explanation. Perhaps there are occasions in which saying that *T* is not the explanation may itself be an explanation; perhaps not. What I wish to argue is that this point is irrelevant to the issue of whether the miracle argument is successful. The question before us is first and foremost a question about *confirmation*. We want to know whether the predictive accuracy of theory *T* is good evidence that *T* is true or approximately true. The issue of explanation matters here only insofar as explanation affects confirmation. My view is that loose talk about abduction has brought these two ideas closer together than they deserve to be.

I'll use Bayes's theorem to illustrate what I have in mind. This theorem says that the probability, $Pr(H/O)$, of a hypothesis (H) in the light of the observations (O) is a function of three other probabilities:

$$Pr(H/O) = Pr(O/H)Pr(H)/Pr(O).$$

We wish to compare the probability of hypothesis (*One*) and hypothesis (*Three*), given that theory *T* has been predictively successful (O). (*One*) is more probable than (*Three*), in the light of this observation, precisely when:

$$Pr(O/One)Pr(One) > Pr(O/Three)Pr(Three).$$

The conditional probabilities in this last expression are called *likelihoods*: the likelihood of a hypothesis is the probability it confers on the observations. Don't confuse this quantity with the hypothesis's posterior probability, which is the probability that the observations confer on the hypothesis. So whether the above inequality is true depends on the likelihoods and the prior probabilities of hypotheses (*One*) and (*Three*).

In this Bayesian format,[5] it is likelihood that represents the ability of the hypothesis to explain the observations. The question of whether the hypothesis explains the observations is interpreted to mean: how probable are the observations, if the hypothesis is true? A hypothesis with a small likelihood says that the observations are very improbable—that it is almost a "miracle," so to speak, that they occurred. Understood in this way, hypotheses (*One*) and (*Three*) are equally explanatory, since they confer the same probability on the observations.

It may be replied, with some justice, that likelihood does not fully capture the idea of explanatory power. Indeed, there are reasons independent of the problem of comparing hypotheses (*One*) and (*Three*) for thinking this. For example, two correlated effects of a common cause may make each other quite probable; given the presence of one effect, it may be more likely to infer that the other is present than that it is absent. Yet neither of the correlates explains the other. All this may be true, but my question, then, is this: When explanatory power diverges from likelihood, why think that explanatory power is relevant to confirmation? In the present case, let us grant that (*One*) is more explanatory than (*Three*). Why is this evidence that (*One*) is more plausible than (*Three*)?

Again, I want to emphasize that my criticism does not reject the idea of inference to the best explanation. Theoretical hypotheses about unobservables – like (*One*) and (*Two*) – have likelihoods. Inference to the best explanation should take those likelihoods into account. The problem, though, is that empirically equivalent theories have identical likelihoods.

I am reluctant to take "explanatory power" as an unanalyzed primitive that conveniently has just the characteristics that realists need if they are to justify their pet discriminations between pairs of empirically equivalent theories. Perhaps a non-Bayesian confirmation theory can make good on this realist idea. I don't know of any proposal that does the trick. So I am reluctant to allow the miracle argument to proceed as a resolution of the problem of choosing between (*One*) and (*Three*).

In the Bayesian inequality stated before, there are other elements besides likelihoods. In addition, there are the prior probabilities of hypotheses (*One*) and (*Three*). If (*One*) were a priori more probable than (*Three*), that would help the realist, although it would be wrong to say that the empirical accuracy of theory *T* was doing any work. If the likelihoods are the same, then the observation is idle. Perhaps we shouldn't call this the miracle argument at all; it isn't that realism is a better explanation of what we observe. Rather, the idea now is that a realist interpretation of a theory is a priori more probable than the alternative.

I am at a loss to see how this idea can be parlayed into a convincing argument for realism. What do these prior probabilities mean? If they are just subjective degrees of belief that some agent happens to assign, we are simply saying that this agent favors realism before any observations have been made. This is hardly an argument for realism, since another agent could have just the opposite inclination.

If prior probabilities are to be used in an argument, it must be shown why hypothesis (*One*) *should* be assigned a higher prior than (*Three*). I know of no way of doing this, though perhaps an a priori argument for realism will someday be invented. Let me emphasize, however, that this is worlds away from Putnam's a posteriori miracle argument. That argument is defective, if explanatory power goes by likelihoods; or it is entirely unclear what the argument says, if explanatory power is to be understood in some other way.[6]

I began this section by rehearsing the standard criticism that empiricism takes an overly narrow view of the scope and limits of scientific inference. Conclusions about unobservables can be blocked only by drastically restricting inferences in a way that seems entirely artificial. But the present discussion of the miracle argument suggests that realism errs in the opposite direction. The idea of inference to the best explanation presupposed by the miracle argument licenses too much, if a roughly Bayesian idea of confirmation is used.

The empiricist wants to show that there is an important sense in which our knowledge cannot go beyond experience. The realist wants to show that our ability to know about quarks is every bit as strong as our ability to find out about tables. The empiricist idea runs into trouble when it artificially limits the power of scientific inference. The realist idea runs into trouble when it artificially inflates that power. It now is time to see how the defensible kernel of each position can be formulated as a single position, one that avoids the excesses of each.

IV

I mentioned before that theory testing is a contrastive activity. If you want to test a theory T, you must specify a range of alternative theories — you must say what you want to test T *against*.

There is a trivial reading of this thesis that I do not intend. To find out if T is plausible is simply to find out if T is more plausible than *not-T*. I have something more in mind: there are various contrasting alternatives that might be considered. If T is to be tested against T', one set of observations may be pertinent; but if T is to be tested against T'', a different set of observations may be needed. By varying the contrasting alternatives, we formulate genuinely different testing problems.

An analogous point has been made about the idea of explanation.[7] To explain why a proposition P is true, we must explain why P, rather than some contrasting alternative Q, is true (Dretske 1973; Garfinkel 1981; Van Fraassen 1980; Sober 1986; but see Salmon 1984 for criticisms). This thesis is nontrivial, since varying the contrasting proposition Q poses different explanatory problems.

A nice example, due to Garfinkel, concerns the bank robber Willi Sutton. A priest once asked Willi why he, Willi, robbed banks. Willi answered that that was where the money was. The priest wanted to know why Willi robbed rather than not robbing. Willi took the question to be why he robbed banks rather than candy stores.

The choice of a contrasting alternative helps delimit what sort of propositions may be inserted into an answer to a why-question. If you ask why Willi robbed banks rather than candy stores, you may include in your answer the assumption that Willi was going to rob something. However, if you ask why Willi indulged in robbing instead of avoiding that activity, you cannot include the assumption that Willi was going to rob something.

Why is it legitimate to insert the statement that Willi was going to rob something into the answer to the first question, but not into the second? Consider the question "Why P rather than Q?" I claim that statements implied by both P and Q are insertable. The question presupposes the truth of such statements, so they may be assumed in the answer. On the other hand, the implications that P has that Q does not cannot be inserted into an answer. It is matters such as these that are at issue, so assuming them in the answer would be question begging.

I have described a sufficient condition for insertability and a sufficient condition for noninsertability. I will not propose a complete account by specifying a necessary and sufficient condition. The modest point of importance here is that the formulation of an explanatory why-question often excludes a statement from being inserted into an answer.

I turn now from explanation to confirmation. Instead of asking "Why P rather than Q?" I want to consider the question "Why *think* P rather than Q?" This is a request for evidence or for an argument showing that P is more plausible than Q. I want to claim here that confirmatory why-questions often exclude some statements from being insertable.

A statement S will not be insertable into an answer to such questions, if it is not possible to know that S is true without already knowing that P is more plausible than Q. What is requested by the question is an independent reason, not a question-begging one.

As in the case of explanation, a statement may be insertable into the answer to one confirmatory why-question without being insertable into the answer to another. One way to change a question so that an answer insertable before is no longer so, is by a procedure I'll call *absorption*. Suppose I ask why Willi led a life of crime rather than going straight? You might answer by citing Willi's tormented adolescence. But if I absorb this answer into the question, I thereby obtain a new question, which cannot be answered by your previous remark. Suppose I ask: Why did Willi have a tormented adolescence and then lead a life of crime, as opposed to having an idyllic adolescence followed by a law-abiding adulthood? The assertion that Willi had a tormented adolescence is not insertable into an answer to this new question.

It is just this strategy of absorption that philosophers have used to generate skeptical puzzles about empirically equivalent theories. The question "Why think P rather than Q?" may have O as an answer. But O cannot be inserted into an answer to a new question: "Why think that P and O are true, rather than Q and *not-O*?" Reichenbach (1958) argued that if I assume a normal physics devoid of universal forces, I can develop experimental evidence for thinking that space is non-Euclidean rather than Euclidean. But if I absorb the assumptions about a normal physics into my question, I obtain a new question that is not, according to Reichenbach, empirically decidable: the reason, he claimed, is that the conjunction of the non-Euclidean hypothesis and normal physics is empirically equivalent to the conjunction of the Euclidean hypothesis and a physics that postulates universal forces.

As noted earlier, empiricists have always maintained that it is not possible to say that one theory has a better claim to be regarded as true than another, if the two are empirically equivalent. Such discriminatory why-questions, the empiricist claims, are unanswerable.

Empiricists have defended this view by claiming that there is a privileged set of statements – formulated in the so-called observation language. Sentences in this special class were supposed to have the following feature – ascertaining their truth required no theoretical information whatever. The empiricist claim then was made that a discrimination can be made between two theories only if they make different predictions about what will be true in this class of sentences.

The standard criticism of this idea was developed by claiming that the distinction between theory and observation is not absolute. The very statements that on some occasions provide independent answers to confirmatory why-questions at other times provide only question-begging answers. A statement that counts as an observation statement in one context can become the hypothesis under test in another. The observation/theoretical distinction, so this criticism of empiricism maintained, is context relative and pragmatic.

The standard empiricist claim about observation has a quantifier order worth noting:

(EA)   There exists a set of observation statements, such that, for any two theories T and T', if it is possible to say that T is more plausible than T', then this will be because T and T' make incompatible predictions as to which members of that set are true.

A weaker thesis, which avoids an absolute distinction between theory and observation, has a different quantifier order:

(AE)   For any two theories T and T', if it is possible to say that T is more plausible than T', then this will be because there exists a set of observation statements such that T and T' make incompatible predictions as to which members of that set are true.

(EA) is committed to an absolute distinction between theory and observation; the required distinction is absolute because what counts as an observation statement is invariant over the set of testing problems. (AE) is not committed to this thesis, because it is compatible with the idea that what counts as an observation is relative to the testing problem considered. Contrastive Empiricism maintains that (AE), rather than the stronger thesis (EA), is correct. What counts as an observation in a given test situation should provide non-question-begging evidence for discriminating between the competing hypotheses.

Contrastive Empiricism makes use of the concept of an observation, as does the very formulation of the problem of empirically equivalent theories, which, recall, is a problem that both realists and empiricists want to solve. I have already mentioned that what counts as an observation may vary from one testing problem to another. But more must be said about what an observation is. I won't attempt to fully clarify this concept here, but again, will content myself with a sufficient condition for empirical equivalence, one with a Quinean cast. Two theories are empirically equivalent if the one predicts the same physical stimulations to an agent's sensory surfaces as the other one does. Observational equivalence is vouchsafed by identity of sensory imput.

Both (*EA*) and (*AE*) mark the special role of experience in terms of a partition of propositions. The scientist testing a pair of theories is supposed to be able to identify a class of sentences in which the so-called observation reports can be formulated. But the empiricist's point about empirically equivalent hypotheses can be made in a quite different way. Consider an analogy: When your telephone rings, that is evidence that someone has dialed your number. But the ringing of the phone when I dial your number is physically indistinguishable from the ringing that would occur if anyone else did the dialing. This is an empirical truth that can be substantiated by investigating the physical channel. The proximal state fails to uniquely determine its distal cause.[8] I don't need to invoke a special class of protocol statements and claim that they have some special epistemological status to make this simple point. Still less does the *telephone* need to be able to isolate a special class of sentences in which it can record its own physical state.

The idea of empirically equivalent hypotheses is parallel, though, of course, more general. It is basically the idea that the proximal state of the whole sentient organism, both now and in the future, fails to uniquely determine its distal cause. Whether two hypotheses are empirically equivalent is a question about the sensory channels by which distal causes can have proximal experiential effects. What engineers can tell us about telephones, psychologists will eventually be able to tell us about human beings. I don't think that the idea of empirical equivalence requires any untenable dualisms.[9]

The main departure of this "engineering" approach to the concept of empirical equivalence from earlier "linguistic‘ formulations is this: In the earlier version, the scientist is viewed as thinking about the world by deploying a certain *language*. The idea of empirical equivalence is then introduced by identifying a set of sentences within that very language; two theories are then said to be empirically equivalent if they make the same predictions concerning the truth of sentences in the privileged class. In the engineering version, we can talk about two theories being empirically equivalent for a given organism (or device) without supposing that the theories are formulable within the organism's language and without supposing that the organism has a language within which the experiential content of the observation is represented without theoretical contamination.[10] It's the sensory state of the organism that matters for the engineering concept, not some special class of statements that the organism formulates.

It is not to be denied that the theories that scientists standardly wish to test do not, by themselves, imply anything about the observations they will make. If neither of two theories has observational implications, then it is only in an uninteresting and vacuous sense that they are empirically equivalent. But what cannot be said of the part can be said of the whole. I take it that two, perhaps large, conjunctions of theoretical claims (including what philosophers like to call auxiliary assumptions) can have observational implications. And what is more, it sometimes can happen that two largish conjunctions can be empirically equivalent. This, I think, is what Descartes wanted to consider when he formulated his evil-demon hypothesis and what Reichenbach had in mind by his conjunction of a physics of universal forces and a geometry. Maybe these hypotheses were short on details, but I do not doubt that there are pairs of empirically equivalent theories. It is about such pairs that empiricism and realism disagree.[11]

The main departure that Contrastive Empiricism makes from previous Empiricisms, including both Logical Empiricism and Constructive Empiricism, is that it is about *problems*, not *propositions*. Previous empiricisms, as I've said, have tried to discriminate one set of statements from another. Van Fraassen, like earlier empiricists, wants to say that science ought to treat some statements differently from others. Contrastive empiricism draws no such distinction. Rather, it states that science is not in the business of discriminating between empirically equivalent hypotheses.

For example, previous empiricisms have wanted to identify a difference between the following two sentences:

(*XI*)   There is a printed page before me.

(*YI*)   Space-time is curved.

I draw no such distinction between these *propositions*. Rather, my suggestion is that there is an important similarity between two *problems*. There is the problem of discriminating between (*XI*) and (*X2*). And there is the problem of discriminating between (*YI*) and (*Y2*):

(*X2*)   There is no printed page before me; rather, an evil demon makes it seem as if there is one there.

(*Y2*)   Space-time is not curved; rather, a universal force makes it seem as if it is curved.

According to Constrastive Empiricism, neither of these problems (when formulated with due care) is soluble.

Although Contrastive Empiricism embodies one part of the empiricist view that knowledge cannot go beyond experience, there is nonetheless an important realist element in this view. Hypotheses about the curvature of space-time may be as testable as hypotheses about one's familiar everyday surroundings. (*X1*) can be distinguished from a variety of empirically nonequivalent alternatives, by familiar sensory means. (*Y1*) can be distinguished from a variety of empirically nonequivalent alternatives, by more recondite, though no less legitimate, theoretical means.

Constrastive Empiricism gives abduction its due. But when the explanations under test are empirically equivalent, it concedes that no difference in likelihood will be found. If we use a rough Bayesian format and claim that there is nonetheless a difference in plausibility between (*X1*) and *(X2)*, or between (*Y1*) and (*Y2*), we therefore must be willing to say that there is a difference in priors. But where could this difference come from? Contrastive Empiricism claims that no such difference can be defended.

In less philosophically weighty problems of Bayesian inference, two hypotheses may have identical likelihoods, but differ in their prior probabilities for reasons that can be defended by appeal to experience. To use an old standby, if I sample at random from emeralds that exist in 1988 and find that each is green, then, relative to this observation, the following two hypotheses have identical likelihoods:

(*H1*)   All emeralds are green.

(*H2*)   All emeralds are green until the year 2000, but after that they turn blue.

In spite of this, I may have an empirical theory about minerals (developed before I examined even one emerald) that tells me that emerald color is very probably stable. This theory allows me to assign (*H1*) a higher prior than (*H2*).[12]

Contrastive Empiricism is not the truism that the likelihoods of empirically equivalent theories do not differ. Rather, it additionally claims that no defensible reason can be given for assigning emprically equivalent theories different priors. A pair of empirically equivalent hypotheses differs from the (*H1*)-(*H2*) pair in just this respect.

This is not to deny that human beings look askance at evil demons and their ilk. We do assume that they are implausible. In a sense, we assign them very low priors, so that even when their likelihoods are as high as the likelihoods of more "normal" sounding hypotheses, we still can say that normal hypotheses are more probable than bizarre evil-demon hypotheses in the light of what we observe. This is how we are, to be sure. But I cannot see a rational justification for thinking about the world in this way. I cannot see that we have any non-question-begging evidence on this issue. Maybe Hume was right that the combination of naturalism and skepticism has much to recommend it.

What does Contrastive Empiricism say about the principles that realists have liked to emphasize? Can't we appeal to simplicity and parsimony as reasons for rejecting evil demons and the like? Won't such considerations count as objective, since they also figure in more mundane hypothesis testing, where the candidates are not empirically equivalent? Simplicity seems to favor (*H1*) over (*H2*) when both are consistent with the observations; so won't simplicity also favor (*X1*) over (*X2*) and (*Y1*) over (*Y2*)? Here the "garden-path argument" threatens to undermine Contrastive Empiricism. If appeals to parsimony/simplicity are permissible when the problem is to discriminate between empirically *non*equivalent hypotheses, how can such appeals be illegitimate when the problem is to discriminate between hypotheses that are empirically equivalent?

Space does not permit me to discuss this issue very much. My view is that philosophers have hypostatized the principle of parsimony. There is no such thing. Rather, I think that when scientists appeal to parsimony, they are making specific background assumptions about the inference problem at hand. There is no abstract and general principle of parsimony, which spans all scientific disciplines like some abductive analog to *modus ponens*. When scientists draw a smooth curve through data points, to use a standard example, they do not do this because smooth curves are simpler than bumpy ones; rather, their preference for curves in one class rather than another rests on specific assumptions about the kind of process they are modeling.

Let me give an example that illustrates what I have in mind. Charles Lyell defended the idea of uniformitarianism in geology. He argued for this view by claiming that a principle of uniformity was a first principle of scientific inference, and that his opponents were not being scientific. However, if you look carefully at what Lyell was doing, you will see that "uniformitarianism" was a very specific theory about the Earth's history. Considered as a substantive doctrine, it is simply not true that uniformitarianism's rivals must be in violation of any first principle of scientific inference (Rudwick 1970; Gould 1985). On the other hand, if one abstracts away from the geological subject matter in the hope of identifying a suitably presuppositionless principle of simplicity or uniformity, what one obtains is a principle that has no implications whatever about whether Lyell's theory was more plausible than the alternatives.

Other examples of this sort could be enumerated. It has recently been popular for biologists to argue that group selection hypotheses should be rejected because they are unparsimonious. Those arguments, I think, are either totally without merit, or implicitly assume that the preconditions for certain kinds of evolutionary processes are rarely satisfied in nature. If parsimony is just abstract numerology, it is meaningless; if it really joins the issue, it does so by making an empirical claim about how evolution proceeds (Sober 1984). The evolutionary problem of phylogenetic inference affords another example: it is an influential biological idea that parsimony can be justified as a principle of phylogenetic inference without requiring any substantive assumptions about how the evolutionary process proceeds. Again, I think this view is mistaken; see Sober (1988) for details.

I grant that a few examples do not a general argument make. I also grant that the three examples I have cited do not involve choosing between empirically equivalent theories. Could one grant my point that appeals to parsimony and simplicity involve contingent assumptions when the competing hypotheses are not empirically equivalent, but maintain that parsimony and simplicity are entirely presuppositionless when the choice is between empirically equivalent theories?

I find this view implausible. It also strikes me as pie-in-the-sky. Within a broadly Bayesian framework, it seems clear that prior probablities are not obtainable a priori.[13] If a plausible non-Bayesian confirmation theory can be developed that says differently, I would like to see it. Although I grant that our understanding of nondeductive inference is far from complete, I simply do not believe that the kind of confirmation theory that realism requires will be forthcoming.[14]

In "Empiricism, Semantics, and Ontology," Carnap (1950) introduced a distinction between internal and external *questions*, which he spelled out by distinguishing one class of *propositions* from another. Quine (1951) and others took issue with this absolute theory/observation distinction, and the rest is history. With Carnap, I believe that the idea that there are two kinds of questions is right; but unlike Carnap, I do not think this notion requires a verificationist semantics or an absolute distinction between observational propositions and theoretical ones.

Contrastive Empiricism reconciles the realist idea that we can have knowledge about unobservables with the empiricist idea that knowledge cannot go beyond experience. The view derives its realist credentials from the fact that it imposes no restrictions on the vocabulary that may figure in testable *propositions*; but it retains an important empiricist element in its claim that science cannot solve discrimination *problems* in which experience makes no difference. Again, the chief innovation of this version of empiricism is its focus on problems, not propositions.[15]

Whether Contrastive Empiricism is more plausible than the thesis and antithesis from which it is fashioned turns on epistemological issues that I have not been able to fully address here. I hope, however, to have at least put a new contrastive why-question on the table: the debate between realism and previous empiricisms — whether of the Logical or the Constructive variety — needs to be enlarged. Detailed work on the theory of hypothesis testing will show whether Contrastive Empiricism is more plausible than the philosophical hypotheses with which it competes.

Notes

1. It is the notion of independence deployed in this realist thesis that I try to clarify in Sober (1982).

2. Reichenbach (1938) is a classic example of this empiricist position.

3. I construe realism and empiricism as theses about how theories should be judged for their plausibility; neither thesis is committed to the claim that scientists *accept* and *reject* the hypotheses they assess. For one view of this controversial matter, see Jeffrey (1956).

4. An example: when biologists argue that the current distribution of living things is evidence that the continents probably were in contact long ago, the argument is not an induction from a sample to a containing population. Biologists did not survey a set of similar planets and see that continental drift accompanied all or most biogeographic distributions of a certain kind, and then conclude that the biogeographic distribution observed here on Earth was probably due to continental drift as well. The inference goes from an observed effect to an unobserved cause. Although the hypothesis that the continents drifted apart is not known by "direct" observation, empiricists nonetheless count it as an observation statement; if we had been present and had waited around for long enough, we could have observed continental drift. So inferences with observational conclusions often require a mode of inference that is neither deductive nor inductive.

5. It will become clear later that this approach is Bayesian only in the sense that it uses Bayes's theorem; the more distinctively Bayesian idea that hypotheses always have prior probabilities is not part of what I have in mind here.

6. Boyd (1980; 1984) advances a form of the miracle argument in which the fact to be explained is the reliability of the scientific method, rather than, as here, the reliability of a given theory. Boyd's version might be viewed as a diachronic analog of the synchronic argument I have discussed. My view is that the diachronic argument faces basically the same difficulties as the synchronic version.

7. The importance of contrasting alternatives has also been explored in connection with the problem of defining what knowledge is; see Johnsen (1987) for discussion and references to the literature.

8. Of course, a probabilistic formulation can be given to this idea: The probability of the phone's sounding a certain way, given that I dial your number, is precisely the same as the probability of its sounding that way if someone else does the dialing.

9. Van Fraassen (1980) rightly emphasizes that whether something is observable is a matter for science, not armchair philosophy, to settle. However, Van Fraassen also claims that an entity whose detection requires instrumentation should not count as "observable;" with Wilson (1984), I find this restriction arbitrary, however much it may accord with ordinary usage (Sober 1985).

10. Skyrms (1984, 117) similarly argues that what counts as an "observation might have a precise description not at the level of the language of our conscious thought but only at the level of the language of the optic nerve."

11. Although Wilson (1984) is properly skeptical about the empirical equivalence of some theory pairs that philosophers have taken to be related in this way, he nonetheless grants that there are such pairs; he cites results due to Glymour and Malament as providing cases in point.

12. There is a more global form of inductive skepticism that blocks this way of discriminating between (*H1*) and (*H2*). If *all* empirical propositions — except those about one's current observations and memory traces — are called into question, what grounds are there for preferring (*H1*) to (*H2*)? This is the "theoretically barren" context in which Hempel (1965) posed his raven paradox. My view is that this skeptical challenge cannot be answered — only in the context of a background theory do observations have evidential meaning (Good 1967; Rosenkrantz 1977; Sober 1988).

13. But see Rosenkrantz (1977) for dissenting arguments.

14. This Bayesian approach to the conflict between realism and empiricism is very much in the spirit of Skyrms's (1984) pragmatic empiricism. Skyrms's main focus is on the idea of confirmation, which the Bayesian understands in terms of a comparison between the posterior and prior probabilities of a hypothesis; my focus has been on the idea of hypothesis testing, which is understood in terms of a comparison of two posterior probabilities. This difference in emphasis aside, we agree that empiricism should not be understood as a semantic thesis; nor should it claim that hypotheses about unobservables cannot be confirmed or tested.

15. Fine (1984) also proposes a compromise between realism and anti-realism, but not, I think, the one broached here. Fine sees the realist and the anti-realist as both "accepting the results of science." The realist augments this core position with a substantive theory of truth as correspondence, whereas the anti-realist goes beyond the core with a reductive analysis of truth, or in some other way.

Fine's idea is to retain the core and reject both sorts of proposals for augmenting it. In contrast, the opposition between empiricism and realism described in the present paper does not concern the notion of truth. What is more, the realism and empiricism with which I am concerned do not in any univocal sense 'accept the results of science," since realism claims that these results include discriminations between empirically equivalent theories, whereas the empiricist denies this.

References

Boyd, R. 1980. Scientific Realism and Naturalistic Epistemology. In *PSA 1980*, vol. 2, eds. P. Asquith and R. Giere. East Lansing, Mich.: Philosophy of Science Association.

———. 1984. "The Current Status of Scientific Realism." In *Scientific Realism*, ed. J. Leplin. Berkeley: University of California Press, 41–82.

Carnap, R. 1950. Empiricism, Semantics, and Ontology. *Revue Internationale de Philosophie* 4: 20–40. Reprinted in *Meaning and Necessity*, Chicago: University of Chicago Press, 1956.

Dretske, F. 1973. Contrastive Statements. *Philosophical Review* 81: 411–37.

Fine, A. 1984. "The Natural Ontological Attitude." In *Scientific Realism*, ed. J. Leplin. Berkeley: University of California Press, 83–107.

Garfinkel, A. 1981. *Forms of Explanation: Rethinking the Questions of Social Theory*. New Haven: Yale University Press.

Good, I. 1967. The White Shoe Is a Red Herring. *British Journal for the Philosophy of Science* 17: 322.

Gould, S. 1985. "False Premise, Good Science." In *The Flamingo's Smile*. New York: Norton, 126–38.

Hempel, C. 1965. *Philosophy of Natural Sciences*. Englewood Cliffs, N.J.: Prentice-Hall.

Jeffrey, R. 1956. Valuation and Acceptance of Scientific Hypotheses. *Philosophy of Science* 23: 237–46.

Johnsen, B. 1987. Relevant Alternatives and Demon Skepticism. *Journal of Philosophy* 84: 643–52.

Maxwell, G. 1962. "The Ontological Status of Theoretical Entities." In *Minnesota Studies in the Philosophy of Science*, vol. 3, *Realism and Reason*, eds. H. Feigl and G. Maxwell, 3–27. Minneapolis: University of Minnesota Press.

Musgrave, A. 1985. "Realism versus Constructive Empiricism." In *Images of Science: Essays on Realism and Empiricism*, eds. P. Churchland and C. Hooker. Chicago: University of Chicago Press, 197–221.

Putnam, H. 1975. *Mind, Language, and Reality: Philosophical Papers*, vol. 2. Cambridge: Cambridge University Press.

———. 1978. "What is Realism?" In *Meaning and the Moral Sciences*. London: Routledge and Kegan Paul. Reprinted in *Scientific Realism*, ed. J. Leplin. Berkeley: University of California Press, 1984, 140–53.

Quine, W. 1951. "On Carnap's Views on Ontology." In *The Ways of Paradox and Other Essays*. New York: Random House, 1966.

Reichenbach, H. 1938. *Experience and Prediction*. Chicago: University of Chicago Press.

———. 1958. *The Philosophy of Space and Time*. New York: Dover.

Rosenkrantz, R. 1977. *Inference, Method, and Decision*. Dordrecht: Reidel.

Rudwick, M. 1970. The strategy of Lyell's *Principles of Geology*. *Isis* 61: 5–55.

Salmon, W. 1984. *Scientific Explanation and the Causal Structure of the World*. Princeton: Princeton University Press.

Skyrms, B. 1984. *Pragmatics and Empiricism*. New Haven: Yale University Press.

Sober, E. 1982. Realism and Independence. *Nous* 61: 369–86

———. 1984. *The Nature of Selection*. Cambridge: MIT Press.

——. 1985. Constructive Empiricism and the Problem of Aboutness. *British Journal for the Philosophy of Science* 36: 11–18.

——. 1986. Explanatory Presupposition. *Australasian Journal of Philosophy* 64: 143–9.

——. 1988. *Reconstructing the Past: Parsimony, Evolution, and Inference.* Cambridge: MIT Press.

Van Fraassen, B. 1980. *The Scientific Image.* Oxford: Oxford University Press.

Wilson, M. 1984. "What Can Theory Tell Us about Observation?" In *Images of Science,* eds. P. Churchland and C. Hooker. Chicago: University of Chicago Press, 222–44.

CONTRIBUTORS

Contributors

Richard Boyd is a professor of philosophy in the Sage School of Philosophy at Cornell University. He is the author of a long series of papers defending scientific realism, one of which appears in this volume. His other published works have been in the philosophy of mind, the philosophy of psychology, semantic theory, the theory of natural kinds, and metaethics.

Arthur Caplan is a professor of philosophy and member of the Center for Philosophy of Science at the University of Minnesota. He is also director of the Center for Biomedical Ethics and a professor of surgery at the University of Minnesota. Caplan received his Ph.D. in philosophy from Columbia University and has taught there and at the University of Pittsburgh. He has written extensively on topics in the philosophy of biology and medicine, medical ethics, and health policy. He is writing a book on the philosophy of medicine.

Paul M. Churchland is professor and chair of philosophy, a member of the Cognitive Science Faculty, and a member of the Institute for Neural Computation, at the University of California, San Diego. His research addresses epistemology and the philosophy of science, perception and the philosophy of mind, and computational neuroscience and connectionist AI. He has authored *Scientific Realism and the Plasticity of the Mind* (1979); *Matter and Consciousness* (1984); and *A Neurocomputational Perspective: The Nature of Mind and the Structure of Science* (1989). He also serves as president of the Society for Philosophy and Psychology.

Ellery Eells is an associate professor of philosophy at the University of Wisconsin, Madison. He received his Ph.D. at the University of California, Berkeley, in 1980. Eells is author of *Rational Decision and Causality* (1982) and a number of papers on decision theory, confirmation theory, probability, and causation. He has recently completed *Probabilistic Causality* (forthcoming from Cambridge University Press).

Adolf Grünbaum is Andrew Mellon Professor of Philosophy, research professor of psychiatry, and chairman of the Center for Philosophy of Science at the University of Pittsburgh. His writings deal with the philosophy of physics, the theory of scientific rationality, and the philosophy of psychiatry. His books include *Philosophical Problems of Space and Time* (2d ed., 1973) and *The Foundations of Psychoanalysis: A Philosophical Critique* (1984). He has contributed over two hundred articles to anthologies and to philosophical and scientific periodicals. Grünbaum is president of the American Philosophical Association (Eastern Division) and of the Philosophy of Science Association, a member of the American Academy of Arts and Sciences, a fellow of the American Association for the Advancement of Science, and a laureate of the international Academy of Humanism. In 1985, he delivered the Gifford Lectures in Scotland, as well as the Werner Heisenberg Lecture to the Bavarian Academy of Sciences in Munich. He received a 1985 Senior U.S. Scientist Humboldt Prize and the 1989 Fregene Prize (Rome, Italy).

Colin Howson is a senior lecturer in the Department of Philosophy, Logic, and Scientific Method at the London School of Economics, from which he graduated in 1967 with first class honours. He has taught at universities in both the United Kingdom and the United States, and is a member of the organizing committee of the British Society for the Philosophy of Science. Howson has published extensively in philosophy of science journals, edited *Method and Appraisal in the Physical Sciences* (1986), and is the author with Peter Urbach of *Scientific Reasoning: The Bayesian Approach* (1989). Much of his recent work has been concerned with promoting the claim of personal probability to be regarded as the foundation of the logic of inductive inference.

Thomas S. Kuhn received his Ph.D. in theoretical physics from Harvard in 1949 and spent the next three years as a junior fellow in the Harvard Society of Fellows, learning to be a historian of that field. Since then he has taught history of science and philosophy of science at Harvard, Berkeley, and Princeton universities, and, since 1979, MIT, where he is Laurance S. Rockefeller Professor of Philosophy. As a historian, Kuhn has published on topics ranging from *The Copernican Revolution* (1957), his first book, to *Black–Body Theory and the Quantum Discontinuity* (1987), his most recent. He is best known, however, for a more theoretical, philosophical volume called *The Structure of Scientific Revolutions* (1970), and for an accompanying volume of essays, *The Essential Tension* (1979).

Henry Kyburg began his academic career as an engineer, with a degree in chemical engineering from Yale University. He did graduate work in philosophy at Columbia University, but on receiving his Ph.D. became an assistant professor of mathematics at Wesleyan University. He has worked on both induction and probability in philosophy of science and epistemology. He is author or editor of twelve books, ranging from the purely mathematical *Probability Theory* to the generally philosophical *Epistemology and Inference*. Kyburg's recent work has concerned the epistemic relation between theory and measurement (*Theory and Measurement*). It is perhaps his engineering background that leads him to take error as inevitable, and therefore to see an intimate connection between theoretical acceptance and observational error.

Larry Laudan is professor and chairman of philosophy at the University of Hawaii. He has written extensively on the problem of scientific change, including *Progress and Its Problems* (1977), *Science and Hypothesis*, and *Science and Values* (1984). He is completing a project that involves a critique of epistemological relativism, out of which his contribution to this volume has grown. That study will be published as *Science and Relativism*. He has taught at London, Pittsburgh, and Virginia Polytechnic Institute and has held visiting positions at Vienna, Konstanz, Melbourne, and Illinois-Chicago Circle.

Alan Nelson is assistant professor of philosophy at the University of California, Irvine. His articles on topics related to "Are Economic Kinds Natural?" have appeared in or will soon appear in *Philosophy of Science*, *Nous*, *Pacific Philosophical Quarterly*, *Ethics*, *Philosophy and Public Affairs*, *Erkenntnis*, and *Midwest Studies in Philosophy*. He is studying the debate about scientific realism and philosophical issues in seventeenth-century science.

Wesley C. Salmon is University Professor of Philosophy at the University of Pittsburgh. Among his books are *Scientific Explanation and the Causal Structure of the World* (1984), *Space, Time, and Motion: A Philosophical Introduction*, and *The Foundations of Scientific Inference* (1967). He edited *Hans Reichenbach: Logical Empiricist* (1979) and *Zeno's Paradoxes*. He is co-editor, with Philip Kitcher, of *Scientific Explanation*, volume 13 of *Minnesota Studies in the Philosophy of Science*. Salmon has taught at UCLA, Washington State College, Northwestern University, Brown University, Indiana University, and the University of Arizona, with visiting appointments at Bristol (England), Melbourne (Australia), Bologna (Italy), and the Minnesota Center for Philosophy of Science. He has served as president of the Philosophy of Science Association and of the American Philosophical Association (Pacific Division). He is a fellow of the American Academy of Arts and Sciences and of the American Association for the Advancement of Science.

C. Wade Savage is professor of philosophy at the University of Minnesota and a member of the Minnesota Center for Philosophy of Science. He was director of the Center from 1980 to 1984 and a codirector of its institute on consensus in philosophy of science during 1985–87. Savage is author of *The Measurement of Sensation: A Study of Perceptual Psychophysics*. He is editor of *Perception and Cognition: Issues in the Foundations of Psychology* (1978) and senior editor of and contributor to *Rereading Russell: Essays on Bertrand Russell's Metaphysics and Epistemology* (1989), both volumes in the Minnesota Studies in the Philosophy of Science series.

Lawrence Sklar received his training in mathematics and physics at Oberlin College and in philosophy of science at Princeton University. Major interests are in the philosophy of physics, the philsosophy of science, and epistemology. Published work includes two books on the philosophy of space and time (*Space, Time, and Spacetime*, 1977, and *Philosophy and Spacetime Physics*, 1985). His current focus of interest is philosophical issues in the foundations of statistical mechanics and in the role of theories in science.

Brian Skyrms is professor of philosophy at the University of California, Irvine. He is the author of *Choice & Chance: An Introduction to Inductive Logic* (1986), *Causal Necessity* (1980), and *Pragmatics and Empiricism* (1984). A new book, *The Dynamics of Rational Deliberation*, is forthcoming.

Elliott Sober is Hans Reichenbach Professor of Philosophy at the University of Wisconsin, Madison. His two books on philosophical issues in evolutionary biology are *The Nature of Selection: Evolutionary Theory in Philosophical Focus* (1984) and *Reconstructing the Past: Parsimony, Evolution, and Inference*. He also is interested in philosophical problems concerning causality, explanation, and confirmation.

John Worrall is reader in philosophy of science at the London School of Economics. He is especially interested in the general phenomenon of theory-change in science and is the author of several articles on rationality, theory-change, and scientific realism. He has made a special study of eighteenth- and nineteenth-century physics, especially optics, and is the author of a forthcoming book on reason and revolution in science, which includes a detailed study of Fresnel's wave revolution in optics. Editor of the *British Journal for the Philosophy of Science* from 1973 to 1982, he also edited (with Elie Zahar) Imre Lakatos's *Proofs and Refutations* (1976) and (with Greg Currie) Lakatos's *Philosophical Papers* (1978).

Consensus Institute Staff

Ned Block, Massachusetts Institute of Technology
Richard Boyd, Cornell University
Robert Butts, University of Western Ontario
Christopher Cherniak, University of Maryland
Paul Churchland, University of California at San Diego
Ellery Eells, University of Wisconsin at Madison
Ronald Giere, University of Minnesota
Clark Glymour, Carnegie-Mellon University
Adolf Grünbaum, University of Pittsburgh
Erwin Hiebert, Harvard University
Colin Howson, University of London
David Hull, Northwestern University
Paul Humphreys, University of Virginia
Thomas Kuhn, Massachusetts Institute of Technology
Henry Kyburg, University of Rochester
Larry Laudan, University of Hawaii
Ernan McMullin, University of Notre Dame
Alan Musgrave, University of Otago
Alan Nelson, University of California at Irvine
David Papineau, Cambridge University
Peter Railton, University of Michigan
Robert Richardson, University of Cincinnati
Merrilee Salmon, University of Pittsburgh
Wesley Salmon, University of Pittsburgh
Lawrence Sklar, University of Michigan
Bryan Skyrms, University of California at Irvine
Paul Smolensky, University of Colorado at Boulder
Elliott Sober, University of Wisconsin at Madison
Stephen Stich, Rutgers University
Zeno Switink, State University of New York at Buffalo
Bas Van Fraassen, Princeton University
John Worrall, University of London

INDEXES

Author Index

Subject Index